D1265742

RECENT ADVANCES IN SCIENCE AND TECHNOLOGY OF MATERIALS

Volume 1

RECENT ADVANCES IN SCIENCE AND TECHNOLOGY OF MATERIALS

Volume 1

Edited by
Adli Bishay
Director, Solid State and Materials Research Center
American University in Cairo
Cairo, Egypt

PLENUM PRESS • NEW YORK AND LONDON

Library of Congress Cataloging in Publication Data

Cairo Solid State Conference, 2d, 1973.
Recent advances in science and technology of materials.

Sponsored by the U.S. National Science Foundation.
Includes bibliographical references.
1. Solids—Congresses. I. Bishay, Adli, ed. II. United States. National Science Foundation. III. Title.

QC176.A1C3 1973 530.4'1 74-17098
ISBN 0-306-37691-1 (v. 1)

The proceedings of the Second Cairo Solid State Conference held in Cairo, Egypt, April 21-26, 1973, will be published in three volumes, of which this is volume one.

A Division of Plenum Publishing Corporation
227 West 17th Street, New York, N.Y. 10011

United Kingdom edition published by Plenum Press, London
A Division of Plenum Publishing Company, Ltd.
4a Lower John Street, London W1R 3PD, England

Printed in the United States of America

SECOND CAIRO SOLID STATE CONFERENCE

REVIEW

The Second Cairo Solid State Conference, sponsored by the U.S. National Science Foundation, was held in Cairo, Egypt, April 21 to 26, 1973. The Conference was organized by The American University in Cairo, in cooperation with Ein Shams University, Cairo.

The Conference, whose theme was "Recent Advances in Science and Technology of Materials," followed the First Cairo Solid State Conference by seven years, and this time combined the fields of Solid State Science and Archaeology. The papers presented dealt with recent discoveries and techniques and their applications in the study of both modern and ancient materials.

From all aspects the Conference was indeed a successful one, attracting over 300 participants from 13 countries. Many of the papers included advances hitherto unreported in several fields. The "Archaeology" sessions included work on analytical techniques and dating as well as restoration and preservation studies, and the sessions on "Modern Materials" attracted papers on technology in addition to studies of the physical and chemical properties of glass, ceramics, metals, and polymers and their composites.

The final session of the Conference was devoted to a review of the papers presented. C. S. Smith (USA), reviewing the papers on Archaeology, noted the

Impact of Physical Science on Archaeological Research

"It is interesting historically to notice that the first laboratory studies of ancient artifacts were in analytical chemistry, the branch of science that first added anything to the ancient empirical knowledge of materials. Later on came petrography and metallography but only recently has the solid state physicist himself become involved. This is precisely the order in which the science of solids developed. For long there was no means of studying structure-sensitive phenomena and the physicist completely ignored them until very recently. But it is, of course, in the fine nuances of structure that the best record of material history lies. At this Conference we have seen examples of superb analytical methods and the most recent tools of the solid state physicist being used to reveal ancient methods of manufacture in more detail than the artisan himself could have conveyed in words, and to provide the subtle but indelible 'fingerprints' revealing date and provenance.

"It was in the field of conservation that physical science first made really significant contributions to archaeology and art history, and it is still in the conservation laboratory where one finds the most sensitive overall view of historical-technical processes. In order to be able to conserve something and sometimes to restore it, one has to understand the material fully. Conservators not only do so in material terms; they are also concerned with all of the art-historical aspects of artifacts, with the individual experience of the artist and the social environment which influenced him. The conservation laboratory therefore provides an excellent arena for cooperative work of mutual interest to artists, historians and scientists."

Professor Smith also noted the

Emphasis on Instrumentation

"It is perhaps inevitable in a conference of this kind that the emphasis should be on instrumentation. The instrumentation is so very impressive, and the techniques so ingenious, that a special effort is needed to remember why the instrumentation is being used in the first place. It is easy for a scientist to think that when he has used his most wonderful gadgetry to get some nice numbers on composition or structure that this is enough in itself. Of course it is not. His figures and patterns are meaningless unless they are put into the fuller context of understanding humanity in the past and provide a new route in humanistic studies. Instead of reading books and manuscripts one now works in a laboratory to read the physical record, but this is done for the same purpose that the scholar in the past gathered information from the verbal or pictorial record. However elegant or expensive the equipment used, and however precise the results, the work is without value as a contribution to historical understanding unless this human meaning is seen."

The papers on "Modern Materials" stressed new applications of solid state science to rapidly advancing technology. Some particularly important and topical subjects reported upon were:

Special Materials for Energy Purposes

Professor J. G. Dienes (USA) pointed out that the field of solid state science and defects in solids faces special challenges from technology. "Higher temperatures and higher exposures are required if we are going to be efficient in our energy production. If we are thinking, in the long run, of thermonuclear reactors -- the so-called 'clean fuel' -- then the materials we are concerned with will have to withstand very severe conditions. A very much altered neutron spectrum will involve transmutation products, impurities, changes in stoichiometry, etc. We are faced with the necessity of using unusual materials, refractory materials, instead of the ones which we are used to. The field is moving essentially from the simple materials to the more complex materials, and a number of fundamental problems will have to be answered."

Glass Fibers for Communications

Professor R. W. Douglas (England) spoke about the situation in the spectroscopy of glasses as revealed by the contributions to this Conference. "At the First Cairo Solid State Conference, a paper was presented which summarized what was then known about the spectroscopy of glasses from the ultra violet to infrared. The spectroscopy of transition metal ions was explored, making use of crystal theory which just ten years before the 1966 Conference had begun to be used by chemists, C. K. Jørgensen in particular. When it was first proposed that glass fibers could be used as transmission lines for a beam of light carrying communications signals, it was at least clear from what was already known about the effect of transmission of metal ions and other absorptions that the proposal should clearly not be considered impossible. Now, the idea is nearly a practicality."

Professor J. Zarzycki (France), reviewing the sessions on "Mechanical Properties," stressed the remaining problems involved in the use of glass fibers for optical communications. "It has been clearly shown that even if we succeed in producing glasses with very small absorption coefficients -- enough for communication purposes -- the problem remains of how to handle these materials. These fibers must withstand hard mechanical treatment, and they must also be stable in time. Static fatigue problems and corrosion will also be involved."

During the Conference, there was considerable discussion and general agreement on the importance of

Cooperation Between Scientists of Different Institutions and Countries, and Between Physical Scientists and Archaeologists

Scientists involved in the study of either ancient or modern materials must, as Professor Zarzycki pointed out, cooperate in their research efforts, if for no other reason than the great expense of the instruments necessary in understanding materials. Professor Smith also referred to the advantages of cooperation, repeating one Conferee's observation that "It is not enough simply to use somebody's machine or to have him make a measurement; you must debate the topic with him and publish jointly. The days have gone by when a physical scientist doing analyses can write an appendix to an archaeological paper, reporting fine measurements that no one bothers to give meaning to. The scientist must become one of the principal authors, if not the principal author, for he must himself ask his own kind of archaeological questions. This is a superb field for cooperative research."

In conclusion, the Conference succeeded in its purpose of bringing solid state scientists and archaeologists together, and has paved the way for further interactions among them in the future. At the same time, scientists have been reminded of the challenges they face in answering the demands of a vastly changing technology, and of the citizen's growing interest in and response to technology.

Adli Bishay

ACKNOWLEDGMENTS

The successful completion of the Conference was largely due to the encouragement and enthusiasm of the late Christopher Thoron, President of The American University in Cairo, the full cooperation of Dr. Ismail Ghanem,* President, and Dr. Nayel Barakat, Dean of the Faculty of Science, Ein Shams University, and the support of the U.S. National Science Foundation and the Egyptian Academy of Science and Technology.

The papers published in the three volumes of these Proceedings went through two screening operations. The abstracts submitted for presentation at the Conference were refereed by a Committee including Professors Arthur Quarrell (Sheffield University), John Thomas (University of Aberystwyth, Wales), Nayel Barakat and Sami Tobia (Ein Shams University), Zaki Iskander (Department of Antiquities, Cairo), Kamal Hussein (Egyptian Academy of Science and Technology), and Hosny Omar, Farkhonda Hassan and James Heasley (The American University in Cairo). The full texts of the papers presented at the Conference were refereed by the Editorial Committee and other specialists whom they appointed.

The coordination between authors and members of the Editorial Committee was due to the skill and hard work of Marjorie Ellen Ekdawi, who also helped in typing the manuscript. Mary Dungan Megalli edited the papers for English language, organized the typing and did the proofreading. Soraya Garas, Mrs. Ekdawi, Victor Yacoub and Kamal Ragy completed these three volumes of difficult and demanding typing in just over two months.

To all the above, my sincere thanks and appreciation. Special thanks are also due to Carl Schieren and Priscilla Blakemore of The American University in Cairo, to my colleagues at the Solid State and Materials Research Center and students at The American University in Cairo, with special reference to members of the Science-Engineering Society.

Adli Bishay

*Currently Minister of Higher Education and Scientific Research.

CONTENTS OF VOLUME 1

DEFECTS IN SOLIDS

ELECTRICAL PROPERTIES

MAGNETIC PROPERTIES

RECENT ADVANCES IN SCIENCE AND TECHNOLOGY OF MATERIALS

Volume 1

KINETICS OF DEFECT FORMATION IN ALKALI HALIDES AT HELIUM TEMPERATURES*

G.J. Dienes

Brookhaven National Laboratory, Upton, New York

The kinetics of defect formation in alkali halides at low temperature was investigated based on the following sequential steps: production of electrons and holes by ionizing radiation, self-trapping of holes to form V_K-centers, recombination of V_K-centers and electrons followed by a collision sequence to form F- and H-centers, ionization of F-centers to form F^+-centers, and capture of electrons by the H-centers to form I-centers. The kinetic equations were solved by computer techniques for steady irradiation and for pulse irradiation. Analytic solutions were found for the steady state approximation, valid after the decay of initial transients. During steady irradiation, once the transients have decayed, the total vacancy concentration (and the concentration of the other centers) increases linearly with irradiation time, in agreement with the available meager experimental data. A specific prediction of the kinetic scheme is the flux (dose rate) dependence of the above growth rate-linear in flux at low flux and gradually approaching flux independence at high flux. When the unstable nature of the close F-H pairs is explicitly taken into account, the response of the system to pulse irradiation simulates the experimental data, at least qualitatively.

INTRODUCTION

It has been shown recently that the direct formation of F and H centers, based on V_K+e recombination followed by a sequence of [110] replacement collisions, is energetically possible.[1] It was

---

*Work supported by the U.S. Atomic Energy Commission.

also suggested that the production of F^+ and I centers is a subsequent event arising from the ionization of F centers and the trapping of electrons by H centers. The next question that arises is the kinetic description of these processes. The available experimental information is rather meager.

Linear growth of total vacancy concentration at liquid helium temperature has been reported by Behr et al[2] for KCl and by Comins[3] for KBr under steady x-ray irradiation at helium temperature at one flux (dose rate). In these experiments the growth rate of the total vacancy concentration is about a factor of 6 to 10 higher than that of the F-center concentration.

The low temperature pulse experiments of Ueta et al[4] showed very clearly that F and H centers are formed first followed by the formation of F^+ and I centers. A high fraction (∿95%) of the F and H centers are unstable and recombine very quickly, presumably by close pair recombination.

A kinetic scheme is proposed in this paper based on the above sequence of events for defect formation. Complete solutions had to be obtained on a computer with some analytical approximations valid in steady state. In Section II the general kinetic scheme is discussed. Steady irradiation and flux dependence are treated in Section III and the response to a pulse of irradiation in Section IV. A brief discussion of the results is given in Section V.

II. THE KINETIC SCHEME

The basic physical steps are: production of electrons (e) and holes (h) by ionizing radiation, self-trapping of a hole to form a V_K center, recombination of a V_K center and an electron followed by a collision sequence to produce F and H, ionization of F to form F^+ and capture of an electron by H to form the I center. In what follows homogeneous kinetics is assumed, i.e. there is no spatial dependence to the rate constants. In the first scheme the immediate F and H recombination is lumped in with the radiative recombination of V_K+e (the scheme is modified for pulsed irradiation, see Section IV). Since the temperature is assumed to be that of liquid helium all ionic motions are neglected except for the above close pair recombination of unstable F and H centers. The kinetic scheme is as follows:

$$\text{Perfect crystal} \underset{K'}{\overset{K\phi}{\rightleftarrows}} e + h \tag{1}$$

$$h \overset{K_1}{\rightarrow} V_K \tag{2}$$

$$V_K + e \xrightarrow{K_2} \text{Perfect crystal (from radiative transition and immediate recombination of close F–H pairs)} \quad (3)$$

$$V_K + e \xrightarrow{K_3} F + H \text{ (non-radiative transition plus collision sequence)} \quad (4)$$

$$F \underset{K_{-4}}{\overset{K_4\phi}{\rightleftarrows}} F^+ + e \quad (5)$$

$$H + e \underset{K_{-5}\phi}{\overset{K_5}{\rightleftarrows}} I \quad (6)$$

where ϕ is the flux of radiation and the K_i are rate constant. The reaction $F + h \rightarrow F^+$ has been neglected. The differential equations describing these reactions are

$$dh/dt = K\phi - K'eh - K_1h \quad (7)$$

$$dF/dt + dF^+/dt = dH/dt + dI/dt = K_3V_Ke \quad (8)$$

$$dV_K/dt = K_1h - (K_2+K_3)V_Ke \quad (9)$$

$$dF/dt = K_3V_KE - K_4\phi F + K_{-4}F^+e \quad (10)$$

$$dH/dt = K_3V_Ke - K_5He + K_{-5}\phi I \quad (11)$$

$$dF^+/dt = K_4\phi F - K_{-4}F^+e \quad (12)$$

$$dI/dt = K_5He - K_{-5}\phi I \quad (13)$$

By adding (10) to (12) and (11) to (13) one immediately finds

$$de/dt = K\phi - K'eh - (K_2+K_3)V_Ke + K_4\phi F - K_{-4}F^+e - K_5He + K_{-5}\phi I \quad (14)$$

and if all the centers are assumed to be at zero concentration at $t = 0$ (not strictly correct) then

$$F + F^+ = H + I \quad (15)$$

If V_K and e have reached steady-state concentration then (14) is immediately integrable to give

$$F + F^+ = H + I = K^*t + A \quad (16)$$

where A is a constant of integration.

Table I

The rate constants chosen for the computer calculations

Run No.	ϕ	K	K'	K_1	K_2	K_3	K_4	K_5	K_{-4}	K_{-5}	$K\phi$	$K_4\phi$	$K_{-5}\phi$
IX-2	10^{13}	10^2	10^{-13}	3×10^1	3×10^{-13}	5×10^{-15}	1.6×10^{-16}	4.5×10^{-17}	2×10^{-17}	10^{-17}	10^{15}	1.6×10^{-3}	10^{-4}
IX-1	10^{14}	"	"	"	"	"	"	"	"	"	10^{16}	1.6×10^{-2}	10^{-3}
IX-4	10^{15}	"	"	"	"	"	"	"	"	"	10^{17}	1.6×10^{-1}	10^{-2}
IX-8	10^{14}	"	"	"	"	"	"	"	"	"	10^{16}	1.6×10^{-2}	0

The non-linear equations (7-13) cannot be handled analytically but were solved by computer techniques for some rather arbitrary sets of rate constants. The basic aim was to obtain some information on the flux dependence of the defect growth rate since flux is an independently variable experimental parameter. There are no systematic experiments on flux dependence. The Behr and Comins experiments indicate that K^* of equation (16) is of the order of 10^{14} cm^{-3} sec^{-1} and the ratio of F^+ to F is of the order of 6 with all concentrations expressed in number per cm^3. The constants chosen for the computer runs are given in Table I. The first three runs explore the dependence on flux and the fourth the influence of setting the rate constants for the reverse reactions at zero. While the nominal time scale is in seconds and the concentrations are in numbers per cc, the units should be considered arbitrary since the rate constants are not known experimentally. Run IX-1 (Table I) was chosen to yield about the observed K^*(later denoted as C_1+C_2) of the Behr and Comins experiments.

III. RESPONSE TO STEADY IRRADIATION

The computer runs showed that rather intricate transient changes occur in e, h and V_K as the flux is turned on. An example for run IX-1 is shown in Fig. 1 where the insert shows the very early time dependence. There is a practically instantaneous production of e, h and V_K as the flux is turned on, followed by a short plateau which leads to maxima and minima on a somewhat longer time scale. Finally h, e and V_K approach an essentially constant value. The corresponding growth of F, F^+ and $(F+F^+)$ is shown in Fig. 2 for the first 500 time units. Clearly F is formed first which is then overtaken by the growth of the F^+ centers. These centers (and also H and I) approach linear growth (i.e. constant growth rate) as e, h and V_K approach steady state values in accordance with equation (16). At $t = 10^4$ (seconds, but in reality arbitrary time units) the computer data (Table II) show that the system is essentially in steady state since e, h and V_K are changing extremely slowly. Note that $e \neq h$. This arises from reactions (5) and (6) - if these reactions were negligible e would be equal to h.

Once the system is in steady state, one can use the following approximate approach, particularly for purposes of extrapolation and interpolation with respect to the flux, ϕ.
Let

$$\dot{F} = C_1 \; ; \; (\dot{F}^+) = C_2$$

From $\dot{V}_K = 0$ (Eq. 9):

$$C_1 + C_2 = K_3 V_K e = K_1 K_3 h / K_2 + K_3 \tag{17}$$

and from $\dot{h} = 0$ (Eq. 7):

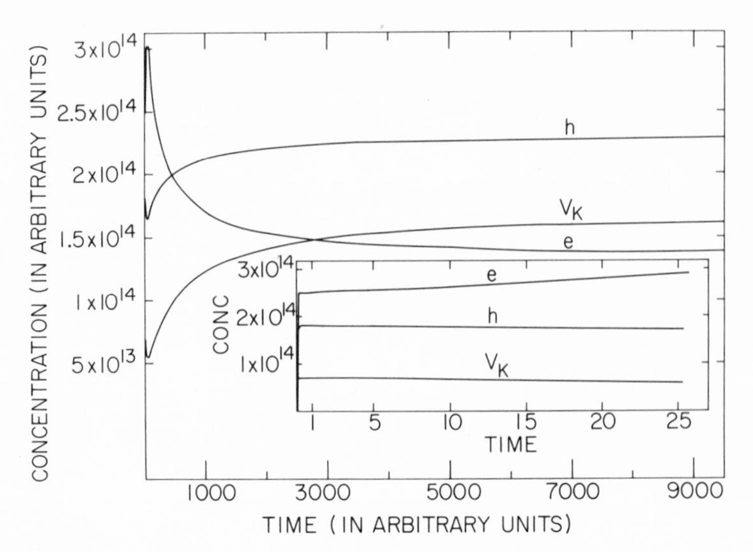

Fig. 1. The concentration of electrons (e), holes (h) and V_K centers as a function of time during steady irradiation. The insert depicts the response at short times. (Rate constants as given in Table I for run IX-1).

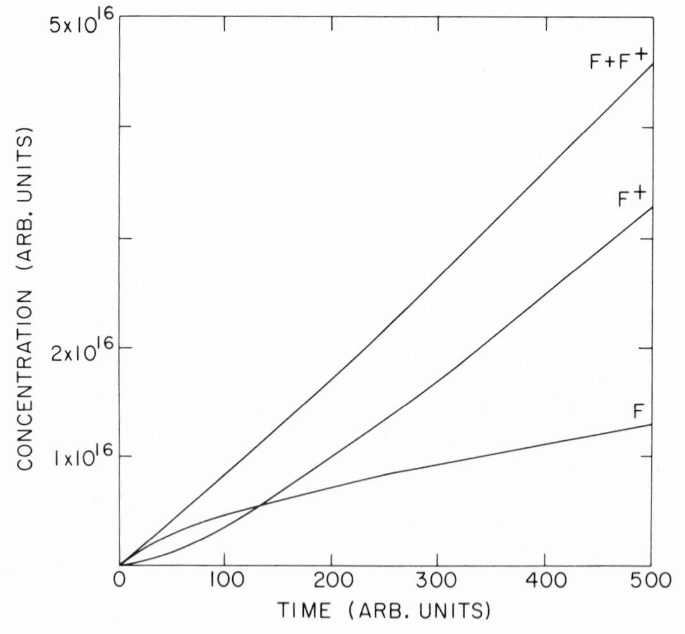

Fig. 2. The growth of F-, F^+- and $(F+F^+)$-centers as a function of irradiation time during steady irradiation. (Rate constants as given in Table I for run IX-1).

Table II

Computer Data at $t = 10^4$

	IX-2 $\phi = 10^{13}$	IX-1 $\phi = 10^{14}$	IX-4 $\phi = 10^{15}$	IX-8 $\phi = 10^{14}$ $K_{-4} = 0$ $K_{-5} = 0$
h	$3.1562\text{x}10^{13}$	$2.2826\text{x}10^{14}$	$6.1102\text{x}10^{14}$	$1.5504\text{x}10^{14}$
e	$1.6833\text{x}10^{13}$	$1.3810\text{x}10^{14}$	$1.3366\text{x}10^{15}$	$3.4500\text{x}10^{14}$
V_K	$1.8443\text{x}10^{14}$	$1.6257\text{x}10^{14}$	$4.4965\text{x}10^{13}$	$4.4202\text{x}10^{13}$
F	$3.3409\text{x}10^{16}$	$1.4972\text{x}10^{17}$	$3.8580\text{x}10^{17}$	$4.7655\text{x}10^{15}$
F^+	$1.1968\text{x}10^{17}$	$8.3248\text{x}10^{17}$	$2.2995\text{x}10^{18}$	$7.5809.10^{17}$
H	$3.3210\text{x}10^{16}$	$1.4947\text{x}10^{17}$	$3.8648\text{x}10^{17}$	$4.9113\text{x}10^{15}$
I	$1.1980\text{x}10^{17}$	$8.3273\text{x}10^{17}$	$2.2988\text{x}10^{18}$	$7.5795\text{x}10^{17}$
$F+F^+$	$1.5309\text{x}10^{17}$	$9.8220\text{x}10^{17}$	$2.6853\text{x}10^{18}$	$7.6286\text{x}10^{17}$
$\dot{h}$	$2.8849\text{x}10^{7}$	$2.7337\text{x}10^{8}$	$1.3549\text{x}10^{8}$	0.0000
$\dot{e}$	$-2.8960\text{x}10^{8}$	$-5.2470\text{x}10^{8}$	$-3.6289\text{x}10^{8}$	0.0000
$\dot{V}_K$	$3.3417\text{x}10^{9}$	$8.1234\text{x}10^{8}$	$2.2179\text{x}10^{7}$	0.0000
$\dot{F}$	$2.3581\text{x}10^{12}$	$1.6067\text{x}10^{13}$	$4.2930\text{x}10^{13}$	$3.2\text{x}10^{3} \simeq 0$
$(F^+)^{\cdot}$	$1.3164\text{x}10^{13}$	$9.6193\text{x}10^{13}$	$2.5757\text{x}10^{14}$	$7.6248\text{x}10^{13}$
$\dot{H}$	$2.3545\text{x}10^{12}$	$1.6065\text{x}10^{13}$	$4.2930\text{x}10^{13}$	$3.2\text{x}10^{3} \simeq 0$
$\dot{I}$	$1.3168\text{x}10^{13}$	$9.6192\text{x}10^{13}$	$2.5757\text{x}10^{14}$	$7.6248\text{x}10^{13}$
$(F+F^+)^{\cdot}$	$1.5522\text{x}10^{13}$	$1.1226\text{x}10^{14}$	$3.0050\text{x}10^{14}$	$7.6248\text{x}10^{13}$
$(F^+)^{\cdot}/F$	5.5825	5.9870	5.9997	$2.38\text{x}10^{10}$
$\dot{h}/h$	$9.1404\text{x}10^{-7}$	$1.1976\text{x}10^{-6}$	$2.2174\text{x}10^{-7}$	0.0000
$\dot{e}/e$	$-1.7204\text{x}10^{-5}$	$-3.7993\text{x}10^{-6}$	$-2.7150\text{x}10^{-7}$	0.0000
$\dot{V}_K/V_K$	$1.8119\text{x}10^{-5}$	$4.9969\text{x}10^{-6}$	$4.9325\text{x}10^{-7}$	0.0000

TABLE III

Comparison of computer results at $t=10^4$ with approximate formulae

Run No.	ϕ	C_1+C_2 Eq.21	C_1+C_2 COMP.	e Eq.19	e COMP.	h Eq.20	h COMP.	V_K Eq.24	V_K COMP.	$(F^+)^\cdot/\dot{F}$ Eq.27	$(F^+)^\cdot/\dot{F}$ COMP.
	10^{11}	1.6393×10^{11}		1.333×10^{11}		3.53×10^{11}		2.459×10^{14}		6.000	
	10^{12}	1.6393×10^{12}		1.333×10^{12}		3.33×10^{12}		2.459×10^{14}		6.000	
IX-2	10^{13}	1.5696×10^{13}	1.5522×10^{13}	1.333×10^{13}	1.6833×10^{13}	3.1915×10^{13}	3.1562×10^{13}	2.355×10^{14}	1.844×10^{14}	6.000	5.5825
IX-1	10^{14}	1.1360×10^{14}	1.1226×10^{14}	1.333×10^{14}	1.381×10^{14}	2.31×10^{14}	2.28×10^{14}	1.70×10^{14}	1.626×10^{14}	6.000	5.9870
	3×10^{14}	2.1080×10^{14}		3.999×10^{14}		4.29×10^{14}		1.05×10^{14}		6.000	
IX-4	10^{15}	3.0100×10^{14}	3.0050×10^{14}	1.333×10^{15}	1.337×10^{15}	6.12×10^{14}	6.11×10^{14}	4.51×10^{13}	4.497×10^{13}	6.000	5.9997
	10^{16}	3.600×10^{14}		1.333×10^{16}		7.33×10^{14}		5.40×10^{12}		6.000	
	10^{17}	3.680×10^{14}		1.333×10^{17}		7.48×10^{14}		5.52×10^{11}		6.000	
	∞	3.690×10^{14}		∞		7.50×10^{14}		0		6.000	
IX-8 ($K_{-4}=0$, $K_{-5}=0$)	10^{14}	7.502×10^{13}	7.625×10^{13}	eq.28 3.556×10^{14}	3.450×10^{14}	1.525×10^{14}	1.550×10^{14}	4.22×10^{13}	4.42×10^{13}	∞	2.38×10^{10}

$$h = K\phi/K_1 + K'e \tag{18}$$

One can obtain an approximate value for the steady state concentration of e as follows.

From Eq. (12):

$$e = -C_2 + K_4\phi F/K_{-4}F^+ = K_4\phi F/K_{-4}F^+ \quad \text{at long times.}$$

The computer runs indicate that $\dot{F} \simeq \dot{H}$ and from equations (10) and (11)

$$-K_4\phi F + K_{-4}F^+e = -K_5He + K_{-5}\phi I$$

and with $F \simeq H$ and $F^+ \simeq I$ (as shown by computer)

$$F^+/F = K_4\phi - K_5e/K_{-4}e - K_{-5}\phi$$

and

$$e = (K_4K_{-5}/K_{-4}K_5)^{1/2}\ \phi \tag{19}$$

Thus, e in steady state is proportional to the flux, ϕ. Substitution into (18) and (17) gives

$$h = K\phi/K_1 + K'(K_4K_{-5}/K_{-4}K_5)^{1/2}\ \phi \tag{20}$$

and

$$d(F+F^+)/dt = (C_1+C_2) = K_1K_3K\phi/(K_2+K_3)[K_1+K'\phi(K_4K_{-5}/K_{-4}K_5)^{1/2}] \tag{21}$$

Calculations based on these approximate equations check quite closely with the computer results for Runs IX-2, IX-1, and IX-4, as shown in Table III. The important behavior is that of the total vacancy production rate, (C_1+C_2), as a function of flux, ϕ. This is given by equation (21). At high flux (C_1+C_2) becomes independent of flux as

$$C_1+C_2 \rightarrow K_1K_3K/(K_2+K_3)K'(K_4K_{-5}/K_{-4}K_5)^{1/2} \tag{22}$$

At low ϕ, (C_1+C_2) becomes linear in ϕ as

$$C_1 + C_2 \rightarrow K_3K\phi/K_2 + K_3 \tag{23}$$

The behavior of $(C_1+C_2) = f(\phi)$ is illustrated in Fig. 3. h in steady state follows a similar course since it is proportional to C_1+C_2 (Eq. 17). V_K in steady state is given by

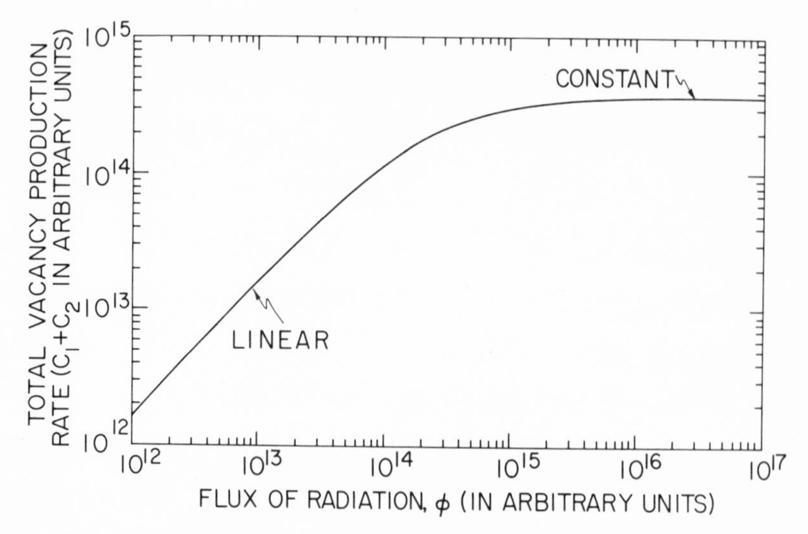

Fig. 3. The flux dependence of the total vacancy ($F+F^+$) production rate. (The rate constants are listed in Table I).

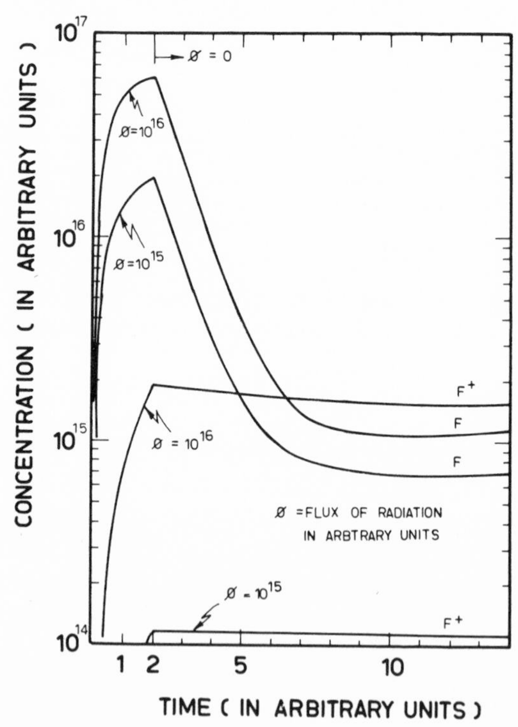

Fig. 4. The response to high flux pulse irradiation. F- and F^+-center concentration as a function of time. Pulse terminated at t = 2.

$$V_K = K_1K/(K_2+K_3)[K_1+K'\phi(K_4K_{-5}/K_{-4}K_5)^{1/2}](K_4K_{-5}/K_{-4}K_5)^{1/2} \quad (24)$$

and at high ϕ

$$V_K \to K_1K\ K_{-4}K_5/K'K_4K_{-5}(K_2+K_3)\phi \quad (25)$$

and at low ϕ

$$V_K \to K/K_2+K_3(K_{-4}K_5/K_4K_{-5})^{1/2} \quad (26)$$

Upon taking derivatives of the approximate expression

$$e \simeq -C_2 + K_4\phi/K_{-4}F^+ \quad \text{with } \dot{e} = 0$$

one obtains

$$(F^+)^{\cdot}/\dot{F} \simeq K_4\phi/K_{-4}e = K_4/K_{-4}(K_{-4}K_5/K_4K_{-5})^{1/2} \quad (27)$$

i.e. at sufficiently long times this ratio is independent of the flux.

If the flux is turned off, after the system has been in steady state, e and h go to zero very quickly, the change in F, F^+, H and I is insignificant, but there can be a significant increase in V_K (by as much as a factor of three). Thus, post-irradiation measurements under these circumstances are satisfactory for all color centers except V_K.

The above approximations cannot be used directly if the back reaction rates are set equal to zero, i.e. $K_{-4} = 0$ and $K_{-5} = 0$. However, in steady state

$$F \to K_3V_Ke/K_4\phi \quad \text{and} \quad H \to K_3V_Ke/K_5e$$

and if $F \simeq H$, which is reasonably well verified by the computer run (Table II), then

$$e \simeq (K_4/K_5)\phi \quad (28)$$

and the rest of the approximate formulas are valid with this value of e. Since $\dot{F} \to 0$ clearly $(\dot{F}^+)/\dot{F} \to \infty$. The comparison between computer runs and the approximate formulas is given in Table III. The agreement is very satisfactory. The flux dependence for (C_1+C_2) is as before.

IV. RESPONSE TO PULSE IRRADIATION

During a high flux pulse irradiation the recombination of unstable F-H pairs cannot be absorbed in the radiative V_K+e reaction (reaction 3) but has to be treated separately with its own decay rate constant, α. The kinetic scheme is, therefore, modified by adding to reactions (1-6) the reactions

$$V_K + e \xrightarrow{K_3'} F^* + H^* \tag{4a}$$

$$F^* \xrightarrow{\alpha} \text{Perfect crystal} \tag{4b}$$

$$H^* \xrightarrow{\alpha} \text{Perfect crystal} \tag{4c}$$

where F^* and H^* are the unstable close pair. Since they are close pairs they disappear by a first order reaction. The differential equations are modified in an obvious way by adding to equations (7-13) the equations

$$dF^*/dt = K_3'V_Ke - \alpha F^* \tag{10a}$$

$$dH^*/dt = K_3'V_Ke - \alpha H^* \tag{11a}$$

In order to have rate constants comparable to the previous ones, K_2 for the pulse calculations was reduced to 3×10^{-14}, K_3' was taken as 2.7×10^{-13} and α was set equal to 1. Two sets of computer runs were made to simulate the response to a pulse, i.e. a pulse of 2 time units at two fluxes (Fig. 4) and a pulse of 10 time units at the same two fluxes (Fig. 5). In these Figures the F concentration plotted is the total observable during and after the pulse (unstable plus stable). The results, particularly for the shorter pulse, are qualitatively very similar to the experimental results of Kondo et al on KBr.[5] The rapid rise of the F center concentration during the pulse is followed by an initially exponential decay. The H-center concentration is equal to the F center concentration for all practical purposes and follows an identical course as a function of time. The F^+-center concentration is highly sensitive to the flux and is essentially unobservable in the lower intensity run. The longer pulses of Fig. 5 are similar, the main difference being that the F-center concentration is past the maximum before the end of the pulse and the F^+-center production is relatively much higher. In these runs, the Decay of F^+ (and the small increase in F past the minimum) after the pulse is due to the capture of electrons by F^+, since the electron concentration is high at these fluxes. h and V_K go to zero very quickly after the pulse.

V. DISCUSSION

The kinetic scheme for defect formation in the alkali halides

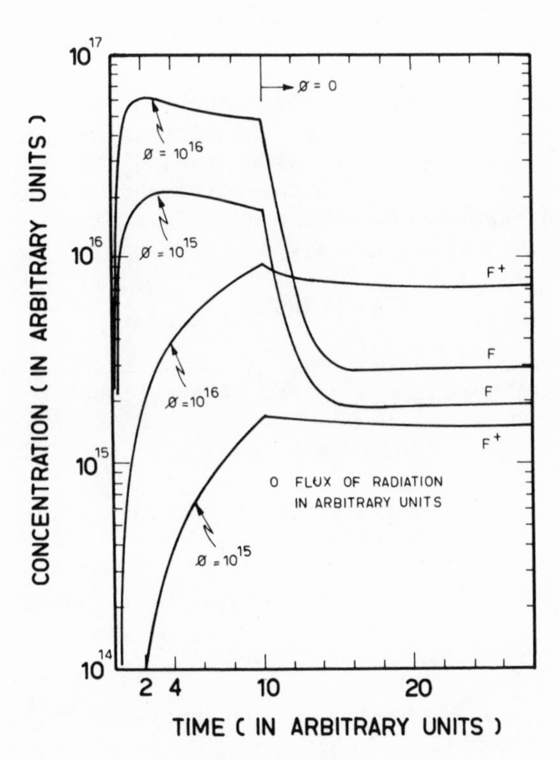

Fig. 5. The response to high flux pulse irradiation. I- and F^+-center concentration as a function of time. Pulse terminated at t = 10.

at helium temperatures proposed in this paper appears to be in qualitative agreement with the available meager experimental data. F and H centers are produced first followed by the formation of F^+ and I centers. In steady irradiation, once the transients have decayed, the growth of total vacancy concentration ($F+F^+$) and F-center concentration is linear with irradiation time. A specific prediction of the kinetic scheme is the flux dependence of the growth rate - linear in flux at low flux and gradually approaching flux independence at high flux. Systematic experiments on flux dependence are needed to test this prediction and to derive some of the rate constants.

When the unstable nature of the close F-H pairs is taken into account, the main features of pulse experiments are also reproduced. The results are sensitive to the intensity and the duration of the pulse. Again, one should be able to derive some of the rate constants as detailed experiments become available with the above variables altered and controlled in a systematic way.

ACKNOWLEDGMENT

It is a pleasure to acknowledge stimulating discussions with Dr. R. Smoluchowski during the course of this study.

REFERENCES

1. R. Smoluchowski, O.W. Lazareth, R.D. Hatcher, G.J. Dienes, Phys. Rev. Lett. 27 1288 (1971).

2. A. Behr, H. Peisl, W. Waidelich, Phys. Stat. Sol. 21 K9 (1967).

3. J.D. Comins, Phys. Stat. Sol. 33 445 (1969).

4. M. Ueta, Y. Kondo, M. Hirai, T. Yoshiuary, J. Phys. Soc. Japan 26 1000 (1969).

5. Y. Kondo, M. Hirai, M. Ueta, J. Phys. Soc. Japan 33 151 (1972).

RELAXATION OF IMPURITY-DEFECT COMPLEXES IN FLUORITE CRYSTALS*

J. H. Crawford, Jr. and E. L. Kitts, Jr.

Department of Physics and Astronomy, University of North Carolina, Chapel Hill, U.S.A.

This paper summarizes recent contributions to the understanding of the structure and the reorientation kinetics of various impurity-defect complexes in alkaline earth fluroides. For the most part the relevant investigations involve relaxation of electric polarization as revealed by the ionic thermo-current (ITC) method on CaF_2, SrF_2 and BaF_2 crystals doped with a variety of trivalent rare earth (RE^{3+}) ions. In the absence of contaminating oxygen the extra positive charge is compensated by an interstitial fluoride ion (F_i^-) which except at elevated temperature forms a complex with the RE^{3+}. In CaF_2 the dominant form of the complex as revealed by EPR and ITC is tetragonal with the F_i^- located in the nearest neighboring interstitial site (Type I complex). In BaF_2 the dominant form is trigonal with the F_i^- situated in a next nearest site along a <111> direction (Type II complex). In SrF_2 both Type I and Type II complexes are clearly evident in the ITC relaxation peak structure. In a crystal of CaF_2:Gd^{3+} treated to introduce oxygen the Type I relaxation at 131 K is replaced almost quantitatively by a new peak at 166 K. Comparison with EPR behavior reveals that the new peak is associated with the relaxation of the CaF_2:$Gd^{3+}O_4{}^{2-}F^-$ complex (Type T_1) which had been previously identified by ENDOR. The reorientation activation energy of 0.49 eV is more nearly comparable to that for Na^+ - fluoride vacancy complexes in CaF_2 (0.5 eV) than to the activation energy for Type I complexes (0.42 eV). The reorientation activation energy for Type I complexes is found to be only slightly dependent upon the size of the rare earth ion, ranging from 0.38 eV for the smallest ion considered (Yb^{3+}), to 0.42 eV for U^{3+}, the largest.

*This work was supported by the U.S. Atomic Energy Commission under Contract AT-(40-1)-3677.

INTRODUCTION

If an ion in a crystal is replaced by an impurity ion with a different charge, compensation for the charge difference must be accomplished. This is usually done by an intrinsic defect, i.e. a lattice vacancy or interstitial host ion, in such a way to balance the excess or deficit of charge. The introduction of cation vacancies along with divalent cation impurities in alkali halides is a familiar example. If these crystals are carefully annealed, nearly all of the compensating defects migrate to the impurity ions to form impurity-defect complexes. It is the behavior of these complexes that are the principal concern in this paper. Almost invariably such complexes are dipolar in character, and when subjected to an electric or mechanical stress field, they will tend to realign to an orientation of lower energy provided the thermal activation is sufficiently great. If the complex consists of an impurity-vacancy couple, reorientation may occur either by the vacancy moving around the impurity or by an exchange of positions of the partners. In general the activation energy for these two distinct reorientation paths is different. If the complex consists of an impurity-interstitial couple, interchange of positions is unlikely and reorientation is considered to occur exclusively by the motion of the interstitial around the vacancy. The structure of the complex is also important; in certain situations of lattice spacing and impurity ion size, the defect located in a next-nearest neighboring position (or one even more remote) may be more stable than a defect occupying a nearest neighbor position. Not only does the structure have an effect on the reorientation process[1] but also on the dipole moment of the defect and hence the degree of coupling to the applied electric field. The degree of alignment of dipoles is governed by the Langevin equation and for a cubic crystal for $kT >> \mu\varepsilon_p$ (μ is the dipole moment and ε_p the external applied field) the polarization is given by

$$P_d = \mu^2 \varepsilon_p N_d / 3kT_p \qquad (1)$$

where N_d is the dipole concentration and T_p is the temperature of polarization. The fractional alignment with the field seldom exceeds 10^{-3} for most experiments of interest.

In this paper we will be principally concerned with complexes of fluoride interstitials (F_i^-) and trivalent rare earth ions (RE^{3+}) in the alkaline earth fluorides. In the fluorite structure compensation of the extra positive charge by a F_i^- is energetically favored over a cation vacancy. Although the relaxational behavior of dipolar complexes in alkali halides is now well understood, having been extensively studied for nearly two decades,[1] only in the past three years has any substantial progress been made in $MF_2:RE^{3+}$. As a result of the recent extensive use of the ionic thermo-current (ITC) or thermal depolarization technique, however, considerable progress has recently been achieved. The purpose of this paper is to summarize these recent results.

EXPERIMENTAL

The ITC technique was developed and first applied by Bucci and co-workers[2] nearly a decade ago. It is more sensitive by at least an order of magnitude than the capacitance bridge measurement of dielectric loss and has the additional advantage that during measurement the specimen is kept at a temperature well below that necessary for migration or break-up of complexes. The method is quite simple and consists of three steps. (a) The polarizing field is applied to the crystal at a temperature high enough for saturation polarization to be reached in a reasonably short time (∼minutes). (b) The crystal is then cooled with the field applied to a temperature low enough to "freeze-in" the polarization. (c) After removing the field, the crystal is warmed at a linear rate and the displacement current caused by the randomization of the aligned dipoles is monitored. Since in reasonably dilute systems each dipole relaxes essentially independently of all the others, the kinetics are first order.

The shape of the ITC or displacement current vs. temperature peak reveals both the activation energy E and the reciprocal frequency factor τ_o for the relaxation process. The area under the curve is equal to the total polarization in the crystal which is proportional to the product of μ^2 and N_d. Therefore, the ITC measurement yields both kinetic parameters and if the dipole moment is known, the concentration of dipoles. The ITC equation may be written as

$$I_d = AP_o/\tau_o \cdot \exp[-E/kT]\exp[-(b\,\tau_o)^{-1}\int_{T_o}^{T} \exp(-E/kT')dT'] \qquad (2)$$

where A is the area of the specimen, P_o is the saturation polarization given by Eq. 1 and b is the heating rate. In the low-temperature shoulder of the peak it can be seen that Eq. 2 reduces to

$$I_d = AP_o/\tau_o \cdot \exp[-E/kT] \qquad (3)$$

and, indeed, in the absence of overlapping processes E can be obtained with reasonable accuracy from the slope of the log I_d vs. 1/T plot. Alternatively, a more elaborate analysis involves the remaining polarization in the crystals:

$$E = kT\,\ell n[A_t^{\infty}/I_d(t)\,\tau_o] \qquad (4)$$

where A_t^{∞} is the area under the ITC curve from t (or T) at which $I_d(t)$ is measured to ∞. Perhaps the most consistently accurate method of analysis is a least squares computer fit to the ITC equation. All of the results tabulated in this paper were obtained using such a computer analysis.

Table I

ITC Data Reported for $MF_2:RE^{3+}$

Crystal	b(K sec^{-1})	T_m(K)	E(eV)	τ_o(sec)	Ref.
$CaF_2:Ce^{3+}$	0.2	147	0.46	6 x 10^{-15}	a
		∿150	0.535	3.9x10^{-17}	b
	0.15		0.45	1 x 10^{-14}	c
$CaF_2:Pr^{3+}$		∿150	0.540	7.8x10^{-17}	b
	0.15		0.47	2 x 10^{-15}	c
$CaF_2:Nd^{3+}$		∿150	0.560	1.1x10^{-17}	b
$CaF_2:Sm^{3+}$		∿150	0.540	7.8x10^{-18}	b
	0.15		0.42	2 x 10^{-14}	c
$CaF_2:Eu^{3+}$		∿150	0.540	2.0x10^{-17}	b
	0.15		0.40	1 x 10^{-13}	c
$CaF_2:Gd^{3+}$		∿150	0.510	1.4x10^{-16}	b
	0.203	139.8	0.395	1.2x10^{-13}	d
$CaF_2:Tb^{3+}$		∿150	0.515	8.0x10^{-17}	b
	0.15		0.39	1 x 10^{-13}	d
$CaF_2:Dy^{3+}$		∿150	0.530	2.2x10^{-17}	b
$CaF_2:Ho^{3+}$		∿150	0.555	3.2x10^{-18}	b
$CaF_2:Y^{3+}$	0.033	131	0.41	2.3x10^{-14}	e
$CaF_2:Er^{3+}$	0.203	∿150	0.550	3.5x10^{-18}	b
		137.9	0.380	2 x 10^{-13}	d
$CaF_2:Yb^{3+}$		∿150	0.560	8.4x10^{-19}	b
$CaF_2:Al^{3+}$	0.15		0.36	10^{-12}	c
$SrF_2:Eu^{3+}$	0.15		0.26	2 x 10^{-9}	c
	0.15		0.28	5 x 10^{-8}	c
	0.15		0.33	2.5x10^{-9}	c

[a] B.S.H. Royce and S. Mascarenhas, Phys. Rev. Lett. 24 98 (1970).
[b] D.R. Stiefbold and R.A. Huggins, J. Solid State Chem. 5 15 (1972).
[c] J. Wagner and S. Mascarenhas, Phys. Rev. B6 4867 (1972).
[d] J.P. Stott and J.H. Crawford, Phys. Rev. B4 668 (1971).
[e] I. Kunze and P. Muller, Phys. Stat. Sol. (A) 13 197 (1972).

The experimental equipment includes a high impedance current measuring instrument such as a vibrating reed electrometer, a programmed control for heating at a prescribed, linear rate and a high impedance specimen chamber whose temperature can be cycled from ∿ 50K to 300K.

RESULTS AND DISCUSSION

The first ITC measurement on a $CaF_2:RE^{3+}$ specimen was reported in 1970 by Royce and Mascarenhas.[3] They found that E = 0.46 eV for a rather heavily doped $CaF_2:Ce^{3+}$ specimen. In view of the fact that all previous values for E based upon dielectric loss and mechanical relaxation measurements were greater than 1 eV,[4,5] this value was considered surprisingly low. However, Franklin and Marzullo,[6] using both dielectric loss and EPR line broadening, obtained a value of 0.38 eV for reorientation in $CaF_2:Gd^{3+}$. Stott and Crawford[7,8] confirmed the low E value from their ITC studies on $CaF_2:Er^{3+}$ (0.38 eV) and $CaF_2:Gd^{3+}$ (0.40 eV). They also observed a peak at a much lower temperature (54K and 60K in $CaF_2:Er^{3+}$ and $CaF_2:Gd^{3+}$) which was enhanced when the crystals were subjected to treatments (uv irradiation or quenching) which enhanced the concentration of isolated RE^{3+}. This relaxation will be discussed in a later section. Additional studies have been made by several other workers and these are summarized in Table I where the values of both E and τ_o are listed. Except for the values reported by Stiefbold and Huggins[9] which are substantially higher than the rest, the values of E seem to cluster near 0.40 eV for most of the ions studied.

In work carried out in our laboratory great care was taken to eliminate any temperature gradient across the specimen during the ITC run. This was done by introducing an exchange gas into the specimen chamber and keeping the heating rate small ($\leq$ 0.1 K sec^{-1}). The result of this precaution was enhanced precision and accuracy in the ITC measurement. A typical curve obtained on CaF_2 containing 0.1 mole% Gd^{3+} is shown in Fig. 1. The circles represent the experimental data and the solid curve represents the computer fit to the ITC equation. The value of E obtained from computer analysis was reproducible to better than 0.01 eV and τ_o was consistent to better than one half order of magnitude for numerous runs on a given sample and for several samples cleaved from two different crystals. This precision technique was applied to CaF_2 crystals doped with a variety of trivalent ions and the results are tabulated in Table II. The values of both E and τ_o are remarkably consistent even though the impurity ion radius ranges from 1.1 to 1.3 Å. There is a small but definite increase in E (increase in τ_o) with decreasing ion size. The values of E range from 0.42 eV for the large ions to 0.38 eV for the smallest, namely Yb^{3+}.

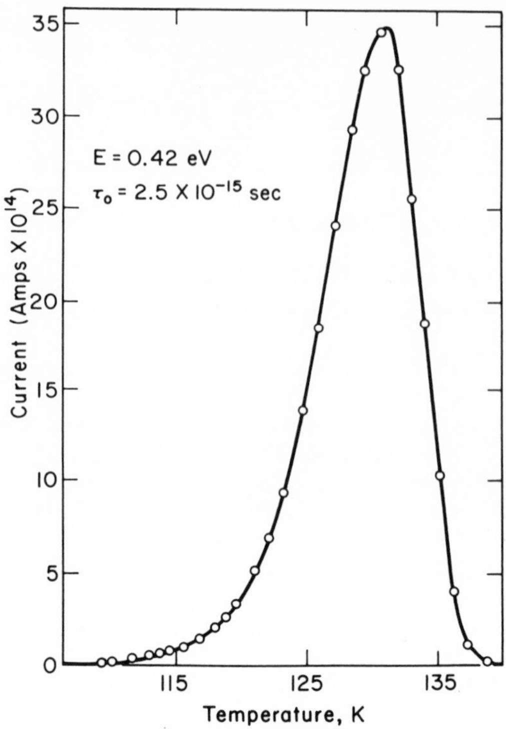

Fig. 1: The principal ITC peak for CaF_2:0.1 mole % Gd^{3+}; T_p = 135 K, ε_p = 3000 V/cm, b = 0.105 K/sec.

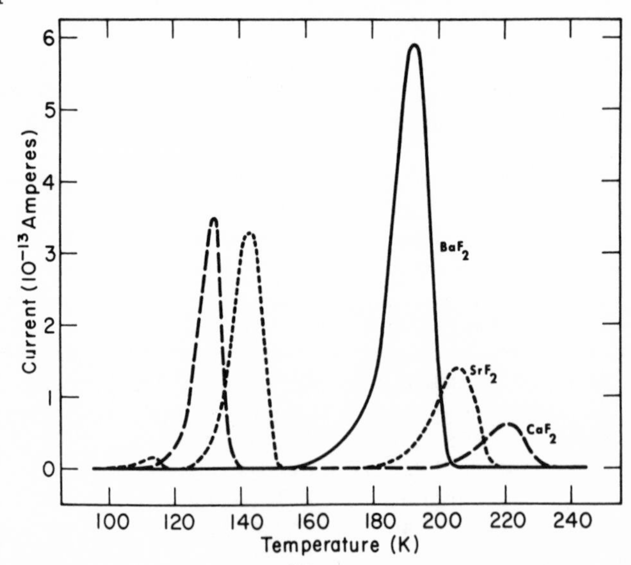

Fig. 2: ITC spectra for MF_2:RE^{3+}. Solid curve BaF:0.05 mole % Gd^{3+}; dashed curve GaF_2:0.1 mole % Gd^{3+}; dotted curve Srf_2:0.1 mole % Gd^{3+}.

Table II

ITC Data for Type I Dipolar Complexes in $CaF_2:RE^{3+}$

Dopant	RE^{3+} Ionic Radius (Å)	Nominal RE Concentration (10^{18} cm^{-3})	T_m(K)**	Dipole Concentration (10^{18} cm^{-3})	E(eV)	τ_o (10^{-15} sec)
U	1.3	12	130	0.83	0.422	3.3
Ce	1.28	12	132	1.9	0.420	6.0
Nd	1.26	12	131	3.8	0.420	5.3
Eu	1.21	49	131	0.46	0.426	1.4
Gd	1.20	24	128	7.0	0.420	2.5
Gd	1.20	12	128	5.6	0.426	1.0
Dy	1.17	12	126	7.3	0.411	2.7
Y	1.155	24	127	5.7	0.412	3.0
Y	1.155	2.4	127	1.5	0.415	2.2
Er	1.14	24	126	3.8	0.406	4.1
Tm	1.13	12	121	4.4	0.384	6.0
Yb	1.12	12	124	4.8	0.390	10

------

*All dopants were trivalent except Eu, which was predominantly Eu^{2+}.

**T_m at a heating rate of 0.05 K sec^{-1}, except 0.1 K sec^{-1} for Eu.

Table III

ITC Data for Type I and Type II Dipolar Complexes in $MF_2:Gd^{3+}$

Crystal	Complex	$T_m(K)$*	Dipole Concentra- (10^{18} cm^{-3})	E(eV)	τ_o(second)
$CaF_2:Gd^{3+}$	I	131	7.0	0.42	2.5×10^{-15}
0.1 mole %	II	221	0.7	0.69	1.0×10^{-14}
$SrF_2:Gd^{3+}$	I	143	8.2	0.45	1×10^{-14}
0.1 mole %	II	206	2.0	0.62	4×10^{-14}
$BaF_2:Gd^{3+}$	I	173	0.6	0.46	1×10^{-12}
0.05 mole %	II	193	5.8	0.60	1×10^{-14}

*T_m at a heating rate of 0.1 K sec^{-1}.

EPR and ENDOR studies have long since established[10,11] that in the absence of oxygen that the dominant complex in CaF_2 has tetragonal symmetry, i.e. the F_i^- is located in the nearest-neighbor interstitial position and denoted as Type I. Thus it is safe to conclude that the peak near 130 K is associated with the relaxation of Type I complexes. However, a second small peak of unknown origin was observed near 220 K. The significance of this peak becomes apparent when the ITC curve of $CaF_2:Gd^{3+}$ is compared with those of $SrF_2:Gd^{3+}$ and $BaF_3:Gd^{3+}$. This is shown in Fig. 2. In $SrF_2:Gd^{3+}$ two peaks are also observed but these are more closely spaced and more nearly the same amplitude. Only one peak is apparent for BaF_2. However, by choosing a polarizing temperature near the low temperature edge of the BaF_2 peak, two peaks could be resolved as shown in Fig. 3. A clue to the origin of this structure can be gained from EPR measurements also. In BaF_2 the trigonal complex (Type II) in which the F_i^- occupies a next-nearest neighbor position is dominant[12–14] whereas in SrF_2 there is a mixture of Type I and Type II with the latter being in the minority.[12,13] Measurements in our laboratory by M. Ikeya indicate a trace of tetragonal structure in the EPR spectrum of BaF_2 and a trace of trigonal in the spectrum of CaF_2. Although these spectral components are too weak to analyse quantitatively, it is reasonable to conclude that they originate from a small admixture of Type I and Type II complexes respectively. On the basis of this correlation we conclude that the low-temperature component of the pair of peaks is associated with Type I dipoles and the upper temperature peak is associated with Type II. This assignment also explains the very large amplitude of the composite $BaF_2:Gd^{3+}$ peak: Because $P_o \propto \mu^2$, each trigonal dipole makes a contribution three times that of each tetragonal dipole. Therefore, even though the nominal doping level for CaF_2, SrF_2 is 0.1 mole % Gd^{3+} as compared to 0.05 mole % Gd^{3+} in BaF_2, the BaF_2 peak is the largest as expected. The relaxation parameters for Type I and Type II Gd^{3+}-F_i^- dipoles for the three materials are summarized in Table III.

Stott and Crawford[7,8] suggested that the very low temperature peaks at 54 K and 60 K in $CaF_2:Er^{3+}$ and $CaF_2:Gd^{3+}$ respectively arises from the reorientation of Type II defects. This tentative identification was made because specimen treatment (uv irradiation[15] of $CaF_2:Er^{3+}$ and quenching of $CaF_2:Gd^{3+}$) which causes an increase in the intensity of the cubic EPR spectrum at the expense of the tetragonal one enhanced substantially the low temperature peaks, and it turns out that a low intensity trigonal spectrum might be missed in the presence of the cubic spectrum. In view of the experiments described above on CaF_2, SrF_2 and BaF_2, this tentative identification now seems untenable and the low temperature ITC peaks must have some other origin. One possible interpretation is that these peaks are indeed attributable to the isolated, cubic RE^{3+} centers which could possess a dipole moment by sitting off-center in the Ca^{2+} site. This hypothesis is not unreasonable because the ions in question are

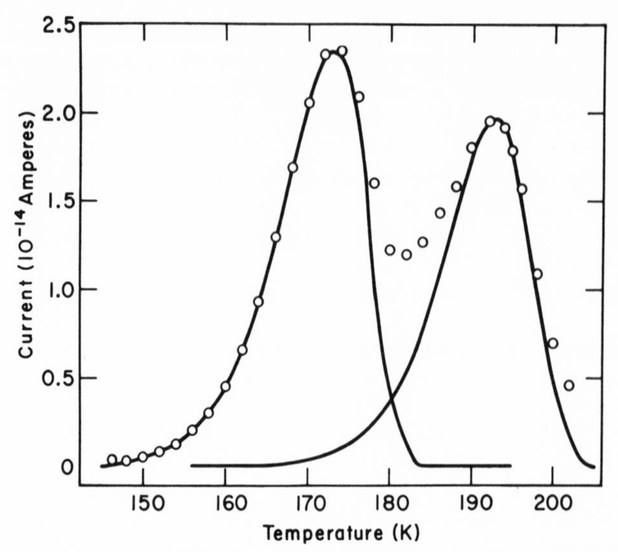

Fig. 3: Resolution of the two ITC peaks in $BaF_2:Gd^{3+}$ by polarizing at $\sim$ 160 K. The 193-K peak under such a treatment has an area only 2% of its saturation area that results from polarization near 190 K.

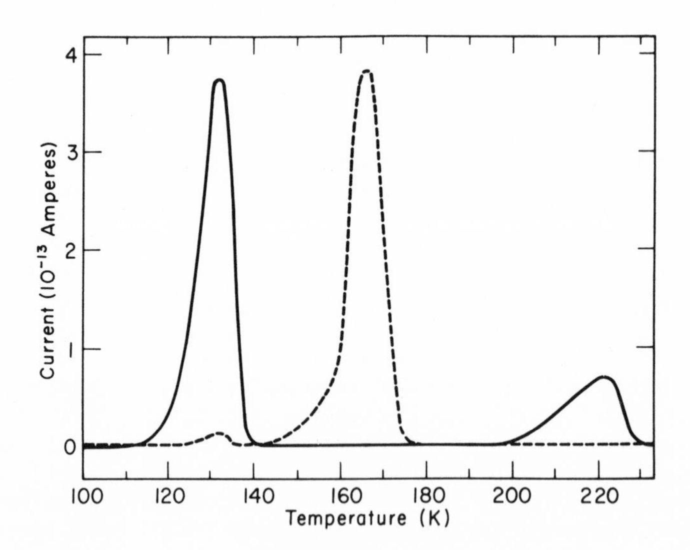

Fig. 4: ITC spectra for $GaF_2:Gd^{3+}$ before (solid curve) and after (dashed curve) hydrolysis.

appreciably smaller than Ca^{2+}. To date there has been no reported EPR evidence that the isolated RE^{3+} ion occupies an off-center site, and on the chance that this may have escaped detection we are currently reinvestigating this possibility.

Extensive studies of hydrogenated crystals using infrared[16,17] and EPR[17] spectroscopy show that the $F_i{}^-$ in the RE^{3+}-$F_i{}^-$ complex can be partially replaced by $H_i{}^-$. Since reorientation of this dipole is also of interest, attempts were made to hydrogenate $CaF_2:Gd^{3+}$ by heating to 800°C in a quartz tube in the presence of flowing hydrogen. The resultant ITC curve (dashed) is compared with the ITC curve (solid) obtained before hydrogen treatment in Fig. 4. The Type I and Type II relaxations are almost completely eliminated and are replaced by a single, well defined peak ($E = 0.49$ eV and $\tau_0 =$ 5×10^{-14} sec.) at 166 K. Although we initially assumed that this peak was associated with the RE^{3+}-$H_i{}^-$ dipole relaxation, an EPR assay[18] revealed that no $H_i{}^-$ was in the crystal. Instead essentially all of the Gd^{3+} was found to be in two trigonal complexes T_1 and T_2 both known to contain oxygen. The structure of these complexes has been determined by the ENDOR method.[19] The T_1 center has the formula $Gd^{3+}\,O_4{}^{2-}F^-$ in which the four O^{2-} ions occupy four F^- sites adjacent to the Gd^{3+} which form the corners of a tetrahedron. The remaining tetrahedron of sites is vacant except for the presence of a single F^- ion. The T_2 center is simply $Gd^{3+}\,O^{2-}F_7{}^-$. Of these two structures only T_1 is able to reorient in an applied field and this reorientation occurs by the interchange of positions of the F^- ion with one of the three vacancies in the tetrahedron of sites adjacent to the impurity. Therefore, the new relaxation peak at 166 K is attributed to the T_1 structure.[20] This is the most complex dipolar imperfection whose relaxation has been identified by ITC.

SUMMARY

To summarize briefly it is evident from the results presented here that ITC is a valuable method for investigating the relaxation of dipolar defects in the fluorite structure. Type I or nearest neighbor RE^{3}-$F_i{}^-$ dipoles originating from a large number of different RE^{3+} ions in CaF_2 have been studied and it is found that there is a small decrease in E and a corresponding increase in τ_0 as the size of these ions decreases. Although Type I dipoles dominate in $CaF_2:Gd^{3+}$, a trace of Type II (next-nearest neighbor) dipoles which relax with a larger activation energy, is also present. In $BaF_2:Gd^{3+}$ the situation is inverted: Type II dipoles are dominant and only a trace of Type I dipoles is detected. In $SrF_2:Gd^{3+}$ the ITC peaks for both types of dipoles are readily detected. Finally, a relaxation peak associated with the trigonal center $T_1(Gd^{3+}\,O_4{}^{2-}F^-)$ as identified by ENDOR has been detected in $CaF_2:Gd^{3+}$ crystals which have been given a hydrolysing heat treatment.

REFERENCES

1. For a review of the relaxation modes and defect structures see A.S. Nowick, in Point Defects in Solids, J.H. Crawford and L.M. Slifkin, eds., Plenum Press, New York (1972) ch. 3.

2. C. Bucci, R. Fieschi, G. Guidi, Phys. Rev. 148 816 (1966).

3. B.H.S. Royce, S. Mascarenhas, Phys. Rev. Lett. 24 98 (1970).

4. J.H. Chen, M.S. McDonough, Phys. Rev. 185 453 (1969).

5. P.D. Southgate, J. Phys. Chem. Solids 27 1623 (1966).

6. A.D. Franklin, S. Marzullo, J. Phys. Chem. 3 L171 (1970).

7. J.P. Stott, J.H. Crawford Jr., Phys. Rev. Lett. 26 384 (1971).

8. J.P. Stott, J.H. Crawford Jr. Phys. Rev. B4 668 (1971).

9. D.R. Stiefbold, R.A. Huggins, J. Solid State Chem. 5 15 (1972).

10. B. Bleaney, P.M. Llewellyn, D.A. Jones, Proc. Phys. Soc. B69 858 (1956) U. Ranon, W. Low, Phys. Rev. 132 1609 (1963).

11. J.M. Baker, E.R. Davis, J.P. Hurrell, Proc. Roy. Soc. A308 403 (1968).

12. J. Sierro, Phys. Lett. 4 178 (1963).

13. U. Ranon, A. Yaniv, Phys. Lett. 9 17 (1964).

14. L.A. Boatner, R.W. Reynolds, M.M. Abraham, J. Chem. Phys. 52 1248 (1970).

15. J.W. Twidell, J. Phys. Chem. Solids 29 1269 (1968).

16. D.N. Chambers, R.C. Newman, Phys. Stat. Sol. 35 685 (1969).

17. G.D. Jones, S. Peled, S. Rosenwaks, S. Yatsiv, Phys. Rev. 183 353 (1969).

18. We are indebted to G.D. Jones and A. Edgar, University of Canterbury, Christchurch, New Zealand, for providing these EPR measurements.

19. T.Rs. Reddy, E.R. Davis, J.M. Baker, D.N. Chambers, R.C. Newman, B. Ozbay, Phys. Lett. 36A 77 (1971).

20. E.L. Kitts Jr., J.H. Crawford Jr., Phys. Rev. Lett. 30 443 (1973).

INTRINSIC POINT DEFECTS IN OXIDES WITH THE RUTILE STRUCTURE*

R.A. Weeks

Solid State Division, Oak Ridge National Laboratory

Oak Ridge, Tennessee

The oxides which are of primary concern here have the rutile structure and are formed from elements of the fourth (Si, Ge, Sn, Ti) group of the Periodic Table. Instrinsic "point" defects of these oxides, oxygen and cation vacancies, cation interstitials and interstitial oxygen molecule ion complexes, have optical and magnetic properties which are a function of their charge state. Properties of the various charge states of oxygen vacancies can be described in terms of linear combinations of s and p states of nearest-neighbor cations, with the exception of TiO_2*, for which a linear combination of d orbitals of the three nearest-neighbor Ti ions is required. Paramagnetic states of oxygen vacancies in these oxides have been observed, and comparison of their properties and rates of production by reactor irradiation exhibits many differences. Electron paramagnetic resonance spectra which have been observed for singly charged oxygen vacancies in glassy* SiO_2 *and* GeO_2 *are compared with those produced in the rutile phases. Configurations of interstitial oxygen ions have several forms, data about which have been derived from electron paramagnetic resonance spectra. Definitive models which have been deduced in a few cases, e.g.,* O_2^- *and* O_3^- *molecule ions, are discussed. Some reasons for the failure to detect either optical absorption bands or paramagnetic states of interstitial cations, with the exception of* Ti^{3+} *and interstitial oxygen ions in the case of* TiO_2 *are discussed.*

---

*Research sponsored by the U.S. Atomic Energy Commission under contract with Union Carbide Corporation.

INTRODUCTION

Diamagnetic insulating oxide solids have widespread technical applications. Physical properties which are important in many applications are frequently determined by defects, either impurities in, or defects of, the crystalline structure such as vacancies. Hence, an understanding of the properties of such defects and the processes by which they are produced is necessary in determining conditions of use. Those oxides formed from elements of the second, third, and fourth groups of the Periodic Table are the most abundant, and hence are the ones which have been most frequently used and whose defect properties have been most extensively studied. Despite these studies, there is a lack of coherence to the knowledge about defects such as vacancies and interstitials and the processes by which they are produced. The purpose of the following discussion will be to show that, in one small class of oxides, intrinsic point defects produced by nuclear reactor radiation have a wide spectrum of types and widely variant production rates.* The data which will be presented will confirm the statement about the lack of coherence but will, perhaps, be suggestive of new directions for research.

The oxides about which the discussion will center are TiO_2, GeO_2, and SnO_2. Each of these may exist at room temperature in several forms, but only the tetragonal or rutile form, common to all three, will be considered. A fourth member of this group, the mineral stishovite (SiO_2), is extremely rare. It has been identified in quartz sands subjected to shock waves produced by nuclear explosions.[1] In order that the data which will be presented about intrinsic point defects in these compounds be viewed in a wider context, reference will also be made to intrinsic point defects in the cubic alkaline-earth oxides. For two types of defects, singly charged anion and cation vacancies, in these compounds, the data base is extensive[2] and models of these defects, based on experiment and theory, are incontrovertible. A similar experiment base and theoretical analysis have not been developed for similar types of defects, or for any other types, in the tetragonal oxides. Thus it will be said that the experimental data to be discussed below are attributed to a certain type of defect. This term means that the available data are consistent with properties of defects deduced from the symmetries of the crystal structure, with properties observed under conditions which exclude defects which are not intrinsic, and that similarities exist between the data and data on defects in other oxide crystals (e.g., the alkaline-earth oxides).

Intrinsic defects frequently have paramagnetic electron states, and investigations of these states by electron paramagnetic resonance (EPR) and electron-nuclear double resonance (ENDOR) spectroscopy have

*Extended defects, e.g. clusters of vacancies or of impurities, will not be discussed.

provided the data upon which their identification often has been based. Excellent examples are the singly charged oxygen (the F^+ center) and cation (the V center) vacancies in the alkaline-earth oxides[2,3] for which EPR and ENDOR spectroscopy have provided extremely detailed models. Theoretical models of the F^+ center are in agreement with experiment, but the observed properties of the V centers are somewhat unusual, and theoretical analysis has yet to provide an explanation. The unusual property of the V center is the cylindrical (axial) symmetry of the paramagnetic state[3,4,5] which is due to the localization of the paramagnetic state on one of the six oxygen ions surrounding the vacancy. (The paramagnetic state does move from one oxygen to another but at room temperature resides on one of the oxygen ions for a time long compared with the reciprocal of spectrometer frequency, typically 10^{-10} sec.) This defect could be considered as an O^- ion at one of the six oxygen sites with the crystal electric field at the ion, due in part to the cation vacancy, partially removing the degeneracy of the oxygen p states.

One means for producing defects in oxide crystalline compounds is irradiation with energetic particles, e.g., electrons, protons, or neutrons, and a convenient source for neutrons is a nuclear reactor. Ions in oxides displaced from their sites, with the exception of ions on the surface of a sample, remain within a sample. These ions may exist as isolated ions in interstitial sites, as molecular species in interstitial sites if they react chemically, as larger defect complexes, or they may recombine with vacancies. Also anion and cation vacancies can combine to form complexes of vacancies. If, for example, defects such as the F^+ center are present after an irradiation, then the displaced oxygen ion, O^- since one of the oxygen electrons is trapped at the F center forming the F^+ center, must be trapped elsewhere in the sample. This ion is paramagnetic and it might be expected that its presence would be detected in the EPR spectrum. The properties of the EPR spectrum of O^- in a variety of crystal electric fields have been discussed.[6] Such spectra have not been observed in any of the alkaline-earth oxides, but spectra due to oxygen ions in molecular complexes have been observed in some oxides[6] and, as will be shown below, in one of the tetragonal oxides.

Intrinsic point defects in the three tetragonal oxides have been produced by reactor irradiations and some of their properties deduced from their EPR spectra. These properties will be discussed. The primary mechanism for producing defects in oxides is displacement of ions by energetic particles*, hence another parameter of interest is the number produced per incident energetic particle (> 0.1 MeV). In the cases discussed here, reactor irradiations were

---

*Photoionization processes such as are effective in producing defects in alkali halides are not of primary importance in oxides.

made to produce defects. It has usually been assumed that a large fraction of the ions displaced by particle irradiations recombine with vacancies, since the number of defects produced per incident neutron, observed by such techniques as EPR spectroscopy, is only a small fraction of the number predicted. However, the production rates in the tetragonal oxides provide interesting and, perhaps, surprising data.

TETRAGONAL OXIDES, TiO_2, GeO_2, AND SnO_2

The rutile structure belongs to the space group D_{4h}^{14} = P4/mnm, with two equivalent cations located in D_{2h} sites and unit cell dimensions in Å for these three oxides are: GeO_2 (a = 4.40, c = 2.86), TiO_2 (a = 4.58, c = 2.95), SnO_2 (a = 4.75, c = 3.17). Each cation is six-fold coordinated and each oxygen is three-fold coordinated with point symmetry C_{2v}. Two of the four oxygen sites per unit cell are related to the other two by the same symmetry operator as the cations. Interstitial sites have point symmetries lower than cylindrical and may be either tetrahedrally or octahedrally coordinated. Hence, the distribution of charge around such sites should be sufficient to remove either p or π orbital degeneracies of either interstitial O^- ions or O_2^- molecule ions, respectively. Interstitial cations would be paramagnetic if their charge state were +3, i.e., Ti^{3+}, Ge^{3+}, or Sn^{3+}. In these cases, the paramagnetic electrons are in 3d, 4s, and 5s orbitals, respectively.

<u>TiO_2</u>. An illustration of some of the characteristics of paramagnetic defect states in rutile structures is found in irradiated TiO_2 single crystals.[10] Single crystals of TiO_2 irradiated in a reactor up to an integrated neutron flux of $\sim 10^{20}$ cm^{-2} at a temperature of $\sim$ 350 K exhibit a slight darkening due to an increase in optical absorption, which increases monotonically with increase in photon energy. No optical absorption bands are resolved and the EPR spectrum is very weak ($< 10^{15}$ spins cm^{-3}).[11] However, crystals irradiated in a reactor at a temperature of 140 K to an integrated flux of 10^{17} fast neutrons cm^{-2} have an intense blue color and an intense EPR spectrum. For such an integrated flux, the concentration of spins in the most intense components of the spectrum is $\gtrsim 10^{18}$ cm^{-3}; i.e., 10 paramagnetic states are produced per incident neutron. In a cubic oxide such as MgO and in an open-network oxide such as α-quartz, $\sim$ 2 per incident neutron are produced with the same irradiation conditions.[12,13] For reasons based upon (i) the principal values of the g-tensor (Table I), (ii) the lack of hyperfine interactions with the nuclei of the titanium isotopes ^{47}Ti and ^{49}Ti, (iii) the concentration relative to the concentration of known impurities, and (iv) the symmetry of the orientation dependence of the EPR components, the defect is attributed to a singly charged oxygen vacancy.

The temperature dependence of the intensity of this component

Table I

Magnetic Resonance Parameters of Radiation-Produced Intrinsic Defects with Spin State S = 1/2 in TiO_2

Center	g-tensor						Sites per unit cell	Model
	$g_{[001]}$	$g_{[110]}$	$g_{[110]}$	g_1	g_2	g_3		
S	1.988	1.968	1.983				2	Singly charged oxygen vacancy
A	1.950	1.950	1.991				2	Perturbed singly charged oxygen vacancy
I	1.971			1.973[1]	1.988[1]		4	Interstitial
X				1.987[2]	1.980[2]	1.966[2]	8	Interstitial

------

1. g_1 and g_2 are in the basal plane and are ±12° with respect to the [010] and [100] directions.
2.

g	Angles (degree) [110]	[110]	[001]
1	36.1	125.3	96.4
2	57.0	98.6	59.0
3	102.9	118.5	31.8

and the effect of subsequent ^{60}Co irradiations at 78 K have been tentatively explained[14] as due to the following effects: 1. Electrons from other defects thermally excited between 113 and 183 K are trapped by singly charged oxygen vacancies converting them to a diamagnetic state. 2. These doubly charged defects lose an electron by thermal excitation between 183 and 225 K. 3. At higher temperatures the displaced oxygen ions recombine with the vacancies and, hence, irradiations at temperatures > 225 K are ineffective for producing vacancies.

In addition to the singly charged vacancies which are observed after low-temperature reactor irradiation, at least three other paramagnetic defects are produced whose concentrations are significant (i.e., $> 10^{16}$ cm^{-3}). These defects (Table I) also have a rather complex thermal dependence, only one of which will be discussed here. This defect (the X center in Table I) is not immediately detected after irradiation, but only after warming to ~ 300 K, cooling to 78 K, irradiating with gamma rays, and measuring without warming above 100 K. The X-center concentration after this sequence of treatments is approximately equal to the concentration of the S center, the singly charged oxygen vacancy, immediately after reactor irradiation. The EPR parameters (Table I) are consistent with those expected for Ti^{3+} in an interstitial site. If the X center is interstitial Ti^{3+}, then it is apparent that reactor irradiation at ~ 140 K will produce approximately equal concentrations of oxygen and titanium vacancies. However, none of the paramagnetic centers which have been detected have properties which are expected[13] for interstitial oxygen ions or titanium vacancies. (Measurements were made on crystals irradiated at ~ 140 K, cooled to 78 K for a first series of measurements and then cooled to 1.2 K for additional measurements). The concentration of interstitial O^- ions and triply charged titanium vacancies should be approximately equal to the concentration of singly charged oxygen vacancies and interstitial Ti^{3+} ions, respectively, which are generated by displacements. In the available low-symmetry interstitial sites the oxygen p states would be nondegenerate. Hence, EPR spectra of such ions should be detectable. If interstitial O^- ions should interact to form O_2^{2-} molecule ions, then no resonance spectra would be detected, since this molecule ion is diamagnetic. Such a solid-state chemical reaction would explain the absence of EPR spectral components due to O^- ions. Furthermore, if this reaction occurs, the effects of annealing indicate that the O_2^{2-} molecule ion dissociates at $T > 225$ K. The triply charged Ti vacancies should also be paramagnetic and have an O^- ion-type EPR spectrum. No explanation of the failure to detect triply charged cation vacancies can be presented at this time.

There are many facets of the TiO_2 defect problems which have not been investigated, such as: (i) the thermal activation energies for the effects described above, (ii) the relation between optical bands and paramagnetic states, (iii) the photoconductive effects of these defects, and (iv) the distinct differences between the defects produced by particle irradiation and those produced when crystals are heated in a reducing atmosphere.[15] Most crystals, for which there are data, have relatively high concentrations of impurities ($\sim 10^{17}$ cm^{-3}) which does make the analysis for intrinsic defects somewhat ambiguous. Preparation of high-purity (impurities < 20 ppm) crystals would reduce the difficulties of such investigations.

Tetragonal GeO_2 and SnO_2. Large ($\sim$ 0.1 cm^3) single crystals of this phase of GeO_2 are difficult to grow and consequently, investigations of intrinsic defects in this structure have been made on crystalline powders.[16] Thus the data on the EPR spectra of paramagnetic defects produced by reactor irradiation are less specific with respect to the defect sites in the crystal structure. The EPR "powder patterns"[17] of paramagnetic defects produced by irradiation can all be described as due to three S = 1/2 states in crystal sites for which the g-tensors are isotropic, axial, and rhombic. These defects are intrinsic, since their concentrations for large integrated neutron fluxes ($> 10^{18}$ cm^{-2}) exceed the concentrations of all impurities which have been detected. The defects with isotropic and axial g-tensors are also produced in low concentrations ($< 10^{15}$ cm^{-3}) by ^{60}Co gamma-ray irradiation; the defect with a rhombic g-tensor is detected only after particle (reactor and electron) irradiations. The EPR parameters of these defects are given in Table II. The models which are given in the table were developed on the basis of many considerations which are given in Ref. 16 and will not be reported here. Data on the effect of irradiating at low temperature (<150K) are not available. These intrinsic paramagnetic defects are stable at room temperature but disappear at temperatures < 350K.[18]

One of the paramagnetic states, the "1" complex in Table II, is attributed to singly charged oxygen vacancies. The symmetry of the oxygen site is C_{2v} and hence one might expect the paramagnetic state of the vacancy to have a g-tensor with approximately axial symmetry as does the E' center in SiO_2 and GeO_2[13,16] for which the symmetry is approximately C_{3v} (i.e., the symmetry of the SiO_3 unit on which the defect electron is localized). However, the atom configurations are quite different, and the defect electron state, through contributions from the 4s electron states of the three Ge ions surrounding the oxygen vacancy in tetragonal GeO_2, differs from the defect electron state of the oxygen vacancy in the alpha quartz structure of GeO_2 and SiO_2. In this latter case, the defect state is approximately a sp^3 hybrid.[13,19] The F-center in MgF_2, which has the same structure as tetragonal GeO_2, has a site symmetry which is also C_{2v} and an isotropic g-value (g = 2.003).[20]

Perhaps the most interesting defect site is that attributed to some form of interstitial oxygen. The identification of this center

Table II

Intrinsic Paramagnetic Defects (S = 1/2) Produced by Reactor Irradiation of GeO_2

Structure	g_1	g_2	g_3	Model
Tetragonal (powder)	2.0034	2.0034	2.0047	Interstitial Ge^{3+}
	2.0020	2.0062	2.0312	Interstitial O_2^- or peroxy radical
	1.9980	1.9980	1.9980	Singly charged oxygen vacancy
Glass	2.002	2.008	2.051	Interstitial O_2^- or peroxy radical
	1.9944	1.9944	2.0010	Singly charged oxygen vacancy

with one of the paramagnetic forms of oxygen has been discussed at some length in Ref. 16, and the reader is referred to this report. This paramagnetic state and its variations, e.g., the defect listed in Table II for GeO_2 glass, are observed in many other oxide glasses and crystals.[13,19,21] An understanding of the varied properties of this type of defect and its role in the structure of oxide glasses and crystalline forms of the glasses, e.g., the tetragonal and alpha-quartz forms of GeO_2, is quite incomplete. There are, for example, a group of crystalline alkali-borate compounds in which solid-state chemical reactions between oxygen ions can produce an amorphous state and an extremely high concentration of O_2^- molecule ion complexes.[22,23] It may be that similar solid-state chemical reactions are induced by irradiation effects in the oxides considered here and in other oxide glasses.

Reactor irradiation of SnO_2 single crystals at low temperature ($\simeq$ 140 K) to integrated fluxes of fast neutrons > 10^{17} cm^{-2} did not produce any EPR spectral components attributable to intrinsic defects.[18] Subsequent irradiations with ^{60}Co gamma rays at 78 K also were ineffective in producing EPR spectral components.

DISCUSSION

Models for three intrinsic paramagnetic defects in tetragonal GeO_2 are based upon less extensive data than those for the defect models in TiO_2, and measurements on single crystals will be necessary to confirm them. Assuming that the models are correct, then two notable similarities between the defects produced in GeO_2 and TiO_2 are evident: (i) singly charged oxygen vacancies and triply charged interstitials are formed by knock-on displacement of ions, (ii) paramagnetic cation vacancies have not been identified in either. The notable differences between GeO_2 and TiO_2 are: (i) intrinsic paramagnetic defects are stable and unstable at 300 K in GeO_2 and TiO_2, respectively; (ii) paramagnetic states associated with displaced oxygen ions are found in GeO_2 but not in TiO_2. In contrast to these similarities and differences in the defects of TiO_2 and GeO_2, no defects are detected in SnO_2, even for reactor irradiations at 140 K.

Table III summarizes the available data on production rates of various types of defects in these three crystals. These defects have all been detected and their concentrations measured by EPR spectroscopy, hence all diamagnetic defects are excluded. Some unknown fraction of the vacancies and interstitials which are produced are probably diamagnetic, and thus these estimates are lower limits. Included in the table for comparison are data on some other oxides. This table shows that these oxides have an extremely wide range of defect production rates, and it is suggested that this wide range is due to solid-state chemical reactions that occur subsequent to displacement events. In the study of such events in metals, it has

Table III

Number of Defects Detected per Incident Fast (> 0.1 MeV) Neutron During Reactor Irradiation

Crystal	T(K) Cation	Vacancies (Interstitial Oxygen)	Oxygen Vacancies	Interstitial Cations	Ref.
Tetragonal					
TiO_2	140	$< 10^{-3}$	> 10	> 10	10
	325	$< 10^{-4}$	10^{-4}	?	11
GeO_2	325	10^{-2}	10^{-3}	10^{-3}	16
SnO_2	140	$< 10^{-4}$	$< 10^{-4}$	$< 10^{-4}$	18
		(not detected)	(not detected)	(not detected)	
α-quartz					
SiO_2	140	2	2	$<10^{-4}$ (not detected)	19
	325	1	1		19
GeO_2	140	1	10	1	16
Cubic					
MgO	140	$< 10^{-4}$	2	$<10^{-4}$ (not detected)	12
	325	$< 10^{-4}$ (not detected)	1	$<10^{-4}$ (not detected)	12

long been recognized that in order to understand similar subsequent reactions irradiation and measurement at very low temperatures (< 20 K) are required. It appears that similar types of experiments are required if defect structures, production rates, and annealing reactions are to be understood in these and other oxides such as the alkaline-earth and silicate oxides.

SUMMARY

It has been shown that the types of defects and their rates of production by reactor irradiation of three oxides with the same crystal structure are different. These effects are also quite different from the effects of reactor or particle irradiation on the cubic alkaline-earth oxides.[2] Although a part of the difference in production rates is due to differences between the particle-scattering cross sections of the ions which comprise the various compounds, most of the difference is probably due to interactions between displaced ions and between displaced ions and vacant ion sites, subsequent to displacement. The failure of experimental techniques heretofore utilized to detect some types of defects may also be responsible for some part of the differences in production rates. At the present time, there is little knowledge about these reactions, and they constitute a large problem to our understanding of intrinsic defects in these oxides and, to some extent, in the alkaline-earth oxides. Before these problems can be resolved, optical absorption bands due to the paramagnetic defects discussed above and to other intrinsic diamagnetic defects must be determined. Spectral techniques such as Raman laser spectroscopy may be quite useful in detecting the presence of molecule ions such as O_2^{2-}.

ACKNOWLEDGMENTS

Discussions with E. Sonder and J. Bates have been of particular value in the preparation of this paper.

REFERENCES

1. W.E. Hagston, J. Phys. C.: Solid St. Phys. 3 1233-1241 (1970).

2. B. Henderson, J.E. Wertz, Adv. in Phys. 17 749-855 (1968).

3. O.F. Schirmer, J. Phys. Chem. Solids 32 499-509 (1971).

4. W.P. Unruh, Y. Chen, M.M. Abraham, Phys. Rev. Lett. 30 446-449 (1973).

5. A.K. Garrison, R.C. Du Varney, Phys. Rev. B7 4689-4695 (1973).

6. J.M. Meese, Solid State Commun. 11 1547-1550 (1972).

7. J.R. Morton, Chem. Rev. 64 453-469 (1964).

8. E.C.T. Chao, J.J. Fahey, J. Littler, D.J. Milton, J. Geophys. Res. 67 419 (1962).

9. R.A. Weeks, T. Purcell, D. Prestel, Proc. 5th Lunar Sci. Conf.

10. T. Purcell, R.A. Weeks, Phys. Rev. 54 2800-2910 (1971).

11. R.A. Weeks, Bull. Am. Phys. Soc. 6 178 (1961).

12. Y. Chen, W.A. Sibley, F.D. Srygley, R.A. Weeks, E.B. Hensley, R.L. Kroes, J. Phys. Chem Solids 29 863-865 (1968).

13. R.A. Weeks, in Interaction of Radiation with Solids, Proc. First Cairo Solid State Conference, A. Bishay, ed., Plenum Press, New York (1967) 55.

14. R.A. Weeks, Abstract 217 in 1971 Int. Conf. Colour Centres in Ionic Crystals, Sept. 6-10, 1971, Reading, U.K.

15. P.F. Chester, J. Appl. Phys. 32 2233 (1961).

16. T. Purcell, R.A. Weeks, Phys. and Chem. of Glasses 10 198-208 (1969).

17. See J. McMillan, Chapters 9 and 10 in Electron Paramagnetism, Rheinhold Book Corp. (1968), for a discussion of "powder patterns" in EPR spectroscopy.

18. R.A. Weeks, unpublished data.

19. R.A. Weeks, in Introduction to Glass Science, L.D. Pye, H.J. Stevens, W.C. La Course, eds., Plenum Press, New York (1971) 137-165.

20. W.P. Unruh, L.G. Nelson, J.L. Lewis, J.L. kolopus, J. Phys. C. (G.B.) 4 2992 (1971).

21. D.L. Griscom, J. of Noncryst. Solids, to be published.

22. J.O. Edmonds, D-L. Griscom, R.B. Jones, K.L. Walters, R.A. Weeks, J. Am. Chem. Soc. 91 1095 (1969).

23. D.L. Griscom, P.J. Bray, private communication.

THE CAPTURE OF ELECTRONS BY NEGATIVELY CHARGED IMPURITY ATOMS IN N-TYPE GERMANIUM

E. F. El-Wahidy, Alexandria University, Egypt

A. G. Mironov, Moscow State University, U.S.S.R.

It is well known that the capture of an electron by an impurity atom is governed by two factors: (1) the electric charge on the impurity atom, and (2) the way in which the electron loses energy in its transition from the conduction band to a captured state. We have studied the dependence of capture rate on the external electric field at low temperature (20°K) for n-Ge containing gold centres as impurities.

The distribution was found by solving the Boltzman equation which takes into account the scattering of electrons by acoustic and optic modes of vibrations and impurity centres, in the presence of an external electric field. The capture rate was calculated using the resulting distribution function.

It was found that the increase of the capture rate with the field is not sufficiently rapid at 20°K to explain electron energy. The life time τ *of electrons changes from* $\tau = 4 \times 10^{-6}$ *sec at* $E = 50$ *volt/cm to the value* $\tau = 4 \times 10^{-7}$ *sec at* $E = 500$ *volt/cm. The life time is obtained for different values of the (small) parameter* ν. *Results show that* τ *varies slowly with* ν *which confirms that the barrier effect is predominant. This agrees with the results of other workers.*

INTRODUCTION

In the present work, we consider the effect of a uniform electric field $\overline{E}$ on the capture rate of electrons by negatively charged gold atoms in n-Ge at low temperatures, T = 20°K. It is assumed that electrons are generated from the three charged gold centers by means of a monochromatic beam of light and are captured on the doubly charged ones.

Under the influence of the applied electric field, electrons with momentum $p < p_o$ are accelerated up to a momentum $p = p_o$ where they will be able to emit optical phonons

$$p_o = \sqrt{2 m \hbar \omega_o}$$

where $\hbar\omega_o = 0.036$ eV, optical phonon energy. The upper limit of the field strength is restricted to the extent that electrons in the passive region (with energy $\varepsilon < \hbar\omega_o$) can make several elastic collisions before reaching $\hbar\omega_o$ and their distribution function thus remains nearly isotropic.[1] This restriction implies that

$$E \ll p_o / e\tau_1(\varepsilon)$$

where $\tau_1(\varepsilon)$ is the momentum relaxation time in the passive region. The emission of optical phonons by electrons with energies $\varepsilon \geq \hbar\omega_o$ is considered to occur so quickly that the distribution function of electrons in the active region ($\varepsilon > \hbar\omega_o$) remains vanishingly small as the electron energy exceeds the optical phonon energy by kT. We assume that

$$\tau_2(\varepsilon) \ll \tau_1(\varepsilon)$$

where $\tau_2(\varepsilon)$ is the relaxation time in the active region. The probability of emission of optical phonons by electrons with energy $\varepsilon \geq \hbar\omega_o$ follows Conwell:[2]

$$\tau_2^{-1}(\varepsilon) = M_o^2 \sqrt{\varepsilon - \hbar\omega_o}\ \theta(\varepsilon - \hbar\omega_o)$$

where M_o is the matrix element of interaction of electrons with optical phonons and $\theta(\varepsilon-\hbar\omega_o)$ is the unit step function:

$$\theta(x) = \begin{cases} 1, & x > 0 \\ 0, & x < 0 \end{cases}$$

THE DISTRIBUTION FUNCTION

The determination of the capture rate is reduced to the problem of finding the distribution function. The usual approach in finding the distribution function is based on an appropriate solution of the Boltzmann transport equation. In our work, the steady state momentum distribution function $f(\overline{p})$ was determined by the following equation:

$$-e\,\overline{E}\,\overline{\nabla}_p\, f(\overline{p}) = (\delta f/\delta t)_{coll.} + \mathfrak{I}\,\delta(\varepsilon - \hbar\Omega_o + E_t) - f(\overline{p})/\tau_r(\varepsilon) + I_{opt} \qquad (1)$$

where $(\delta f/\delta t)_{coll.}$ describes the rate of change of $f(\overline{p})$ due to scattering by acoustical phonons and by impurity centers. The second and third terms at the right-hand-side of eq. (1) describe the generation and recombination processes:

$$\mathfrak{I} = \pi^2\,\hbar^3\, Q\, I\, (N_o - N) \,/\, m^{3/2}\,\sqrt{\hbar\Omega_o - E_t}$$

where Q is the ionisation cross section for the impurity center, I is the intensity of the incident photon flux, N_o and N are the concentrations of all gold atoms and doubly charged ones respectively, $\hbar\Omega_o$ is the energy with which electrons are generated from the impurity level E_t, m is the effective mass of the electron ($m = 0.25\, m_o$),

$$f(\overline{p})/\,\tau_r(\varepsilon) = f(\overline{p})\,\sigma(\varepsilon)\,v(\varepsilon)\,N$$

$v(\varepsilon)$ is the velocity of the electrons, and $\sigma(\varepsilon)$ is the capture cross section of the negatively charged gold centers. $\sigma(\varepsilon)$ is given by the well known formula:[3]

$$\sigma(\varepsilon) = \sigma_o\,\varepsilon^{\nu-1}\,[\exp.\,\{2\pi Z e^2\}/\{\varepsilon\hbar v(\varepsilon)\} - 1]^{-1}$$

where σ_o is a slowly varying function, ν is a small parameter (the actual form of σ_o and the value of the parameter ν depend on the energy-loss mechanism operative), Ze is the charge on the impurity atom, ε is the dialectric constant. $I_{opt.}$ is the current of electrons entering the passive region after their emission for optical phonons.

For spherical constant energy surfaces and isotropic scattering the momentum distribution function can be expanded in Legendre polynomials in θ, the angle between the momentum vector $\overline{p}$ and the direction of the applied electric field $\overline{E}$. The standard approximation is to neglect all but the two lowest order Legendre polynomials. Thus, we can write

$$f(\overline{p}) = f_o(\varepsilon) + f_1(\varepsilon)\cos\theta \qquad (2)$$

The insertion of (2) in (1) leads to two equations, one for the symmetric part of the distribution function $f_o(\varepsilon)$ and the other equation for the antisymmetric part of $f_1(\varepsilon)\cos\theta$, as follows:

$$\frac{d}{dx}\left[x^2 \left(1 + \frac{\alpha}{\lambda} x\right) f_o'(x) + x^2 f_o(x) \right] + G\,\delta(x-x_o) - A\,R(x)\,f_o(x) + I_{opt} = 0 \qquad (3)$$

and

$$f_1(\bar{p}) = \frac{e(\bar{E}.\bar{p})}{m}\,\tau_1(\varepsilon)\,\frac{df_o(\varepsilon)}{d\varepsilon} \qquad (4)$$

where

$$x = \frac{\varepsilon}{kT}$$

$$x_o = \frac{\hbar\,\Omega_o - E_t}{kT}$$

$$\alpha = \frac{3\pi}{16}\left[\frac{\mu_{ac}\,E}{v_s}\right]^2 = 1.12 \cdot 10^4 \left(\frac{20}{T}\right)^3 \left(\frac{E}{50}\right)^2$$

$$\lambda = 6\,\frac{\mu_{ac}}{\mu_{im}} = 5.5 \cdot 10^2 \left(\frac{20}{T}\right)^3 \left(\frac{N_{im}}{10^{16}}\right)$$

$$G = \frac{\pi^2\,\hbar^3\,Q\,I\,\ell_{ac}\,(N_o-N)}{2\,m^2\,v_s^2\,kT}$$

$$A = \frac{\sigma_o\,(kT)^\nu\,\ell_{ac}\,N}{2\,m\,v_s^2}$$

$$R(x) = x^\nu \exp(-\gamma/\sqrt{x})$$

$$\gamma = \frac{2\,\pi\,Ze^2}{\varepsilon\,\hbar\,v_T} = 38.5\sqrt{\frac{20}{T}}$$

μ_{ac} and μ_{im} are the electron drift mobilities (in vanishingly small field) due to electron scattering by acoustical phonons and impurity atoms respectively.

v_s is the longitudinal velocity of sound in the medium,
ℓ_{ac} is the mean free path of the electron,
v_T is the thermal velocity of the electron,
N_{im} is the concentration of the equivalent single charged impurities:

$$N_{im} = \sum_t Z_t\,N_t$$

The parameters G and A are related to each other by the boundary condition:

$$G = A \int_0^\infty R(x)\ f_o(x)\ dx \tag{5}$$

To solve equation (3) we have first to examine I_{opt} which is formulated by the solution of the transport equation in the active region, where scattering by optical phonons predominates.

In the active region, the distribution function $\phi(x)$ can be determined from the solution of the equation:

$$\frac{1}{\sqrt{x}}\frac{d}{dx}\left[\frac{-2}{3}\ \frac{(eE)^2}{m}\ x^{3/2}\ \tau_1(x)\ \frac{d\phi(x)}{dx}\right] + M_o^2\ (kT)^2 \sqrt{x-\frac{\hbar\omega_o}{kT}}\ \phi(x) = 0$$

Its solution is given by

$$\phi(x) = \text{const}\ K_{2/5}\left[\frac{4}{5}\sqrt{\zeta}\ \frac{(kT)^{3/2}}{\hbar\omega_o}\ (x - \frac{\hbar\omega_o}{kT})^{5/4}\right]$$

where $K_{2/5}(x)$ is the Macdonald function

$$I_{opt} = \sum_{\varepsilon'} M_o^2\ (n_o-1)\ f_o(\varepsilon')\ \delta\ (\varepsilon' - \varepsilon - \hbar\omega_o)$$

and n_o is the equilibrium number of optical phonons,

$$n_o = [\ \exp\left(\frac{\hbar\omega_o}{kT}\right) - 1\ ]^{-1}$$

Using the asymptotic expression for the Macdonald function we get

$$I_{opt} = C\ M_o^2\ kT \sqrt{x+\frac{\hbar\omega_o}{kT}}\ \ \exp - \mu(x - \frac{\hbar\omega_o}{kT})^{3/2}$$

where

$$C = \phi(\omega_o) \text{ and}$$

$$\mu = \frac{\sqrt{2m}}{3}\ \frac{M_o^2\ kT}{eE} \sqrt{\frac{kT}{\hbar\omega_o}} = 1.53\ \frac{50}{E}$$

Finding I_{opt} equation (3) becomes:

$$\frac{dJ(x)}{dx} + G\delta(x-x_o) - AR(x)\ f_o(x) + \phi(\omega_o)\ \frac{\ell_{ac}}{2\sqrt{m}}\ \frac{M_o^2 kT}{v_s^2} \sqrt{\frac{\hbar\omega_o}{kT}}\ \sqrt{x}\ e^{-\mu x^{3/2}} = 0 \tag{5}$$

where $J(x) = [x^2(1+\frac{\alpha}{\lambda}x)\ f_o'(x) + x^2\ f_o(x)\ ]$

is the electronic current in the energy space.

We seek solutions of equation (5) in the form

$$f_o(x) = \eta\, \phi(x) + \xi(x) \tag{6}$$

that satisfy the boundary conditions

$$f_o(\infty) = 0, \quad J(o) = 0 \tag{7}$$

where η is a free parameter; $\phi(x)$ is the solution of equation (5) in the absence of generation and recombination, i.e. $\phi(x)$ is the solution of the equation

$$\frac{dJ_\phi(x)}{dx} + \phi(\omega_o)\frac{\ell ac}{2\sqrt{m}}\frac{M_o^2\ kT}{v_s^2}\sqrt{\frac{\hbar\omega_o}{kT}}\ \sqrt{x}\ e^{-\mu x^{3/2}} = 0 \tag{8}$$

Equating the solution of (8) to the solution for $\phi(x)$ in the active region at $x = \hbar\omega_o$, we get:

$$\phi(x) = \phi(\omega_o)\{1+(\alpha/\lambda)x\}^{-\lambda/\alpha}\Big[\{1+(\alpha/\lambda)(\hbar\omega_o/KT)\}^{\lambda/\alpha} + \\ + \beta\int_x^{\hbar\omega_o/kT}\frac{\{1+(\alpha/\lambda)x'\}\{1-e^{-\mu x'^{3/2}}\}}{x'^2\{1+(\alpha/\lambda)x'\}}dx'\Big] \tag{9}$$

where

$$\beta = (eE\ell_{ac}/2mv_s^2)(\hbar\omega_o/kT) = 1.73\cdot10^4(E/50)$$

It is clear from (9) that $\phi(x)$ tends to the Maxwellian distribution function as E tends to zero. The parameter η is chosen such that

$$J(\omega_o) = J_\phi(\omega_o)$$

Then, it follows that $J_\xi(\omega_o) = 0$ (10)

Substituting (6) in (5) we get for the $\xi(x)$ function:

$$dJ_\xi(x)/dx - AR(x)\xi(x) - \eta\, AR(x)\,\phi(x) + G\delta(x-\{\hbar\omega_o/kT\}) = 0 \tag{11}$$

Using the boundary condition (10), we found the value of the parameter η together with the $\xi(x)$ function by using the variational principle.

After numerical computation the net result for the symmetric part of the distribution function can be written in the form

$$f_o(x) = G/A\left[f_\phi(x) + (\lambda/\alpha)AF(x,x_o)\right] \tag{12}$$

where

$$f_\phi(x) = \phi(x)/\mathfrak{J}_{R\phi}\left[1-(\lambda/\alpha)A\{a_1-a_1a_2R(x_o)(1+(\sqrt{x_o}/\gamma)+3(x_o/\gamma^2))\}\right]$$

and

$$F(x,x_o) = \left[\frac{\theta(x-x_o)}{2}\left(\frac{1}{x^2}-\frac{1}{x_o^2}\right)+b_1R(x)+b_2\frac{R(x)}{x^2}+\left(\frac{1}{2x_o^2}-b_3\right)\right]$$

$a_1 = 2.48\cdot10^{-8}$ $\quad a_2 = 10^6$

$b_1 = 2.0\cdot10^2$ $\quad b_2 = 1.27\cdot10^5$ $\quad b_3 = 2.45\cdot10^{-4}$

$$\mathfrak{J}_{R\phi} = \int_0^\infty R(x)\,\phi(x)\,dx \quad .$$

THE CAPTURE RATE

The capture rate per center is given by

$$C_n = \frac{\int_0^\infty \sqrt{\varepsilon}\, f_o(\varepsilon)\, v(\varepsilon)\, \sigma(\varepsilon)\, d\varepsilon}{\int_0^\infty \sqrt{\varepsilon}\, f_o(\varepsilon)\, d\varepsilon} \tag{13}$$

If C_o is the measured value for the capture rate at vanishingly small fields, when the distribution function of electrons is Maxwellian, we can approximate a value for the parameter σ_o. Using for C_o the value 2.10^{-12} cm^3/sec^{-1} we get $\sigma_o = 10^{-23}$ (KT). Then, putting in (13) the obtained distribution function (12), we get, after numberical calculations, for $\nu = -1$, the results given in Table I.

Table I

E V/cm	C_n cm^3 sec^{-1}	$\tau = 1/NC_n$ sec
50	$0.3\cdot10^{-9}$	$3.3\cdot10^{-6}$
250	$1.1\cdot10^{-9}$	$0.9\cdot10^{-6}$
500	$2.2\cdot10^{-9}$	$0.45\cdot10^{-6}$

From the table it is clear that as the electrons become hotter the capture rate increases and consequently the life time decreases.

Numerical calculations for the life time for different negative values of the parameter ν have shown that

$$\tau\ (E) = \tau\ (50)\ (50/E)^{\rho}$$

The values of ρ corresponding to different values of ν are given in Table II.

Table II

ν	-1	-2	-3
ρ	.84	0.8	0.75

We see that the life time varies slowly with ν, confirming that the barrier effect is predominant, in agreement with other reported results.[4,5]

REFERENCES

1. E.E. Vasilios, E.B. Levinson, JETP 50 1660 (1966).

2. E.M. Conwell, High Field Transport in Semiconductors and Solid State Physics, Suppl. 9, Academic Press, New York (1967).

3. V.L. Bonch-Bruevitch, Fiz. Tverdogo Tela, Suppl. 182 (1959).

4. R.G. Pratt, B.K. Kidley, Proc. Phys. Soc. (1963).

5. V.L. Bonch-Bruevitch, Fiz. Tverdogo Tela 6 2047 (1964).

RADIATION DEFECTS IN HEAVILY IRRADIATED GaAs

E.Yu. Brailovskii, V.N. Broudnyi,
I.D. Konozenko, and M.A. Krivov

Institute of Nuclear Research Academy of Sciences
of the Ukrainian SSR, Kiev

Siberian Physico-Technical Institute at Tomsk
University, Tomsk, USSR

Electrical properties and IR-absorption in GaAs irradiated at 80°C with various doses of 2.0 MeV electrons were investigated. Heavily irradiated specimens of n- and p-type GaAs are characterized by very low conductivity, carrier density and Hall mobility and large optical absorption in the interval of 0.5-1.5 eV. With increasing dose, the hole mobility, in contrast to electron mobility, decreases initially and passes through a minimum that is caused by recharging of E_v + (0.20 - 0.25) eV donor centres. The absorption spectra of n- and p-type irradiated crystals are identical. We observed structureless absorption and some absorption bands, 0.96-0.98 eV being the dominant one. The temperature dependence of conductivity and thermo-stimulated current measurements indicate the presence of E_v + (0.45 ± 0.05) and E_v + (0.25 ± 0.05) eV centres which are very much like those introduced by heat treatment. It is found that upon annealing of heavily irradiated p-GaAs, the hole mobility exhibits very large reverse annealing at 100-200°C, caused by the creation during the annealing process of additional E_v + (0.20 - 0.25) eV centres. Results of isochronal annealing on carrier concentration in p-GaAs suggest that a large concentration of point defects agglomerate, forming the clusters whose annealing takes place at a temperature above 400°C.

INTRODUCTION

Heavily irradiated semiconductors are of great current interest, both for the investigation of structural defect behaviour in extreme conditions and for the practical importance of this question in ion implantation technique.

In the case of large primary radiation damage, concentration interactions are expected which would result in essential changes of the configuration of radiation damage and in certain properties characteristic of the heavily irradiated materials.

There is relatively little well defined experimental data on structural, electrical and optical properties of the GaAs samples after heavy irradiation with fast neutrons and electrons.[1,5]

In this paper we report experimental observations of electrical and optical properties of the n- and p-type GaAs single crystals irradiated with large doses of 2.0 MeV electrons and the annealing behaviour of these samples.

EXPERIMENTAL

The samples were cut from n- and p-type GaAs single crystals doped with Te and Zn respectively and having carrier concentrations of 1.10^{16} - 1.10^{18} cm^{-3} at 300°C. Sample faces were oriented perpendicular to the <111> direction.

The cross-shaped sample geometry was used for electrical measurements. In+Te(1%) and In+Zn(5%) alloys were used to solder contacts for n- and p-GaAs respectively.

Thermally stimulated current measurements were carried out at a heating rate of β = 0.12°C/sec. The trap level energies were found upon the assumption of equilibrium between the traps and valence band.

Optical transmission measurements were performed at 300 and 80°K using the spectrometer model IKS-12 with optical slit-widths not more than 0.02 eV in the range 0.5 - 1.5 eV. The absorption coefficient was computed from transmission measurements. A value of 0.30 for reflection coefficient was used in our analysis of electron-irradiated samples.[6]

Irradiation was performed with electrons of 2.0 MeV energy from a Van de Graff accelerator with doses ranging from 10^{17} to 10^{19} el/cm^2. The sample temperature during irradiation was 80°C.

The 30 min isochronal annelaing of the irradiated samples of p-type GaAs was carried out in steps of 30°C in the temperature range 80 to 600°C.

RESULTS

Electrical measurements. The effect of electron irradiation on the conductivity of n- and p-type GaAs for doses up to 10^{19} el/cm^2 is shown in Fig. 1. In all cases we observed a decrease of conductivity as a result of irradiation. For the samples 1 and 2 (starting carrier concentration 1.10^{16} cm^{-3}) the conductivity reaches the value $4\text{-}6.10^{-10}$ ohm^{-1} cm^{-1} at $\phi = 1.10^{18}$ el/cm^2, and for further irradiation up to $\phi = 10^{19}$ el/cm^2, changes little. The Fermi level moves up to the middle of the gap and reaches the value E_c - 0.61 and E_v + 0.65 for n- and p-type GaAs respectively.

The electron mobility was found to decrease monotonously with increasing dose. The change of hole mobility versus carrier concentration during irradiation is shown in Fig. 2 (curves 1,2). It is seen that the hole mobility first decreases with irradiation and has a minimum at the hole concentration value in the range 10^{14} - 10^{15} cm^{-3}. The observed hole mobility minimum may be caused by recharging of the scattering centres introduced as a result of irradiation, as the Fermi level position changes during irradiation. In this case such centres must be donors because they become neutral or weakly scattering when the Fermi level moves up in p-type GaAs. The Fermi level value corresponding to the hole mobility minimum, calculated from

$$E_f = kT \ln p/Nv$$

where N is the density of states in v-band, was found to be E_v + (0.20 - 0.25) eV. Curve 3 of Fig. 2, which describes the hole mobility change as a result of isochronal annealing, lies under curve 2, occurring during irradiation of the same sample, and has the deeper minimum at the same value of E_f. These results indicate that defects having a level of E_v + (0.20 - 0.25) eV and causing the anomalous hole mobility behaviour during irradiation, are also formed by the annealing process.

Figure 3 shows the dark conductivity changes versus reciprocal temperature for two heavily irradiated samples of p-GaAs. It is seen that after a dose of 8.10^{17} el/cm^2 (Fig. 3a) there are two slopes, corresponding to the levels with energies E_v + (0.45 ± 0.05) eV and E_v + 0.22 ± 0.03) eV. At the dose rising to 3.10^{18} el/cm^2 (Fig. 3b) the E_v + 0.25 eV level is observed at toom temperature and at the lower temperatures we find the E_v + 0.10 eV level.

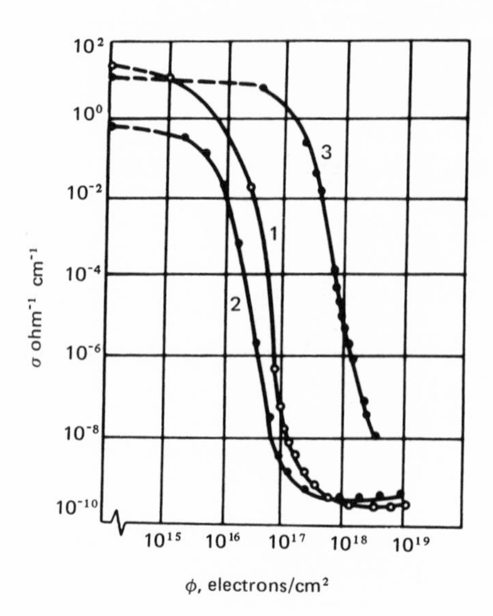

Fig. 1: Conductivity at 300°C vs. 2M eV electron fluence for GaAs.

1. n-GaAs, $n_o = 1 \cdot 10^{16} cm^{-3}$
2. p-GaAs, $p_o = 1 \cdot 10^{16} cm^{-3}$
3. p-GaAs, $p_o = 7 \cdot 10^{17} cm^{-3}$

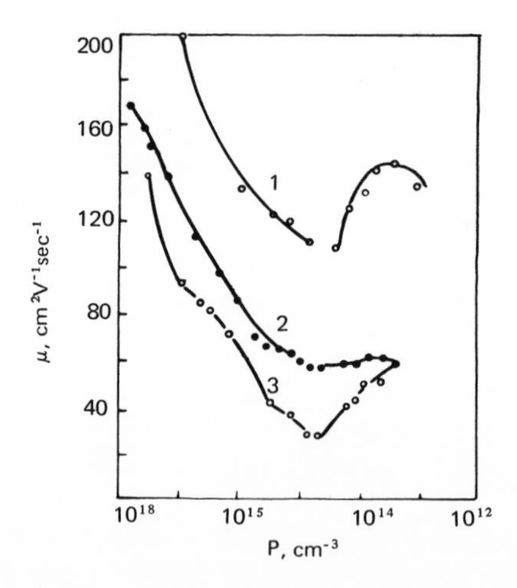

Fig. 2: Hole mobility vs. hole concentration during electron irradiation. Curve 1: $p_o = 1 \cdot 10^{17} cm^{-3}$. Curve 2: $p_o = 6 \cdot 10^{17} cm^{-3}$. Same, during annealing, Curve 3: for same sample as Curve 2.

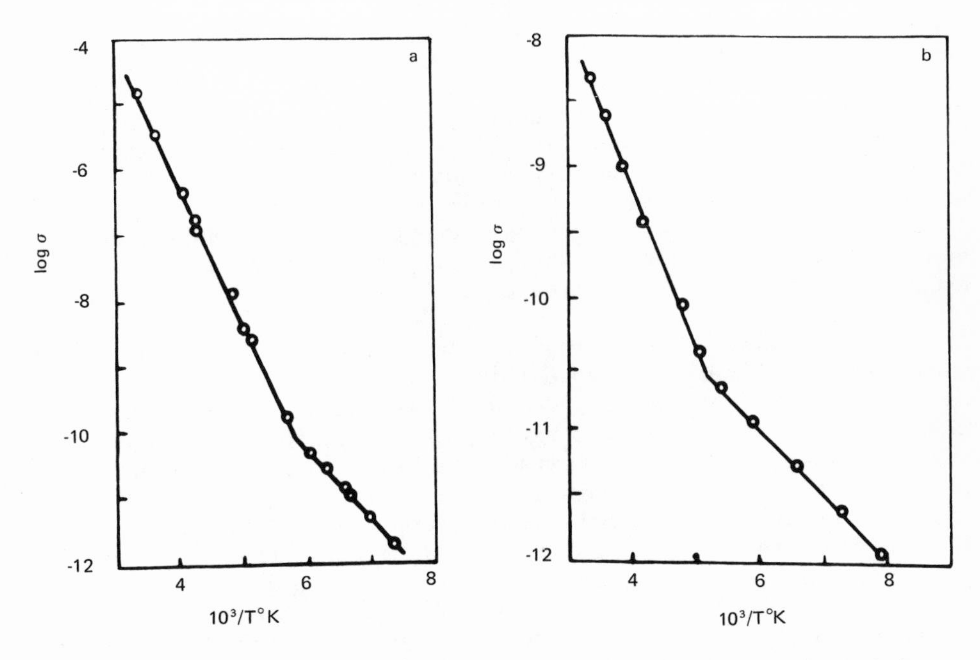

Fig. 3: Temperature dependence of dark conductivity in p-GaAs after irradiation:

a; $\phi = 8\cdot10^{17}el/cm^2$

b: $\phi = 3\cdot10^{18}el/cm^2$

Thermally stimulated current measurements were carried out for the same irradiated sample of p-GaAs. Two TSC peaks are found in these samples, occurring at about 120 and 245°C (Fig. 4) and corresponding to traps having an energy of roughly E_V + (0.25 ± 0.05) and E_V + (0.46 ± 0.05) eV, which agrees well with the results presented above.

The results of isochronal annealing of the hole concentration and mobility for the samples of p-GaAs ($p_o = 6.10^{17}$ cm^{-3}, μ_o = 160-170 cm^{2}v^{-1} sec^{-1}) irradiated with various electron doses are shown in Fig. 5. For the least dose ($\phi = 1.2 \cdot 10^{17}$ el/cm^2) the concentration almost completely anneals to 300°C, with a predominant annealing stage at 100-200°C (Fig. 5a, curve 1). As the dose increases we find another, high temperature, stage at 400-600°C, whose importance in the total annealing rises with increasing irradiation. Finally, in the case of the largest dose, we observed two annealing stages almost overlapping and a continuous recovery of carrier concentration in the temperature range 100-600°C. After 600°C, annealing does not reach the starting carrier concentration value, and unannealated fraction increases with dose.

We found that the hole mobility annealing behaviour also depends essentially on irradiation dose (Fig. 5b). At small doses we have a single annealing stage at 150-250°C. As the dose increases, as in the case of carrier concentration annealing, the high temperature annealing stage appears. Simultaneously with this, the pronounced hole mobility decrease appears in the temperature range 100-200°C (the reverse of annealing mobility). Comparison with Fig. 2 (curve 3) data shows that the reverse of annealing mobility in the heavily irradiated samples is associated with the E_V + (0.20 - 0.25) eV centres. As seen from Fig. 5a, the influence of these centres on the hole mobility at annealing temperatures up to 150°C is insignificant. The carrier scattering change as a result of irradiation and as a result of reverse annealing $(\Delta\, 1/\mu)_{irr}$ and $(\Delta\, 1/\mu)_{r.a.}$ respectively, may be written as follows:

$$(\Delta\, 1/\mu)_{irr} = 1/\mu_\phi - 1/\mu_o \quad ;$$

$$(\Delta\, 1/\mu)_{r.a.} = 1/\mu_{min} - 1/\mu_\phi$$

where μ_o , μ_ϕ and μ_{min} are the mobility values before irradiation, after irradiation, and the minimum values during annealing. For the cases 3 and 4 (Fig. 5b) the ratio $(\Delta\, 1/\mu)_{r.a.} / (\Delta\, 1/\mu)_{irr} \sim 3$, i.e. the increase in carrier scattering by the centres, formed in the process of reverse annealing, is larger than by the defects introduced immediately by irradiation. This shows that E_V + 0.20 eV defects are highly effective carrier scattering centres.

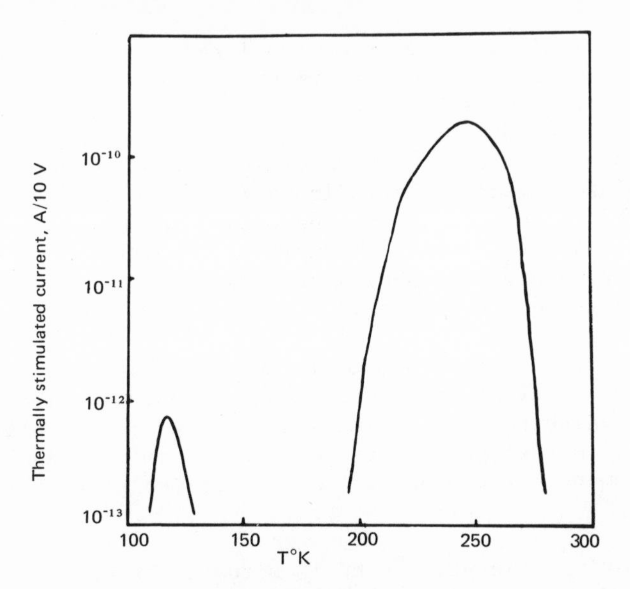

Fig. 4: Thermally stimulated current in sample p-GaAs after irradiation with dose $\phi = 3\cdot10^{18}$el/cm^2.

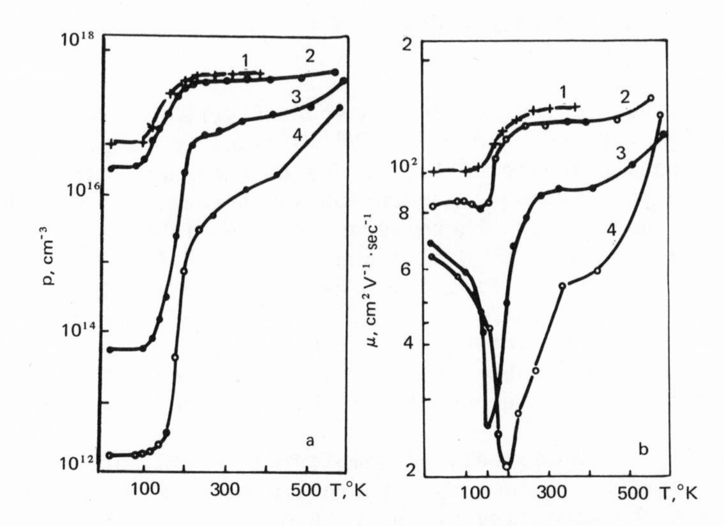

Fig. 5: Isochronal annealing of hole concentration (a) and hole mobility (b) in p-GaAs ($p_o = 6\cdot10^{17}$cm^{-3}) after irradiation.

1. $\phi = 1.2\cdot10^{17}$el/cm^2 3. $\phi = 4.4\cdot10^{17}$el/cm^2

2. $\phi = 1.5\cdot10^{17}$el/cm^2 4. $\phi = 8.5\cdot10^{17}$el/cm^2

Some optical properties. 2 MeV electron irradiation with sufficiently large doses has been found to produce a strong absorption on the low-energy side of the band gap. The growth of this absorption depends linearly on the fluence over the entire spectral range from 0.5 to 1.5 eV. However, the more rapid growth occurs in the region near the fundamental absorption edge. Some more or less well defined absorption bands have been found in the absorption spectra of the heavily irradiated samples. Comparison of Fig. 6 and 7 shows that absorption spectra of irradiated samples of n- and p-type GaAs are quite similar.

The integrated absorption coefficient is at least $3.10^2 cm^{-1}$ eV between 0.5 and 1.4 eV after the irradiation dose of 1.10^{19} el/cm^2. Assuming the f-value to be 1, we calculated the lower limit of the concentration of absorbing species to be $8.10^{18} cm^{-3}$. Thus, reasonably good agreement between optical and electrical measurements is observed.

The dominating absorption band is situated at photon energy 0.96-0.98 eV. The absorption coefficient in this band maximum increases in proportion to irradiation dose over the entire range of exposure. Careful absorption spectra measurements in the range 0.5-1.5 eV reveals additional absorption bands at 0.84-0.86 eV and 1.08-1.10 eV as shoulders on the 0.96-0.98 band. In addition we have found an absorption band at 1.18-1.20 eV in the spectra of samples irradiated with a dose $\phi = 4.10^{17}$ el/cm^2 (fig. 6, curve 1).

The analyses of the absorption spectra were carried out assuming a Gaussian form for the 0.96-0.98 eV band. Accurate determination of the band slope is difficult due to the fact that the band is superposed on continuous absorption which rises towards shorter wave lengths, and because there is an additional shoulder in the spectra. The 0.96-0.98 eV band measured at 80^oK is broad and has a half-width of 0.24 eV (Fig. 6). The 0.84-0.86 eV step may be interpreted as an absorption band at 0.76-0.78 eV. As the temperature decreases the intensity of the 0.96-0.98 eV band increases, whereas its integrated absorption practically does not change. A weak temperature displacement of this band occurs.

Annealing of the irradiated samples has shown that the region of spectra near the fundamental absorption edge recovers most quickly while the annealing of the 0.96-0.98 eV band begins at temperatures above 300^oC. Complete recovery of this band does not occur even after heating up to 450^oC (Fig. 6, curves 4,5).

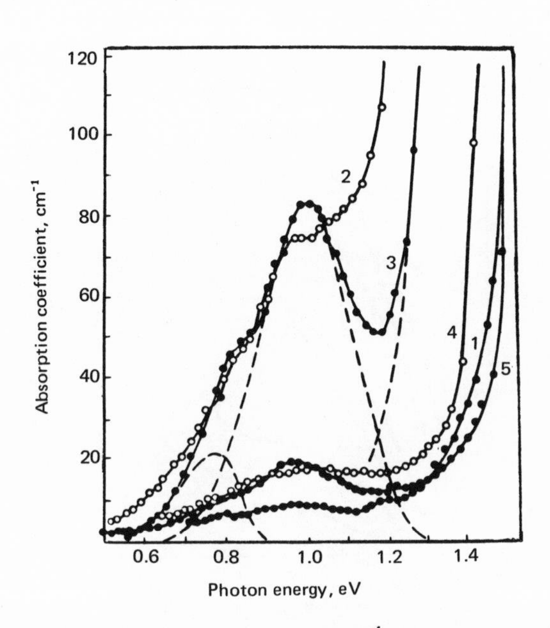

Fig. 6: Variation of absorption coefficient as a function of photon energy in n-GaAs after irradiation and annealing.

1. $\phi = 4 \cdot 10^{17} el/cm^2$, T = 80°K

2. $\phi = 3.8 \cdot 10^{18} el/cm^2$, T = 300°K

3. $\phi = 3.8 \cdot 10^{18} el/cm^2$, T = 80°K

4. After 30 min. annealing at 450°C, T = 300°K

5. After 30 min. annealing at 450°C, T = 80°K

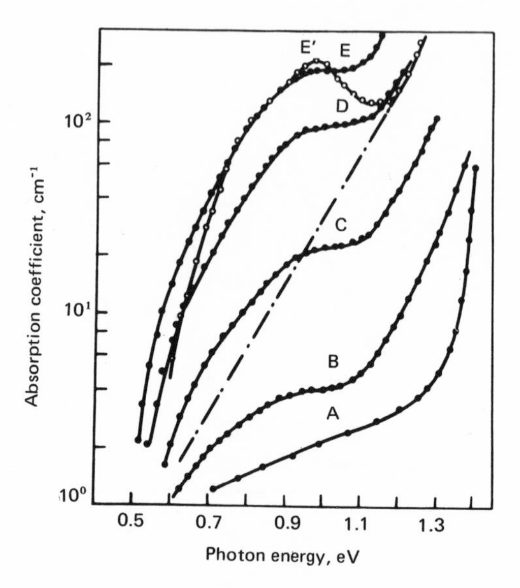

Fig. 7: Effect of 2M eV electron irradiation on absorption coefficient of p-GaAs ($p_o=1\cdot10^{16}cm^{-3}$).

A: Before irradiation

B,C,D,E,E': After irradiation, $\phi = 1\cdot10^{17}$, $1\cdot10^{18}$, $1\cdot10^{19}$, and $1\cdot10^{19}$ el/cm^2, respectively

A,B,C,D,E, 300°K; E', 80°K

DISCUSSION AND CONCLUSIONS

The results obtained in this work show that specimens of either n- or p-type GaAs, after sufficiently heavy electron irradiation, are characterized by the following properties: very low values of conductivity, carrier density and Hall mobility; and very strong optical absorption at wavelengths extending well beyond the fundamental absorption edge.

Coates and Mitchell[3] have found anomalous behaviour of conductivity in GaAs irradiated with large doses of fast neutrons ($\phi > 10^{18}$ neutrons/cm^2). The temperature dependence of resistivity indicates that the mechanism of conduction at high neutron doses is tunnel-assisted hopping.[3]

In our case of electron irradiation, the conductivity value reaches $4\text{-}6.10^{-10}$ ohm^{-1}cm^{-1} at doses of 1.10^{18}el/cm^2 and hardly changes after that for doses up to 10^{19}el/cm^2. However, it should be noted that a very small rise of conductivity occurs in the dose range 10^{18} - 10^{19}el/cm^2 (Fig. 1). It is possible that for still higher doses of electron irradiation ($\phi > 10^{19}$el/cm^2) the conductivity may behave similarly to the case of neutron irradiation. The temperature dependence of conductivity measured for the samples irradiated with doses of 3.10^{18}el/cm^2 (Fig. 3) gives no evidence that the mechanism of conduction in these samples is tunnel-assisted hopping.

The isochronal annealing results show that irradiation dose has an important influence on the defect annealing behaviour in p-type GaAs. The change in annealing behaviour for hole concentration at high doses is very similar to that observed previously in our work[7] for n-type GaAs. The high temperature annealing stage observed in heavily irradiated p-GaAs coincides with disordered-region annealing in GaAs irradiated with fast neutrons.[8] Proceeding from these facts and taking into account the results obtained by Vook[2], we may conclude that, as for n-type GaAs, annealing of heavily irradiated p-type GaAs can cause agglomeration of point defects, forming clusters whose annealing occurs at temperatures higher than 400°C.

The electrical measurements show that in samples irradiated to high electron dose there are defects with levels at E_V + (0.2-0.25) and E_V + (0.45 ± 0.05) eV. The centres at E_V + 0.2 eV having a strong influence on carrier mobility have also been found in GaAs irradiated with heavy particle (proton or fast neutron).[9,10] The pronounced increase in the concentration of defects located at E_V + (0.2-0.25) eV during annealing of samples containing a large concentration of point radiation defects, taken together with the results of heavy particle irradiation[9,10] lead to the conclusion that the responsible defects leave a complex structure. It must be noted that earlier authors did not observe similar defects in electron irradiated GaAs. Blanc el al[11] found that as a result of heat treatment, defects with levels at E_V + 0.5 and E_V + 0.2 were introduced, and that the latter has a large carrier scattering cross section. This led them to distinguish between electron-induced and thermally-induced defects.

Our measurements have shown that heavy electron irradiation creates the radiation defects with levels of E_V + (0.2-0.25) and E_V + (0.45 ± 0.05) eV, which are very much like centres introduced in GaAs by heat treatment.

REFERENCES

1. F.L. Vook, J. Phys. Soc. Japan 18 Suppl. II, 190 (1963).

2. F.L. Vook, Phys. Rev. 135 A1742 (1964).

3. R. Coates, E.W. Mitchell, J. Phys. C: Sol. State Phys. 5 L 113 (1972).

4. E. Yu. Brailovskii, V.N. Broudnyi, M.A. Krivov, V.B. Red'ko, Fiz. i Tekh. Poluprovodnikov 6 2075 (1972).

5. K.V. Vaidyanathan, L.A.K. Watt, M.L. Swanson, Phys. stat. sol. A10 127 (1972).

6. V.C. Burkig J.L. McNickols, W.S. Ginell, J. Appl. Phys. 40 3268 (1969).

7. E. Yu. Brailovskii, V.N. Broudnyi, Fiz. i Tekh. Poluprovodnikov 5 1248 (1971).

8. L.W. Aukerman, R.D. Graft, Phys. Rev. 127 1576 (1962).

9. K. Wohlleben, W. Beck, Z. Naturforsch. 21a 1057 (1966).

10. L. Borghi, P. De Stepano, P. Mansherretti, J. Appl. Phys. 41 4665 (1970).

11. J. Blanc, R.H. Bube, L.R. Weisberg, J. Phys. Chem. Solids 25 225 (1964).

THE PROBLEM OF CHEMICAL INHOMOGENEITIES IN ELECTRONIC MATERIALS

Harry C. Gatos and August F. Witt

Massachusetts Institute of Technology

Cambridge, Massachusetts

Chemical inhomogeneities are present in semiconductors (and generally in all solids) solidified under destabilizing thermal gradients. Their origin is associated with temperature fluctuations at the growth interface (leading to flucutations in the microscopic rate of growth) and/or with fluctuations of the boundary layer thickness. The formation of such inhomogeneities in semiconductors, such as InSb, Ge and Si, is discussed in the light of results obtained with the aid of "rate striations" which allow the determination of the microscopic growth rate and of the morphological characteristics of the growth interface. The quantitative relationship between impurity dopant concentration and the microscopic growth rate is presented.

INTRODUCTION

To achieve desired properties (chemical, physical or electronic) "impurity" components usually must be added to solids. These components are incorporated either in solid solution or as distinct phases. In either case, for optimum performance, the distribution of these components must be accurately controlled, within a functional scale, throughout the solid. Compositional inhomogeneities, gradual or abrupt, often cause a significant degradation of the sought-after properties. They are, to a varying degree, always present in solids prepared from melts or solutions. Their origin is associated with inhomogeneities in these liquids brought about primarily by density (gravity) effects and by segregation effects at the solid-liquid interface. Although the adverse effects of compositional inhomogeneities have been always recognized, their identification, study and elimination, still present major difficulties.

In the case of electronic materials, and especially semiconductors, dopant inhomogeneities have pronounced effects on electronic properties. In view of the recent development of semiconductor device technology on a micron or even submicron scale, precise control of the dopant distribution on such a scale has become of increased importance.

The difficulties in studying dopant inhomogeneities in semiconductors and quantitatively determining their origin can be summarized as follows: a) variations of the average concentration, (which is the order of parts per million) must be determined with a linear resolution on a micron or submicron level; b) the microscopic rate of solidification (which strongly affects the dopant segregation behavior) must be determined; this rate is sensitive to small temperature fluctuations ($<\pm 0.01^{o}C$) at the solidifying interface: c) in highly doped melts, the morphological characteristics of the interface associated with constitutional supercooling effects must be taken into consideration.

In a number of instances semiconductor materials are prepared from the vapor phase. The origin of inhomogeneities in this case, to a large extent, is similar to that of solidification from liquids. However, only solidifcation from liquids will be considered here.

In this paper an attempt is made to outline the present status of experimental and theoretical techniques for studying the nature and origin of compositional inhomogeneities in semiconductors and to discuss some of the recent experimental results (obtained on InSb, Ge, and Si) in the light of the prevailing theoretical models. It will also be shown that compositional inhomogeneities, in turn, constitute a unique tool in studying the basic aspects of the solidification process on a microscale.

ORIGIN OF COMPOSITIONAL INHOMOGENEITIES

Inhomogeneities in Czochralski Growth under Rotation. The Czochralski technique has been most commonly employed for the growth of semiconductor cyrstals ever since the discovery of the transistor in 1948, the starting point of the solid state electronics era. This technique has been found most suitable, among all others, since it provides for crystal and/or crucible rotation and, thus, minimizes thermal and compositional inhomogeneities in the melt, which is under destabilizing thermal gradients. It further allows the growth of large crystals with very high degree of crystalline perfection.[1]

"Periodic" dopant inhomogeneities in Czochralski grown Ge crystals were reported as early as 1953.[2] They were identified by autoradiography using crystals containing radioactive dopant. (Typical periodic inhomogeneities such as those referred to in the 1953 study are shown in Fig. 1.)

Variations in resistivity and minority carrier lifetime by a factor of up to five were found to be associated with the originally observed inhomogeneities (Fig. 2). Thus, the adverse effects of dopant inhomogeneities on the electronic characteristics of semiconductors were unambiguously established. The origin of these inhomogeneities was attributed to non-steady state growth conditions associated with thermal fluctuations at the solid-melt interface.[4] These investigators arrived at a theoretical expression relating the effective distribution coefficient, $k_{eff} = C_S / C_L$ (C_S and C_L are the concentrations of the dopant in the solid and in the bulk liquid, respectively), to the equilibrium distribution coefficient k_o, the rate of growth, V, and the thickness of the boundary layer, δ, for steady state conditions under diffusion control:

$$k_{eff} = k_o / (k_o + (1-k)\exp(-\delta V/D)) \qquad (1)$$

where D is the diffusion coefficient of the dopant in the melt. This relationship has become fundamental in considering segregation phenomena.

Dopant striations in semiconductor crystals grown under rotation have been detected and studied by a number of investigators. Various techniques have been employed in these studies including electrical resistivity,[2,5] autoradiography,[2,6] infrared absorption,[2] electroplating,[7,8] and chemical etching.[9] The semiconductors primarily investigated in these studies have been Ge and InSb, and to a lesser extent, Si. The observed periodic striations were attributed to the fact that the temperature in melt is not radially symmetric about the rotational axis. Consequently, as the crystal rotates the microscopic growth rate and thus the concentration of the dopant incorporated into the crystal vary with angular position.[10] In fact, it was shown that, in melts with pronounced radial asymmetry, remelting can take place during the rotation of the crystal through the "hot" part of the melt, leading to "remelt" type of inhomogeneities.[3]

Core Formation. In 1959 it was observed[11] that in InSb crystals grown in the <111> direction the inner part (core) exhibits a different dopant concentration that the outer (off-core) part. This phenomenon was independently observed[8] in germanium crystals grown in the <111> direction. The core formation has been shown to be associated with a facet present at the growth interface. It is

Fig. 1. Typical rotational striations in an InSb crystal grown in the <111> direction at various rotational rates[3] (section along the growth axis). 210X

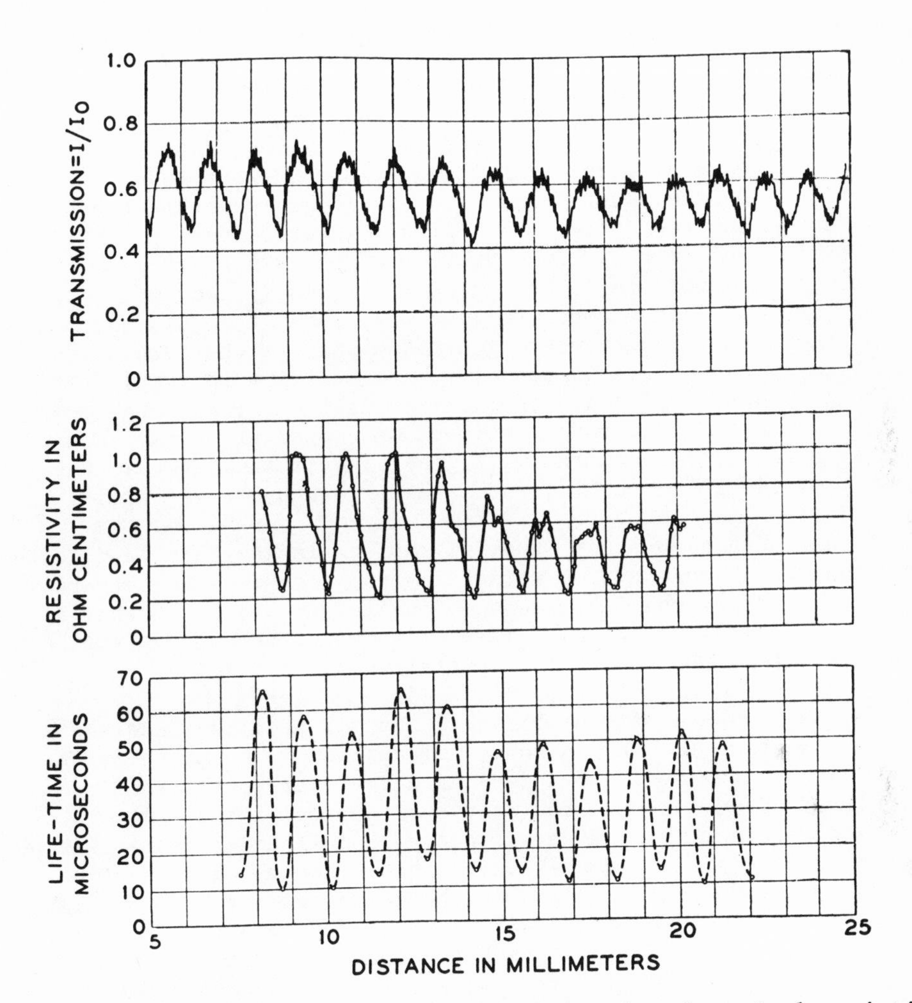

Fig. 2. Optical transmission of autoradiograph, electrical resistivity and hole lifetime for a germanium crystal exhibiting striations.[2]

a common characteristic of semiconductor crystals grown from the melt (Fig. 3) even in directions other than the <111>.[13] Its size varies with experimental conditions. Although it has been extensively studied, the detailed mechanism of core formation has not as yet been established. The "core", as a compositional inhomogeneity will not be further discussed in the present paper.

Other Compositional Inhomogeneities. As advances were made in the techniques for detecting dopant inhomogeneities, additional types other than the rotational striations were observed.[14] Thus, in InSb, within successive rotational striations various types of inhomogeneities were found, random and periodic in nature (both in the "on-core" and the "off-core" region). These types of striations are primarily due to thermal convection which is not overcome, at low rotational rates, in low temperature growth (for example, in InSb and Ge). In high temperature growth, such as in Si, thermal convection persists even under high rotational rates (forced convection). A typical example of inhomogeneities in an Si crystal is shown in Fig. 4. The density, intensity and other characteristics of such random striations depend primarily on the impurity concentration in the melt, the size of the crystal, the rate of rotation, and pulling rate and temperature gradients.

It has been also shown[3] that vibrations in the melt lead to the formation of inhomogeneities reflecting the frequency and other characteristics of the vibrations. Finally, it was shown that in the absence of crystal rotation additional types of inhomogeneities can be incorporated into the pulled crystals.[9,14]

In the presence of destabilizing thermal gradients and when forced convection is not dominant, compositional inhomogeneities (other than those due to rotation) have been attributed to direct or indirect effects of the thermo-hydrodynamics of the melt. According to thermo-hydrodynamic theory, a melt under destabilizing thermal gradients, can successively exhibit turbulent convection, oscillatory thermal instability or thermal stability as its height decreases. A number of studies[16] have been systematically addressed to the cause and effect relationship between such thermo-hydrodynamic characteristics of the melt and the resulting inhomogeneities in the solid. The results have been consistent with such a relationship.

Regardless of their specific nature and/or morphology the compositional inhomogeneities constitute variations of k_{eff} brought about by variations in the growth rate (caused by temperature fluctuations of the growth interface) and the fluid dynamics of the melt.

Theoretical Studies. In parallel to the experimental investigations theoretical models were developed dealing with the stationary and dynamic characteristics of the solid-melt interface and their role

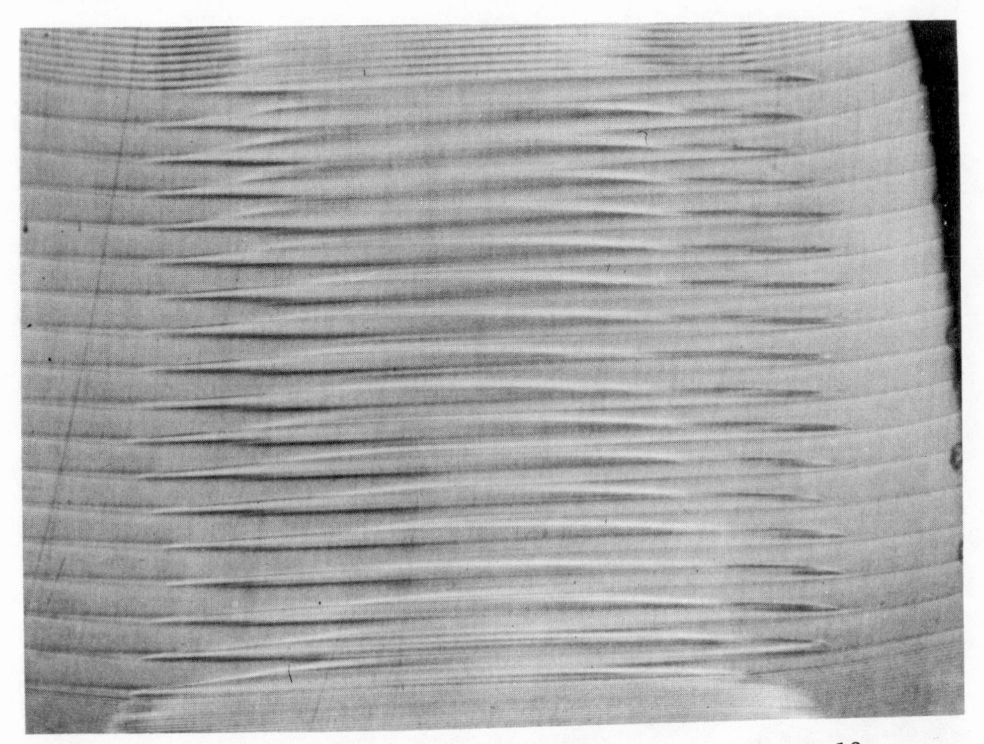

Fig. 3. "Core" in an InSb crystal grown under rotation.[12] 14X

Fig. 4. Non-rotational striations between successive rotational striations in a Si crystal (section along the growth axis)[15] 675X

in the solidification process. Pertinent parameters were considered in these models including diffusion under dynamic boundary conditions,[4] heat transfer,[10,17] and forced convection.[18]

Attempts to compare the theoretical models with experiment have met with limited success.[4,5] This result is not surprising in view of the fact that the idealized conditions assumed in developing these models are not generally encountered in the complex process of crystal growth.

SEGREGATION BEHAVIOR AND MICROSCOPIC GROWTH RATES

Growth under Forced Convection - Czochralski Growth under Crystal Rotation. As pointed out above, in the presence of radial asymmetry about the rotational axis, the microscopic growth rate, V, at any point at the crystal-melt interface varies continuously, so that the concentration of the dopant incorporated into the solid varies with angular position leading to rotational striations.[10] An expression has been derived relating V to the conditions of thermal asymmetry:[3]

$$V = V_o - [2\pi R\,\Delta T / G]\cos 2\pi Rt \qquad (2)$$

where V_o is the pulling rate, R is the rate of rotation, ΔT is the temperature variation that a given point at the interface undergoes during a 360° rotation, G is the temperature gradient in the melt adjacent to the growth interface and t is the time. Taking $2\pi R\Delta T/GV_o = \alpha$ then eq. (2) becomes:

$$V = V_o(1 - \alpha\cos 2\pi Rt) \qquad (3)$$

According to eq. (3) the minimum value of V is

$$V_{min} = V_o(1 - \alpha) \qquad (4)$$

It is apparent that when $\alpha > 1$, then V_{min} becomes negative, i.e., remelting should take place during part of each rotation leading to "remelt" rotational striations pointed out above.

The quantitative determination of the microscopic growth rate, V, during each rotational cycle has become possible by introducing in the growing crystal "rate striations", i.e. compositional inhomogeneities of short duration and known frequency. Rate striations have been originally introduced by coupling to the melt vibrations of known frequency[3] and more recently by applying current pulses[19] across the growing system. By measuring the distance between rate striations in the crystal one can determine, from their known frequency, the microscopic growth rate at any position in the crystal. The rate striations were shown not to interfere in either case with the overall growth process.

The microscopic growth behavior, as determined by means of rate striations, is shown in Fig. 5 for a crystal grown under thermal asymmetry leading to "remelt" striations. It is clearly seen that the "off-core" (off-facet) part of the crystal grew at an average rate greater than that of the "core" (facet) which, in this case, is equal to the pulling rate.

Recently an experimental approach was developed combining differential etching, rate striations, and spreading resistance measurements (from which dopant concentrations can be determined).[21] Thus, it was possible to determine quantitatively the dopant concentrations (with a linear resolution of less than 5 μm) and the precisely corresponding growth rates on a microscale. A typical case is shown in Fig. 6. An analysis of the effect dopant segregation according to the Burton, Prim and Slichter theory[4] indicated the presence of steady state conditions (constant boundary layer thickness) only for the segments of increasing growth rates within each rotational cycle. The "effective diffusion layer thickness" was found to be smaller than computed from a Cochran analysis[22] and decreased significantly in the presence of remelting.

Growth under Thermal Convection. In the absence of forced convection (no rotation), if the dopant rejected during growth (for $\kappa < 1$) is too slowly removed from the interface by thermal convection (transport through diffusion is too slow to be effective at the commonly encountered solidification rates), then excessive dopant accumulation at the growth interface can lead to constitutional supercooling conditions and interface breakdown.

However, even if interface breakdown does not take place, the problem of compositional inhomogeneities is far more complex than in the presence of forced convection. As pointed out above, the melt under destabilizing thermal gradients may exhibit turbulent flow, oscillatory thermal instability or thermal stability. The nature of the convective flow depends, primarily, on the magnitude of the vertical and lateral gradients, the aspect ratio and viscosity of the melt.

In a recent study of solidification, direct time correspondence between crystal growth behavior and the thermal characteristics of the melt was established by means of "time markers"; they were introduced (by current pulses) into the growing crystals and simultaneously registered on the continuous recording of the thermal behavior of the melt.[23] It was, thus, unambiguously shown that the thermo-hydrodynamics of the melt control the growth and dopant

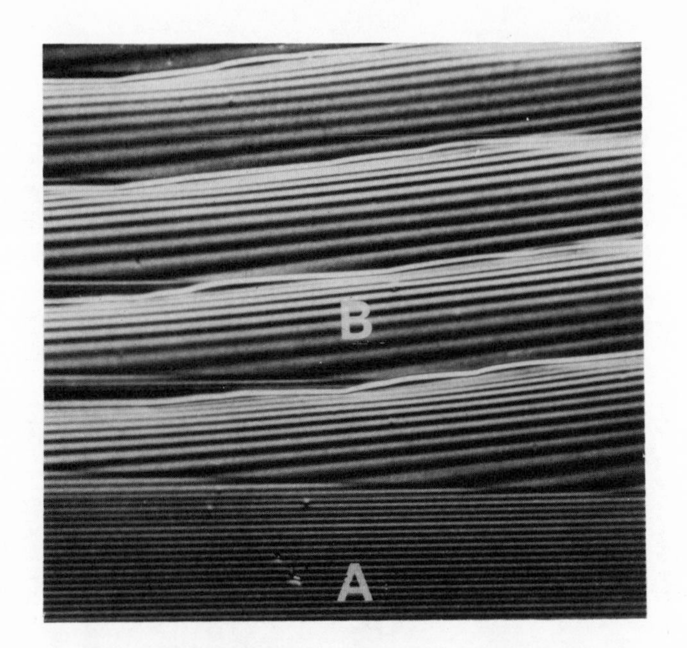

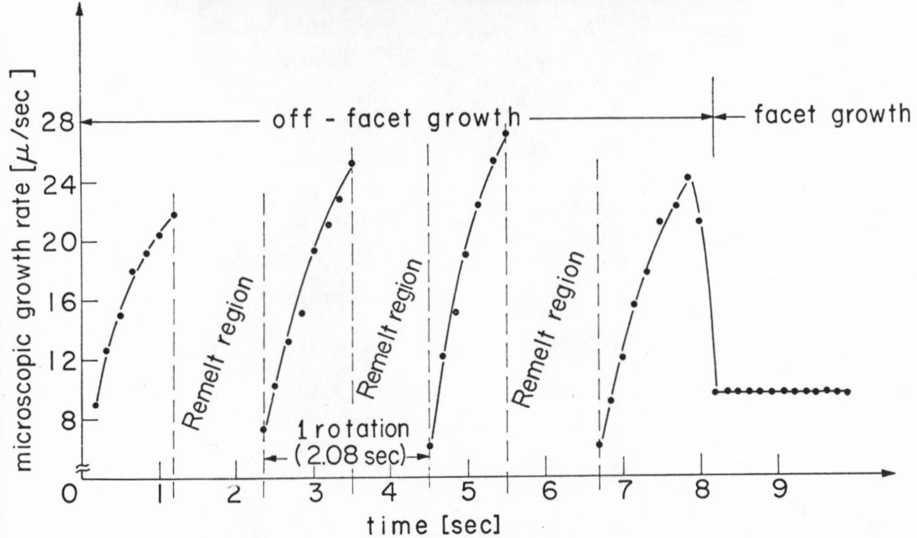

Fig. 5. Off-facet region, A, and, facet region, B, of an InSb crystal; the instantaneous microscopic growth rates computed from the spacing of the rate striations is also shown. In the upper part of the photomicrograph pronounced remelt lines are observed. Portions of the core can be observed on the extreme left (B). In the lower part, rotation was stopped (the core expanded)[20]- pulling rate 35 mm/hr; rate of rotation 29 rmp; frequency of rate striations 6/sec. ca. 350X

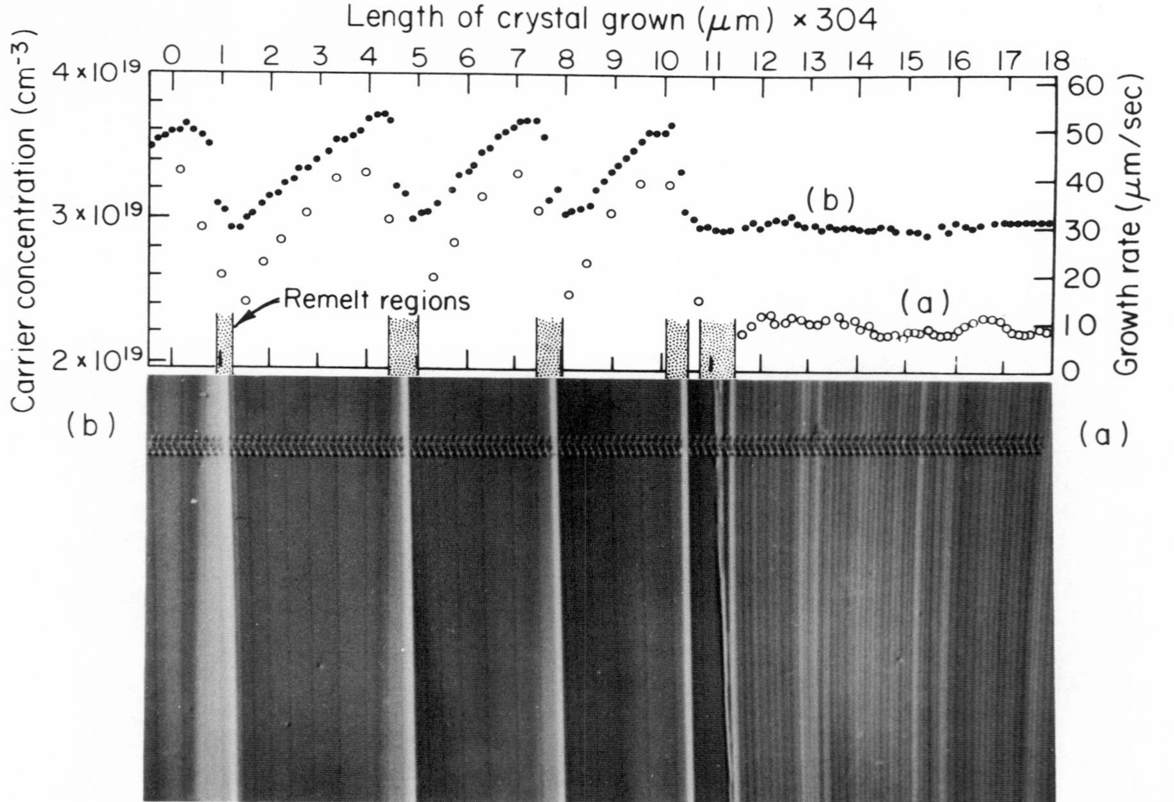

Fig. 6. Differentially etched segment of gallium doped germanium single crystal pulled with seed rotation (left part) and without seed rotation (right part) and dopant distribution analysis: curve (a) growth rates as obtained from the spacings of the rate striations introduced at 0.5 sec intervals. Growth rate analysis could not be carried out in the immediate vicinity of the remelt striations (remelt regions). Curve (b) carrier (dopant) distribution as obtained from spreading resistance measurements at spacings of 5 μm. 200X

segregation behavior. Turbulent convection, oscillatory instability and finally thermal stability were successively present during solidification with decreasing melt height. A typical example of compositional inhomogeneities in the crystal and the corresponding thermal fluctuations in the melt (thermal oscillatory instability in this case) are shown in Fig. 7. The numbered time markers permit a direct comparison between the thermal characteristics of the melt and the simultaneous growth behavior.

The Rayleigh number, which is often used as a criterion for hydrodynamic behavior, ranged from 3×10^5 to about 4×10^3 for turbulent convection, from 3×10^3 to 2×10^3 for oscillatory thermal instability and from 10^3 to 0 for thermal stability. The transition from thermal instability to stability occurred at Rayleigh number value of about 2×10^3 which is reasonably close to the value of 1.7×10^3 predicted from theoretical considerations, although the assumed boundary conditions in the theory are not quite the same as those in the present experiments.

Interface Instability. As pointed out above, excessive accumulation of the dopant at the growth interface leads to conditions of interface instability and eventually interface breakdown. Numerous theoretical studies have been reported on the parameters and relationships associated with interface instability.[24] In a recent theoretical treatment of interface instability during Czochralski growth from stirred heavily doped melts, oscillatory behavior of the growth interface (for convection controlled solute transport) was predicated with a defined frequency and phase velocity.[25] Such oscillatory interface instability has been observed[26] during Czochralski growth of germanium from a Ga-doped melts. A typical instance of such instability is shown in Fig. 8. It can be seen that complex segregation effects are associated with the oscillatory instability. It should be pointed out that the experimentally determined wavelength and phase velocity of this oscillatory instability were found to be in reasonable agreement with the theoretical predictions.

SUMMARY

Compositional inhomogeneities (striations) are generally present in semiconductor crystals grown from doped melts under destabilizing thermal gradients; periodic inhomogeneities (rotational striations) are formed in crystals pulled under rotation (Czochralski growth) in the presence of thermal asymmetry. The characteristics of such inhomogeneities reflect the thermal conditions at the growth interface associated with forced convection, thermal asymmetry and thermal convections; they reflect also changes in the characteristics of the boundary layer.

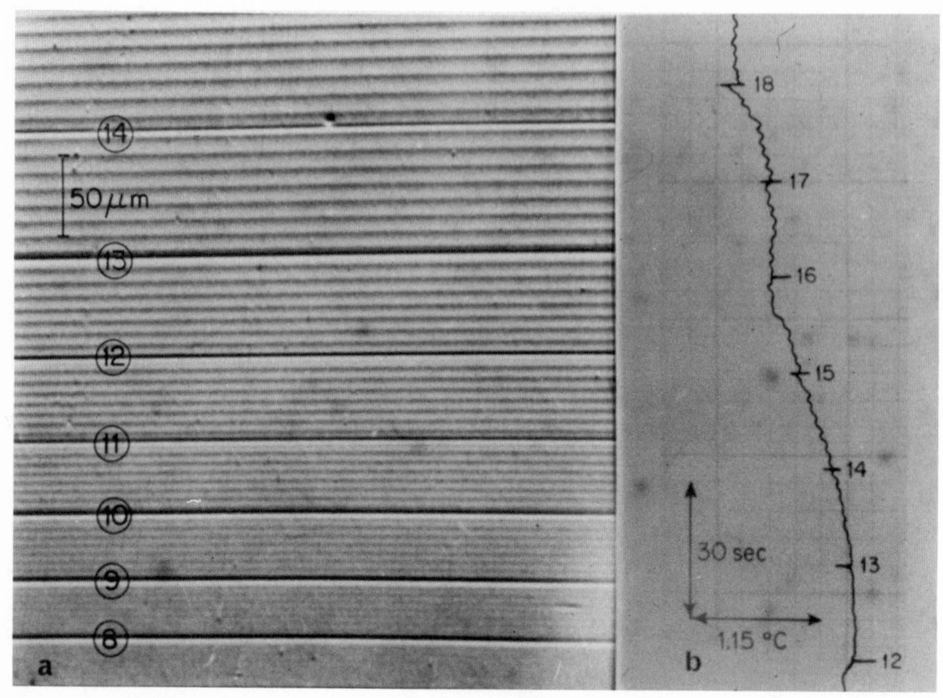

Fig. 7. (a) Section of crystal grown in the presence of thermal oscillations in the melt. The dopant heterogeneities caused by the thermal oscillations are clearly visible. Note that their spacing decreases from top to bottom. The time markers make it possible to establish that the frequency of the thermal perturbations is constant over the depicted region (the number of dopant heterogeneities between successive time markers is constant) and that the microscopic growth rate decreases from top to bottom (the spacing of successive time markers decreases). (b) Recording of the temperature in the melt during growth of the crystal segment depicted in Fig. 7(a).

Fig. 8. Rotational striations reflecting the interface oscillatory instability during growth of a germanium crystal from a Ga-doped melt. 600X

The direct correspondence of compositional inhomogeneities in the crystal to the prevailing thermal or compositional conditions at the interface during growth is preserved because diffusion of dopants (with minor exceptions) in semiconductor crystals is negligible.

This characteristic of the compositional inhomogeneities to reflect accurately the conditions prevailing at the interface can be employed in studying growth and segregation phenomena. Thus, inhomogeneities of known frequency and duration ("rate striations" and "time markers") have been successfully employed in determining microscopic rates of growth and in relating growth and segregation behavior to thermal and compositional parameters of the melt and the growth interface.

Dopant inhomogeneities in semiconductors have pronounced adverse effects on their electronic characteristics such as minority carrier lifetime and carrier mobility. In the case of semiconductor devices dopant inhomogeneities can affect the breakdown voltage, the saturation current in a p-n junction, the current ratios in tunnel diodes and can cause undesirable built-in fields.

ACKNOWLEDGMENTS

The authors are grateful to the National Science Foundation and to the National Aeronautics and Space Administration for financial support and encouragement.

REFERENCES

1. R.A. Laudise, The Growth of Single Crystals, Prentice-Hall (1970).

2. J.A. Burton, E.D. Kolb, W.P. Slichter, J.D. Struthers, J. Chem. Phys. 21 1991 (1953).

3. K. Morizane, A.F. Witt, H.C. Gatos, J. Electrochem. Soc. 114 738 (1967).

4. J.A. Burton, R.C. Prim, W.P. Slichter, J. Chem. Phys. 21 1987 (1953).

5. J.R. Carruthers, Canad. Met. Quart. 5 55 (1963).

6. M.D. Banus, H.C. Gatos, J. Electrochem. Soc. 109 829 (1962).

7. P.R. Camp, J. Appl. Phys. 25 459 (1954).

8. J.A.M. Dikhoff, Solid-State Electron. 1 202 (1960).

9. H.C. Gatos, A.J. Strauss, M.C. Lavine, T.C. Harman, J. Appl. Phys. 32 2057 (1961).

10. E. Billig, Proc. R. Soc. A229 346 (1955).

11. K.F. Hulme, J.B. Mullin, Phil. Mag. 4 1286 (1959).

12. K. Morizane, A.F. Witt, H.C. Gatos, J. Electrochem. Soc. 115 747 (1968).

13. A.J. Strauss, Solid-State Electron. 5 97 (1962).

14. A.F. Witt, H.C. Gatos, J. Electrochem. Soc. 113 808 (1966).

15. A.F. Witt, H.C. Gatos, in Semiconductor Silicon, R.R. Haberecht and E.L. Kern, eds., Electrochemical Society, Inc., New York (1969) 146.

16. A. Müller, M. Wilhelm, Z. Naturforsch. 19a 254 (1964).

17. W.R. Wilcox, in Fractional Solidification, Vol. I, Zief and W.R. Wilcox, eds., Arnold, London (1967).

18. J.R. Carruthers, J. Electrochem. Soc. 114 959 (1967); W.S. Robertson, Brit. J. Appl. Phys. 17 1047 (1966).

19. R. Singh, A.F. Witt, H.C. Gatos, J. Electrochem. Soc. 115 112 (1968).

20. A.F. Witt, H.C. Gatos, J. Electrochem. Soc. 115 70 (1968).

21. A.F. Witt, M. Lichtensteiger, H.C. Gatos, J. Electrochem. Soc. (in press).

22. W.G. Cochran, Proc. Camb. Phil. Soc. 30 365 (1934).

23. K.M. Kim, A.F. Witt, H.C. Gatos, J. Electrochem. Soc. 119 1218 (1972).

24. R.F. Sekerka, J. Cryst. Growth 3,4 71 (1968).

25. R.T. Delves, J. Cryst. Growth 3,4 562 (1968).

26. R. Singh, A.F. Witt, H.C. Gatos, J. Appl. Phys. 41 2730 (1970).

A REVIEW OF THE ELECTRICAL PROPERTIES AND MICROSTRUCTURE OF VANADIUM PHOSPHATE GLASSES

D.L. Kinser and L.K. Wilson

Vanderbilt University

Nashville, Tennessee, U.S.A.

Studies of the electrical properties of glasses in the VO_2-V_2O_5-P_2O_5 system are reviewed. Electron microstructural studies on glasses throughout the system revealed glass-glass immiscibility over a wide area. Correlation of the electrical properties and the corresponding microstructures indicates that the previously observed conductivity maxima in this system is a consequence of the microstructure as well as the electronic behavior of this system. There is evidence for the existence of a second conductivity maxima near the V^{4+}=V^{5+} composition although glasses in this composition range are difficult to prepare. Of these two maxima one is the result of microstructural segregation while the second is a consequence of the hopping conduction mechanism. It is thus concluded that the anomalous behavior of this system is in fact not anomalous in an electronic sense, but is a consequence of microstructural features of the system.

LITERATURE SURVEY

Electrical Properties. Vanadium phosphate glasses have been known since the early work of Roscoe[1] and Tammann and Jenckel.[2] The electrical and other physical properties of the vanadium phosphate glasses were initially examined by Denton, Rawson and Stanworth.[3] This work was later extended to a systematic examination of BaO-V_2O_5-P_2O_5 glasses by Baynton, Rawson and Stanworth.[4] Munakata conducted an extensive study of the electrical property, composition and oxidation state inter-dependencies in this system.[5,6] He concluded that the maximum electrical conductivity of the high vanadium phosphate glasses occurred at a V^{4+}/V^{total} ratio between 0.1 and

0.2 independent of the BaO content, and that conduction in these glasses was similar to conduction in crystalline magnetite, and involved valency exchange between V^{4+} and V^{5+} ions. Munakata appeared to recognize that this implied a maxima in conductivity at $V^{4+} = V^{5+}$ although he did not attempt to relate his observed maxima with this expectation.

Several Soviet investigators[7,8] have examined the electrical behavior of vanadium phosphate glasses with ternary additions of numerous other oxides. Nador[9] has reported upon a binary series of vanadium-phosphate glasses and noted evolution of oxygen during the melting of V_2O_5-P_2O_5 batches. The compositional dependence of the softening point was also reported with DTA analyses of the series. Nador also noted a minor change in conduction activation energy with extended thermal treatments at 220°C. The work of Kitaigorodskii, Frolov, and Kuo-Cheng,[10] is notable in that they examined a 75.7 V_2O_5-24.3 P_2O_5* glass melted under conditions which modified the V^{4+}/V^{total} ratio. They noted a maxima in conductivity at V^{4+}/V^{total}= 0.11.

Hamblen, Weidel and Blair[11] examined a series of vanadium pentoxide phosphate glasses with V_2O_3 as well as numerous other ternary additions. It was noted that the presence of reduced V^{4+} ions appreciably modified the electrical properties. These authors noted that devitrification of some of the ternary glasses gave specific resistivities in the 10 ohm-cm range from glasses with resistivities of 10^3 ohm-cm at room temperature. Janakirama-Rao[12] reported the electrical conductivity of several binary vanadium phosphate glasses in a paper which also included a study of the infrared spectra of the glasses. The conductivities were generally 10 to 10^3 times greater than those found by other workers.[3,9,11] No rationalization of this disparity has been presented to the authors' knowledge.

An extensive investigation of the AC and DC electrical and optical properties of vanadium phosphate glasses has been reported by Nestor and Kingery[13]. Their results indicate that the DC conduction observations can be analyzed on the basis of narrow band or polaron conduction model.[14,15] Their AC measurements at low temperatures indicated a relaxation process which they attributed to the resonance of electrons. Their results support the conduction model proposed by Munakata.[5,6] Hench and Jenkins[16] examined the AC conductivity glass and reported that the AC and DC conduction are indistinguishable at frequencies below 5 x 10^4 Hz over the temperature range 25°K.

---

* All composition quoted and discussed in this paper are on a molar composition basis.

At higher frequencies a pronounced frequency dependence of conductivity was observed at temperatures below 343°K. This transition behavior was attributed to the predominance of the hopping mobility term below 343°K which arises in a small polaron treatment[17,18,19] of conduction.

Brown[20] has conducted a study of glasses prepared from the V_2O_3, V_2O_5 and P_2O_5 oxides with somewhat surprising results. His conclusions indicate that the oxidation state sensitivity of DC conductivity is quite pronounced but he notes that glasses prepared from the V_2O_3 oxide come to equilibrium with air with a higher concentration of V^{4+} ions than glasses prepared from V_2O_5. He concludes that this is a result of retention of some structural features of the V_2O_3 oxide in the liquid state or a memory effect. It will be noted later that some of his results for the glasses prepared from V_2O_5 do not agree with those of other workers.

Allersma and Mackenzie[21] have examined the Seebeck coefficient of a series of vanadium phosphate glasses in order to gain further insight into the conduction mechanism. Their analysis indicated that the ratio of high to low valence ions plays a significant role in the Seebeck coefficient behavior in transition metal phosphate glasses; the properties of vanadium glasses were amenable to analysis using "simple" theories but glasses based on iron oxide were not.

Kennedy, Hakim, and Mackenzie[22] have reported the preparation of amorphous V_2O_5 by a vapor deposition and the conductivity of this material. Linsley, Owen and Hayatee[23,24] made an extensive investigation of the dependence of the electrical properties on composition and oxidation; the conductivity goes through a maximum with changing V^{4+}/V^{total}. A maximum is expected but at 0.5 and not at 0.1 to 0.2 as found; however Munakata found similar behaviour[5,6]. These glasses are unlike the iron phosphate glasses[25] where the maxima in conductivity occurs at $Fe^{2+}/Fe^{total} = 0.5$. These workers have suggested that this behavior is a consequence of complexing of some of the V^{5+} leading to its inactivity in the conduction process. Two additional papers[26,27] which appeared about the same time as those of Linsley indicated that phase separation occurs in these glasses. The thermal history of samples examined under the microscope was quite different from those used in the electrical studies, but phase separation of a glass-glass type was clearly indicated. Schmid[28,29] has studied AC and DC conductivity at lower temperatures than most workers except Hench and Jenkins[16]. His results and analyses indicate that a small polaron model accurately describes the AC and DC behvaior of the glasses examined. This model explains the tendancy towards temperature independence of the conductivity at low temperatures. Schmid also noted that changes in the melting temperature altered the DC resistivity of the glasses. He noted that these changes were consistent with a higher V^{4+}/V^{total} ratio in the glasses melted at higher temperatures.

Caley and Krishna Murthy[30] made a systematic survey of the effect of WO_3, GeO_2, SiO_2, B_2O_3, and TeO_2 additions upon the electrical conductivity of the 80 V_2O_5 -20 P_2O_5 glass. They found that the vanadium glass is relatively unaffected by the replacement oxides. They noted that binary WO_3 - P_2O_5 glasses exhibited a pronounced change in DC conductivity with a small V_2O_5 addition.

Magnetic Properties. Recently several workers[31,32,33] including the present authors[34] have studied magnetic resonance in vanadium phosphate glasses. From nuclear magnetic resonance (NMR) studies, Landsberger and Bray[35] postulated that the site symmetry of the V^{4+} and V^{5+} ions is identical in the glass. This implies that electron hopping can occur without structural rearrangement of the glass. The electron paramagnetic resonance (ESR) work of Sayer and the present authors indicates that an antiferromagnetic exchange interaction exists between V^{4+} ions. This indicates an increased V^{4+} - V^{4+} interaction greater than would be expected from a random distribution of V^{4+} and V^{5+}ions. Thus the electron exchange probability between aliovalent ions is effectively reduced. In addition, the ESR line width was found to be temperature independent for all V^{4+}/V^{total} ratios. Since the conductivity is nonlinear over the same temperature range, this observation suggests that the mobility rather than the charge carrier density is temperature dependent. These observations have also indicated considerable order in first nearest neighbours although structural order beyond this point is uncertain. A more recent work[36] has suggested that, in most transition metal phosphate glasses, not all of the aliovalent ions are magnetically coupled, and in fact some behave as isolated ions. The effect of compositional modification of microstructure has also been examined in a series of vanadium-phosphate glasses by the authors[39]. This work has clearly indicated the presence of compositional segregation of vanadium ions in these glasses.

The dielectric behavior and transport properties have been examined in considerable detail in the work of Sayer[31,32,33]. This work showed that the carrier concentration calculated from dielectric relaxation behavior and DC conductivity using a thermal diffusion model agrees well with electron paramagnetic resonance measurements.

In summary, the electrical and magnetic observations made on vanadium phosphate glasses indicate that the conduction mechanism involves an electron exchange between V^{4+} and V^{5+} ions but several inconsistencies appear. First, the maxima in conductivity versus V^{4+}/V^{total} should appear at equal concentrations of V^{4+} and V^{5+}; but the available experimental data gives the maxima at V^{4+}/V^{total}= 0.1 to 0.2. There also are several observations which indicate that these glasses are microstructurally heterogeneous and subject to thermal history. The microscopic compositional segregation has not been considered in any of the models advanced to this point. The magnetic results suggest that some V^{4+} ions behave differently from others and the role of the antiferromagnetically coupled ions is effectively reduced in electrical conductivity.

A REINTERPRETATION

As a new approach to presenting the extensive electrical property observations of numerous workers on vanadium phosphate glasses, consider the ternary glass system VO_2 - V_2O_5 - P_2O_5. If a glass is made from a batch calculated to give $90V_2O_5$ $10P_2O_5$, the composition may change by loss of oxygen:

$$V_2O_5 \rightarrow 2\ VO_2 + 1/2\ O_2 \uparrow$$

In practice, reduction to VO_2 is never complete and the actual glass "composition" lies between $90V_2O_5$ - $10P_2O_5$ and 94.7 VO_2 - 5.3 P_2O_5. The resulting glass is therefore a mixture of the two vanadium oxides and phosphorous pentoxide. The three component diagram in Fig. 1 indicates the course of the reaction and shows the possible positions on the ternary diagram where the glass made from a 90 V_2O_5 - 10 P_2O_5 batch can lie. A series of "binary" V_2O_5 - P_2O_5 glasses, melted in air for 1 hour at 1100°C, were found to have compositions as indicated in Fig. 1. It is thus evident, but of course not a new observation, that the glasses richer in P_2O_5 are considerably more reduced. Several isolated observations have confirmed that this curve will shift upward for higher melting temperatures.[37]

A resistivity composition diagram at constant temperature for the ternary system for all the available resistivity observations for which oxidation state data are available, as given in Fig. 2. Much of the data in the literature is collected on this one plot and much of it is in poor agreement. It appears that the most consistent data are those of Linsley et al.[23,24] while that of Brown[20] appears to be consistently different. This could be the result of different melting temperatures, although our own data agree well with Linsley even though we melted at 1100°C while he melted at 800 to 900°C. It is also probable that the quenching and annealing treatments are different among the various investigators.

In Fig. 2 the location of the minima in resistivity according to Linsley et al. is shown. This line is almost parallel to the base of the triangle thus suggesting some structural origin of the maximum conductivity.

We have noted that the data plotted in this manner do not place the minima in resistivity or the maxima in conductivity in the same

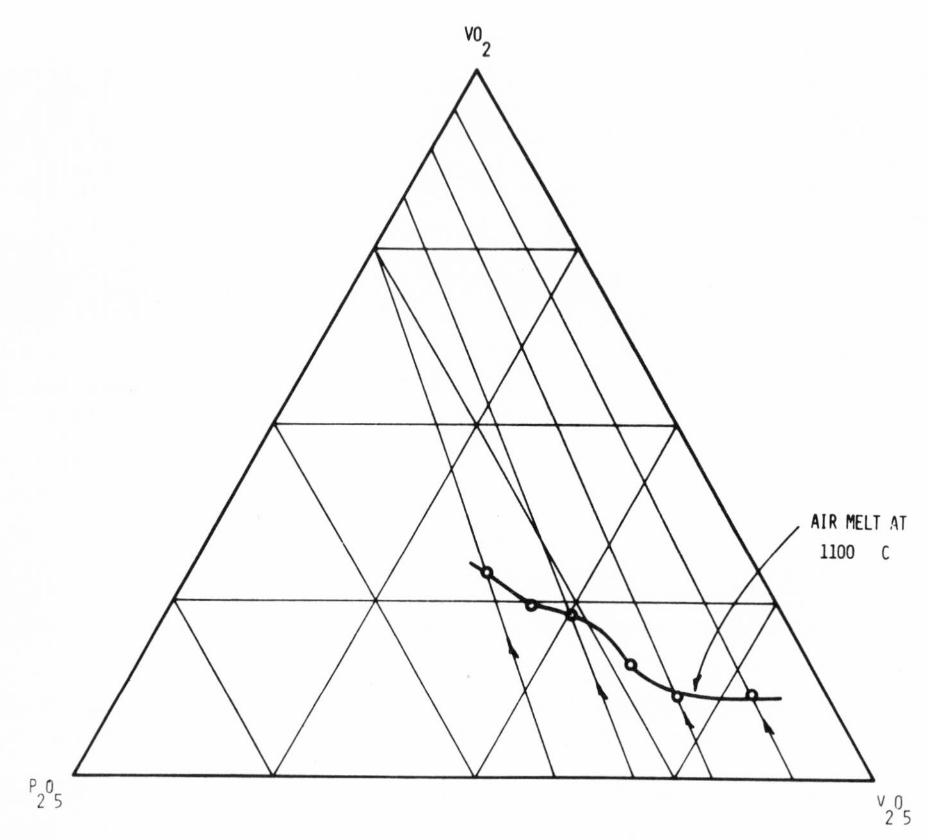

Fig. 1: Ternary plot of the VO_2-V_2O_5-P_2O_5 glass system. The lines from the Binary V_2O_5-P_2O_5 side of the diagram to the VO_2-P_2O_5 side of the diagram are the loci of possible glass compositions resulting from melting of binary V_2O_5-P_2O_5 glasses. The curve denotes the actual compositions of glasses melted one hour in air at 1100°C.

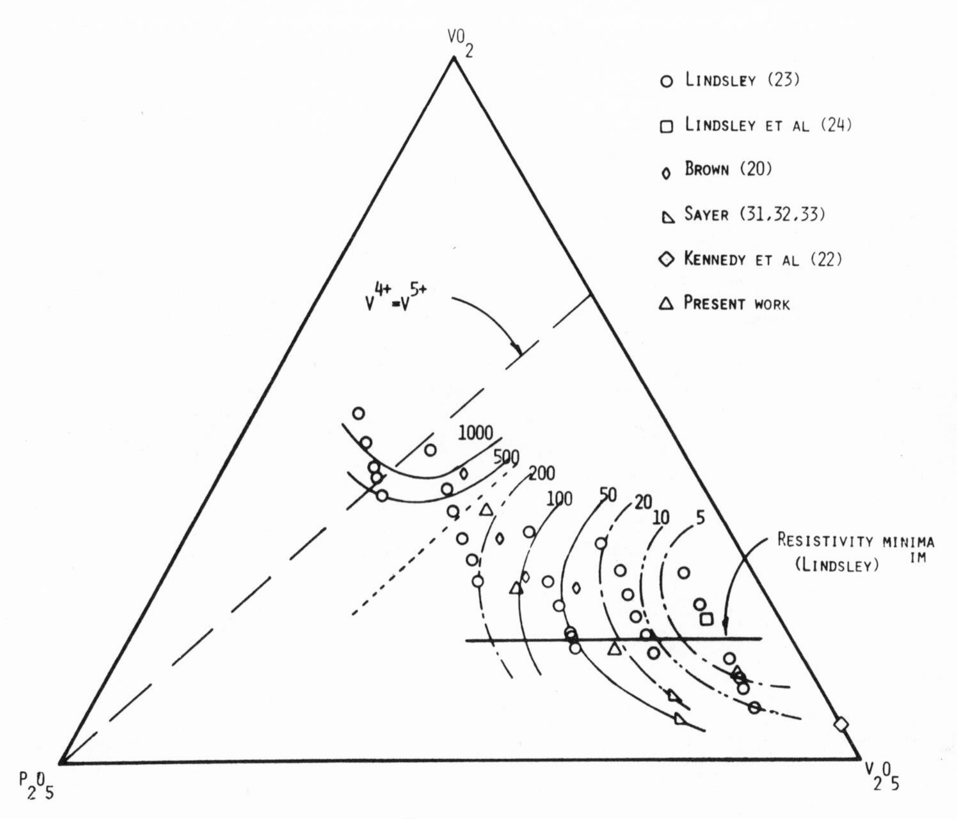

Fig. 2A: Ternary plot of 10^3 times the electrical resistivity at $10^3/T = 3.0$ (65°C). Data of the various authors are smoothed and some points from the results of Brown[20] were discarded because of poor agreement.

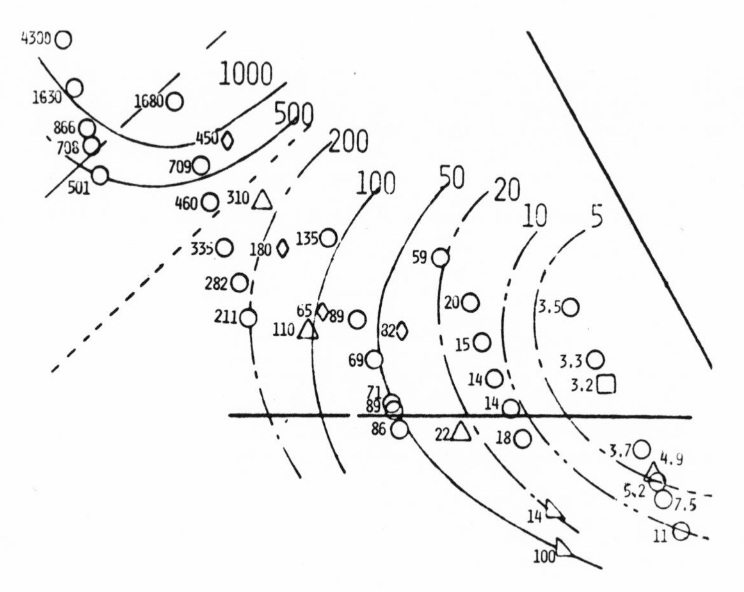

Fig. 2B: Detail from part A (Fig. 2) with actual data inserted for purposes of analysis.

location as that of Linsley et al. It appears from this composite that the minima in resistivity should be located somewhat more toward the reduced side of the diagram although the indicated upward shift has a small influence upon the value of the V^{4+}/V^{total} ratio at the conductivity maxima. Kitaigorodskii et al.[10] found the maxima in conductivity to be at $V^{4+}/V^{total} = 0.12$, which is somewhat lower than the Linsley value; unfortunately these observations could not be compared at the same temperatures and this may be the origin of the difference. Kitaigorodskii found a more complex behavior with composition than other workers. This inconsistency is perhaps a consequence of different thermal history as melting temperatures are approximately equivalent.

There is another important observation to be made from this composite plot. The behvaior of the contours in the neighborhood of $V^{4+}/V^{total} = 1/2$ suggests that a second minimum in resistivity may occur at this theoretically predicted position. It is unforunate that data are not available for this composition but the data which are available suggest, but do not prove, the existence of a minima at the predicted position. These glasses are of course most difficult to prepare because of the extreme reducing conditions required and because of the increasing tendency toward crystallizations in the VO_2 corner of the system.

An investigation of the microstructural features of the ternary vanadium phosphate glasses[38] showed extensive glass-glass phase separation. Such immiscibility was suggested by Anderson and Luehrs[26,27] and Janakirama-Rao[12], but a systematic investigation has not been reported. A typical series of micrographs is shown in Figs. 3, 4, 5 and 6, which show how the microstructure varies from the V_2O_5 rich corner of Fig. 1 along the "S" shaped curve toward the more reduced side of the diagram. These micrographs indicate a clear trend from the nominal $60V_2O_5$-$40P_2O_5$($29VO_2$-$37V_2O_5$-$34P_2O_5$) glass of Fig. 3 to the other extreme at the nominal $90V_2O_5$-$10P_2O_5$ ($12VO_2$-$79V_2O_5$-$9P_2O_5$). It is evident that the heterogeneity is maximum at each extreme of the compositional series examined; those glasses with compositions near the maximum conductivity line are practically homogeneous, while those removed from that composition line are clearly heterogeneous. The maxima in conductivity could therefore be the result of a maximum miscibility of V^{4+} and V^{5+}. Compositions removed from this line have a segregated or immiscible phase containing either V^{4+} or V^{5+} depending upon their position with respect to the observed "homogeneous" glasses.

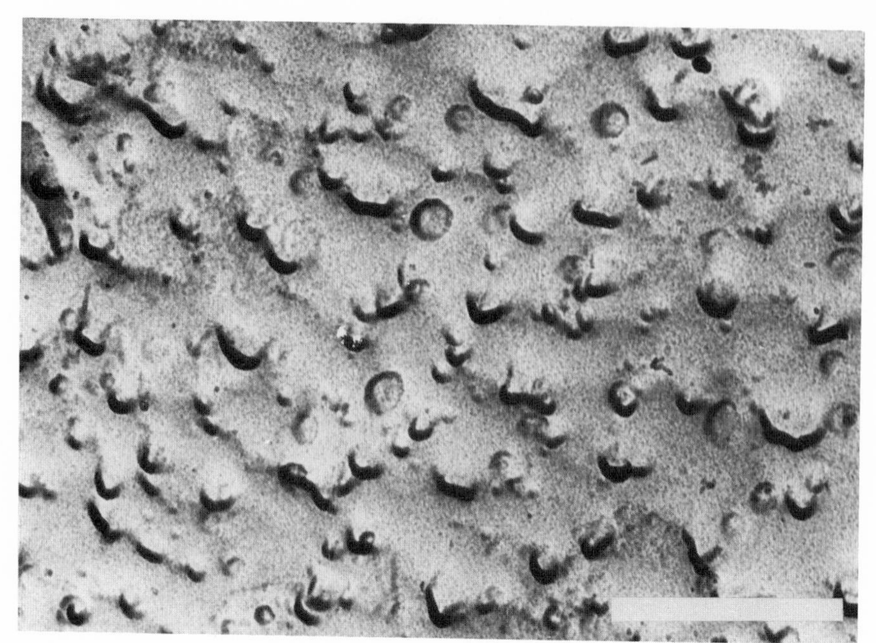

Fig. 3: Replica electron micrograph of the $29VO_2$-$37V_2O_5$-$34P_2O_5$ glass (Batch $60V_2O_5$-$40P_2O_5$). Etched 30 seconds in 1% aqueous HCl. (Bar = 1μ)

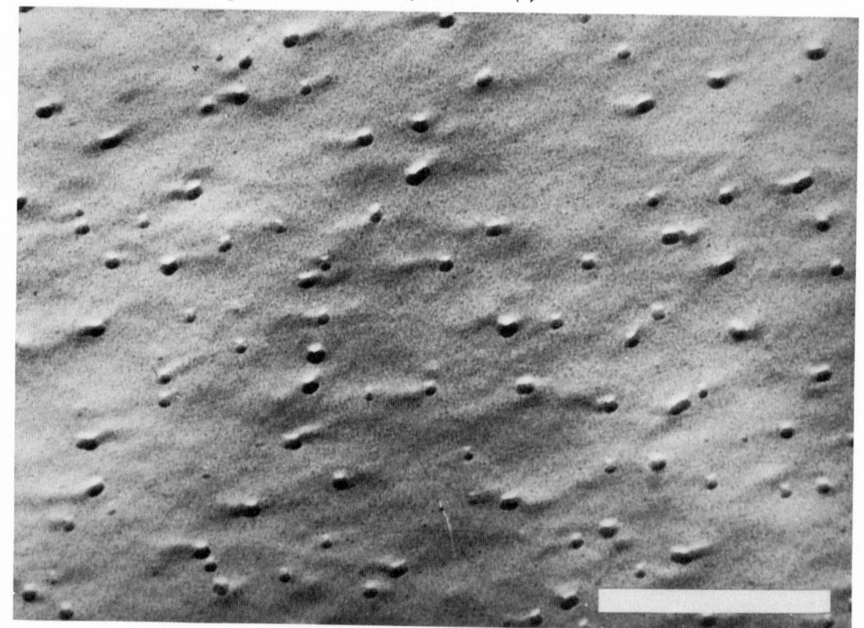

Fig. 4: Replica electron micrograph of the $23VO_2$-$51V_2O_5$-$26P_2O_5$ glass (Batch $70V_2O_5$-$30P_2O_5$). Etched 30 seconds in 1% aqueous HCl. (Bar = 1μ)

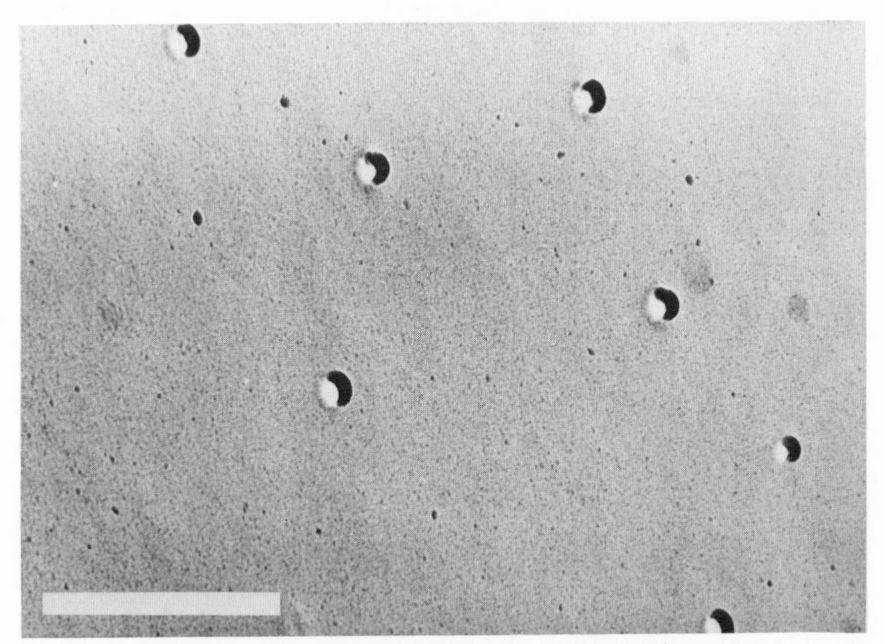

Fig. 5: Replica electron micrograph of the $12VO_2$-$69V_2O_5$-$19P_2O_5$ glass (Batch $80V_2O_5$-$20P_2O_5$). Etched 30 seconds in 1% aqueous HCl. (Bar = 1μ)

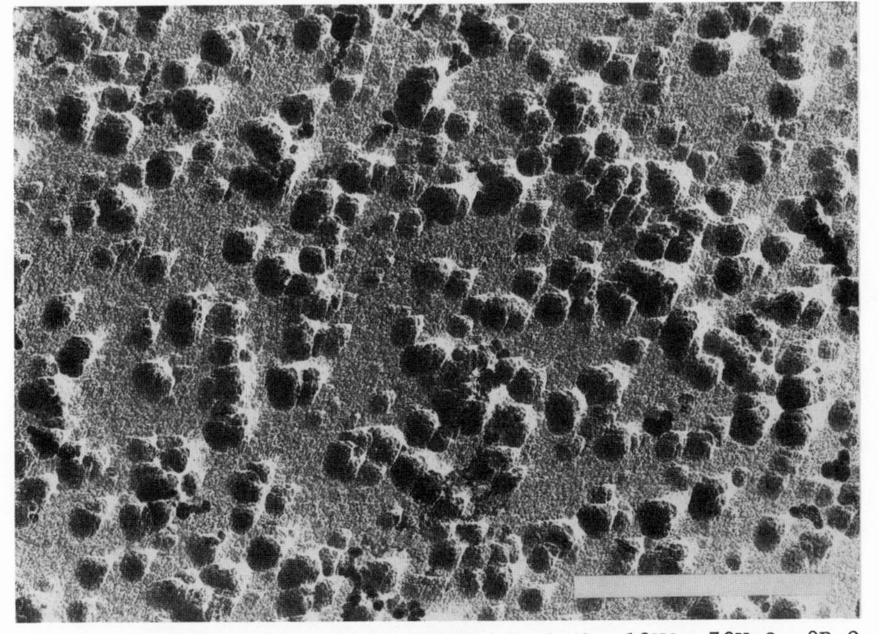

Fig. 6: Replica electron micrograph of the $12VO_2$-$79V_2O_5$-$9P_2O_5$ glass (Batch $90V_2O_5$-$10P_2O_5$). Etched 30 seconds in 1% aqueous HCl. (Bar = 1μ)

ACKNOWLEDGEMENTS

The authors wish to thank A. W. Dozier, E.J. Friebele and J. G. Vaughan for their help in collecting the experimental data and calculations. The financial support of the U.S. Army Research Office, Durham, under contract DAHC04-70-C-0046 is gratefully acknowledged.

REFERENCES

1. H.E. Roscoe, Phil. Trans. Roy. Soc. 158 1-27 (1867).

2. G. Tammann, E. Jenckel, Z. Anorg. Allgem Chem. 184 416-420 (1929).

3. E.P. Denton, H. Rawson, J.E. Stanworth, Nature 173 1030-1032 (1954).

4. P.L. Baynton, H. Rawson, J.E. Stanworth, J. Elchem Soc. 104 237-240 (1957).

5. M. Munakata, S. Kawamura, J. Asahara, M. Iwamoto, J. Ceram. Assoc. Japan 67 344-353 (1959).

6. M. Munakata, Sol. St. Electronics 1 159-163 (1960).

7. V.A. Ioffe, I.B. Patrina, I.S. Poberouskaya, Solid State 2 609-614 (1960).

8. L.A. Grenchanik, N.V. Petrouykh, V.G. Karpechenko, Soviet Physics - Solid State 2 1908-1915 (1960).

9. B. Nador, Steklo i Keramika 17 18-21 (1960).

10. I.I. Kitaigorodskii, V.K. Frolov, Kuo-Cheng, Steklo i Keramika 17 5-7 (1960).

11. D.P. Hamblen, E.A. Weidel, G.E. Blair, J. Amer. Ceram. Soc. 46 499-504 (1963).

12. Bh. V. Janakirama-Rao, J. Am. Ceram. Soc. 49 605-609 (1966).

13. H.H. Nestor, W.D. Kingery, Proc. Intern. Congress du Verre, 106-110.

14. R.R. Heikes, R.W. Ure, Thermoelectricity: Science and Engineering, Interscience Publishers, New York (1961) 75-82.

15. N.F. Mott, R.W. Gurney, Electronic Processes in Ionic Crystals, Oxford, London (1950).

16. L.L. Hench, D.A. Jenkins, Phys. Stat. Sol. 20 327-330 (1967).

17. H.R. Killias, Phys. Letters 20 5-6 (1966).

18. T. Holstein, Am. Phys. (USA) 8 343-389 (1959).

19. L. Friedman, Phys. Rev. 135A 233-246 (1964).

20. R.M. Brown, PhD Thesis, University of Illinois (1966).

21. T. Allersma, J.D. Mackenzie, J. Chem. Phys. 47 1406-9 (1967).

22. T.N. Kennedy, R. Hakin, J.D. Mackenzie, Mat. Res. Bull. 2 193-201 (1967).

23. G.S. Linsley, PhD Thesis, University of Sheffield (1968).

24. G.S. Linsley, A.E. Owen, F.M. Hayatee, J. Non-Cryst. Solids 4 208-219 (1970).

25. K.W. Hansen, J. Electrochem. Soc. 112 994-6 (1966).

26. G.W. Anderson, F.V. Luers, J. Appl. Phys. 39 1634-38 (1968).

27. F.V. Luers, G.W. Anderson, Proceedings, 26th Annual Electron Microscopy Soc. of America, C.J. Arceneaux, ed. (1968) 422.

28. A.P. Schmid, J. Appl. Phys. 39 3140-49 (1968).

29. A.P. Schmid, J. Appl. Phys. 40 4128-36 (1969).

30. R.H. Caley, M. Krishna Murthy, J. Am. Ceram. Soc. 53 254-57 (1970).

31. M. Sayer, A. Mansingh, J.M. Reyes, G.F. Lynch, Second Int. Conf. on Low Mobility Materials, Taylor and Francis Ltd., London (1972) 115-123.

32. G.F. Lynch, M. Sayer, S.L. Segel, G. Rosenblatt, J. Appl. Phys. 42 2587-91 (1971).

33. M. Sayer, A. Mansingh, J.M. Reyes, G. Rosenblatt, J. Appl. Phys. 42 2857-64 (1971).

34. E.J. Friebele, L.K. Wilson, D.L. Kinser, J. Am. Ceram. Soc. 55 164-68 (1972).

35. F.R. Lansberger, P.J. Bray, J. Chem. Phys. 53 2757-68 (1970).

36. L.K. Wilson, E.J. Friebele, D.L. Kinser, in Amorphous Magnetism, H.O. Hooper and A.M. de Graaf, eds., Plenum, New York (1973).

37. A. Kato, R. Nishibashi, M. Nagano, I. Mochida, J. Am. Ceram. Soc. 55 183-85 (1972).

38. D.L. Kinser, E.J. Friebele (submitted to J. Am. Ceram. Soc.).

ELECTRICAL SWITCHING IN CALCIUM PHOSPHATE GLASSES CONTAINING IRON

M. H. Omar and M. N. Morcos

Solid State and Materials Research Center

The American University in Cairo

Electrical switching was studied in a series of calcium phosphate glasses containing iron. The samples were prepared by melting and evaporating of the glasses. The switching phenomena was studied using a.c. and d.c. voltages. The samples showed switching, from a high resistance state (off-state $\geq 10\ M\Omega$) to a low resistance state (onstate $\leq 1\ K\Omega$) at certain threshold voltages. The switching was invariably of the memory type and a high current pulse was required to switch the samples to the off-state. In most cases, no negative resistance was observed. The threshold switching voltage decreased with higher iron content and with decreasing thickness of the samples. The most significant parameter appeared to be the temperature; a marked decrease in threshold voltage was observed with increasing temperature of the sample.

INTRODUCTION

Switching effects have been observed in many amorphous materials. Most of the research work carried out in this field deals mainly with chalcogenide glasses,[1,2,3] and only a few papers have been published on switching effects in transition-metal-oxide semiconducting glasses.[4,5] In these glasses, electrical conduction is believed to occur between ions of the transition metal in their different valence states. They are characterised by their high melting points, cheap constituent materials and the fact that their electrical properties are not sensitive to the presence of impurities.

In almost all published work on switching in semiconducting

glasses, there has been a general feeling that these devices suffer from lack of reproducibility, and that they show erratic behaviour. In our previous study on switching in borate glasses containing iron, we observed a wide scatter in the values of the threshold for bulk samples.[5] The switching behaviour of samples which were melted in vacuum showed some improvement. The purpose of the present investigation is to study another transition metal oxide glass[6] and to try other methods of sample preparation which may show better reproducibility and performance. Attempts have also been made to reduce the threshold voltage and to manufacture very thin devices of known chemical composition by means of sputtering techniques.

EXPERIMENTAL

Preparation of the Samples. The series of glasses investigated in this work are of the molar composition CaO P_2O_5 x Fe_2O_3, where x is the concentration of iron oxide which ranged between 10 and 30 mole%. These glasses (from which the switching devices were made) were prepared by melting under normal conditions in zircon crucibles at 1200°C "Melted samples" were prepared by melting a small glass fragment on a molybdenum sheet heated electrically under a vacuum of about 10^{-4} mm Hg. "Evaporated samples" were prepared by flash evaporation of a glass fragment on a cold metal substrate.

A few samples were prepared by sputtering in an argon gas atmosphere. However, the films prepared by sputtering were too thin and too fragile to give reliable results.

Measurements. Different types of electrodes were tried. In one type, the metal substrate acted as one electrode while a tungsten whisker with controlled pressure was used as the other electrode. The contact area for such whiskers was of the order of 10^{-4} mm^2. Larger contact areas up to about 1 mm^2 were also used. Evaporated silver electrodes on the surface of the samples, by shadow masking, were made with different contact areas. Other types of electrodes such as mercury-indium droplets or colloidal silver paint were also tried.

Electrical measurements were carried out using a.c. (50 c/s), d.c. and pulsed d.c. voltages applied across the samples. A high resistor connected in series with the devices served to limit the current when switching to the on-state took place. The current in the samples was measured by the voltage across a standard resistor connected in series with the samples. I-V characteristics were displayed on a dual-trace oscilloscope equipped with a polaroid camera. The samples were placed in a holder which enabled measurements to be carried out under vacuum. Provision was made to vary the temperature of samples between liquid nitrogen up to about 500°C.

RESULTS

Melted thick films. Freshly prepared samples were found to require relatively high voltage (about 800 volts) to switch for the first time. In the following successive cycles a much lower voltage (about 200 volts) was needed for switching. The application of a high "forming" voltage during the first switching was found to be necessary for all the melted samples, and was in general three to four times greater than subsequent threshold voltages. After the initial forming process, the device generally could be switched repeatedly in the usual way between the two stable states. A typical behaviour of a sample of this type (prepared from a glass containing 30 mole % Fe_2O_3) is shown in Fig. 1. The threshold voltage for the sample is about 250 volts. The same characteristics were obtained using 50 c/s a.c., d.c. or pulsed d.c. of up to 10 Kc/s. In the case of a borate glass containing the same Fe_2O_3 concentration a negative resistance was obtained[5] instead of the I-V behaviour shown in Fig. 1.

With lower iron content (less than 20 mole%), melted samples of phosphate glass showed the same switching characteristics with a memory similar to that shown in Fig. 1. However, the threshold voltage for these films was found to vary over a large range of values and in many cases it was difficult to revert the devices back to the off-state by the usual current pulse. A histogram showing the behaviour of 3 devices (30 mile % Fe_2O_3 melted film) is shown in Fig.2.

Evaporated thin film. Off-state resistance for such devices was of the order of 1 - 10 Ω and the on-state resistance was not more than a few hundred ohms. The low resistance state was stable even when the applied voltage was reduced to zero (memory switching). The application of the high current pulses immediately reverted the sample to its original high resistance state. No higher forming voltage was required in the first switching event as in the case of melted films. Again the I-V characteristics are identical to those shown in Fig. 1, but with low threshold voltage values. Threshold voltage histograms for three points on an evaporated sample (30 mole% Fe_2O_3) are shown in Fig. 3. The threshold voltage at which switching occurred for the evaporated thin films appeared to be lower and more reproducible than that of melted films. Several resistance measurements were performed on the devices in both the off-state and the on-state. An example of such measurements for a film evaporated from a glass containing 10 mole % Fe_2O_3 is shown in Fig. 4.

In order to compare melted and evaporated films, data on the distribution of the threshold voltage for 28 successive switching cycles is shown in Fig. 5. As the samples were not of the same thickness, comparison here is generally based on the distribution of the threshold voltage only.

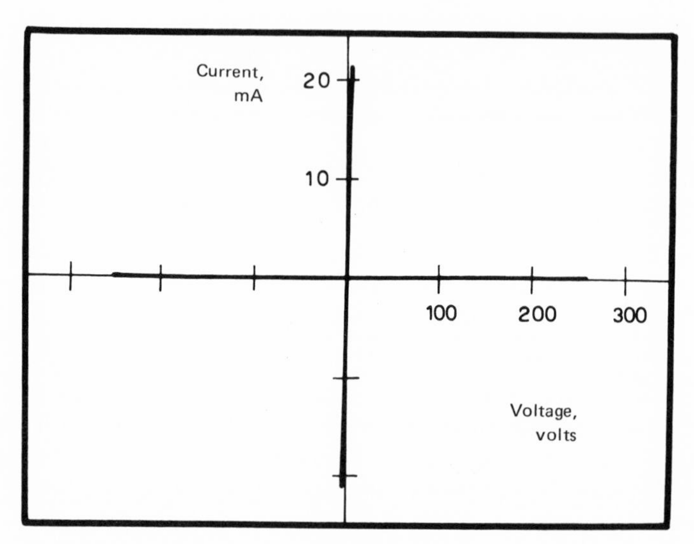

Fig. 1: I-V characteristic of melted film of calcium phosphate glass containing 30 mole% Fe_2O_3.

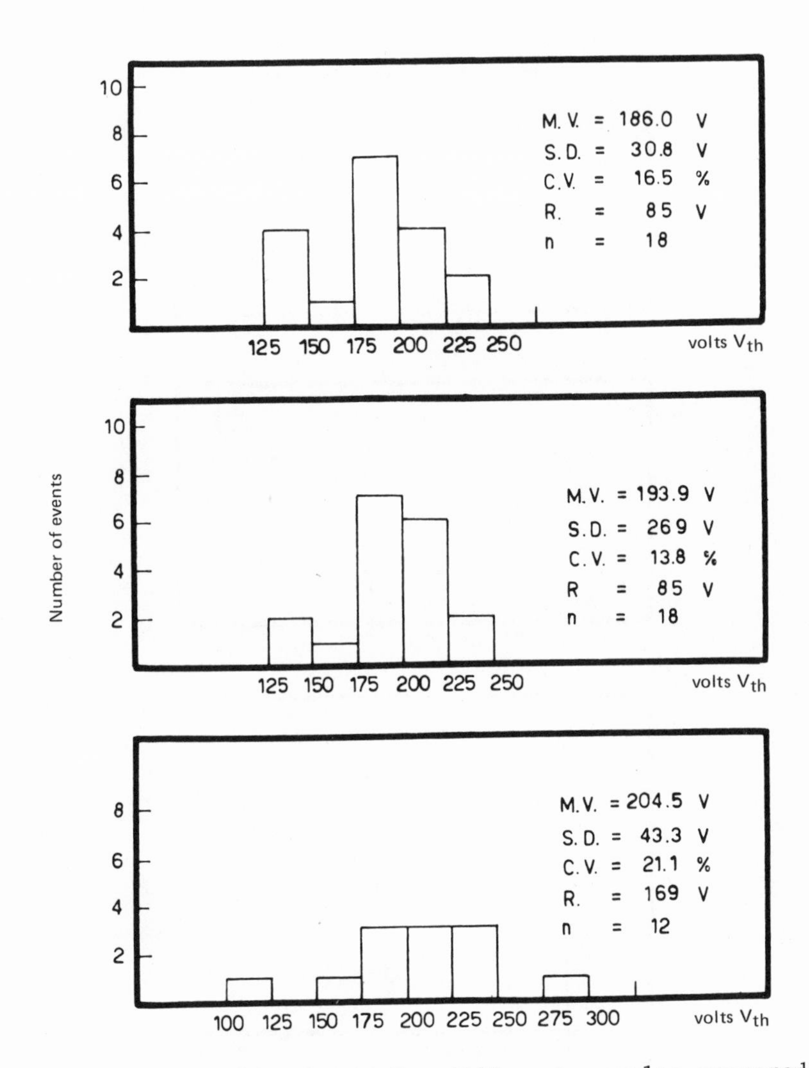

Fig. 2: Threshold voltage for different samples prepared by melting from a glass containing 30 mole% Fe_2O_3.

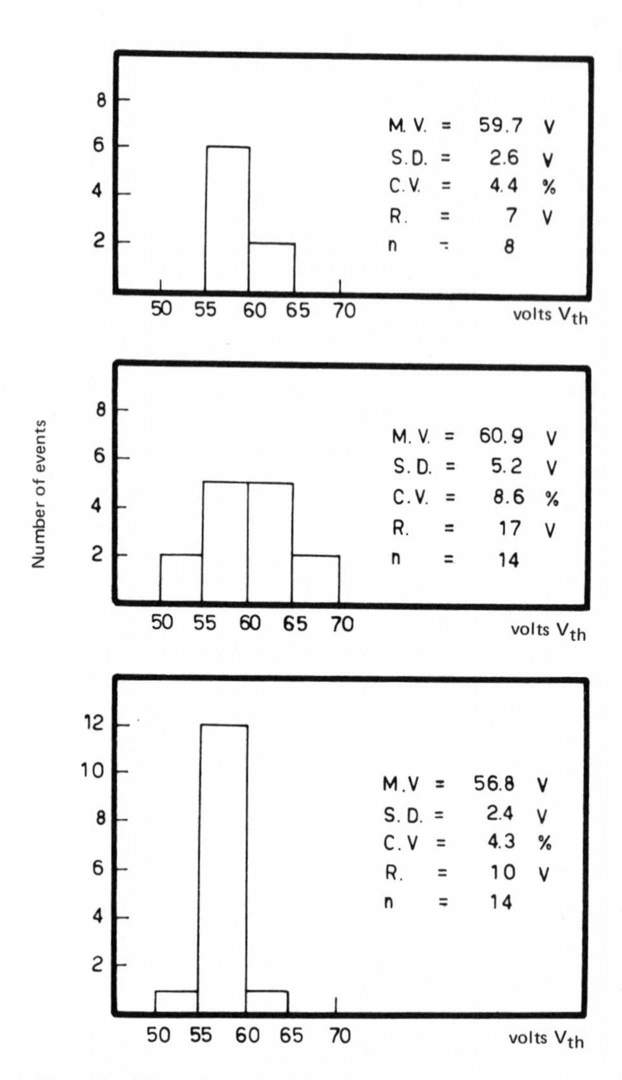

Fig. 3: Threshold voltages for different samples prepared by evaporation from a glass containing 30 mole% Fe_2O_3.

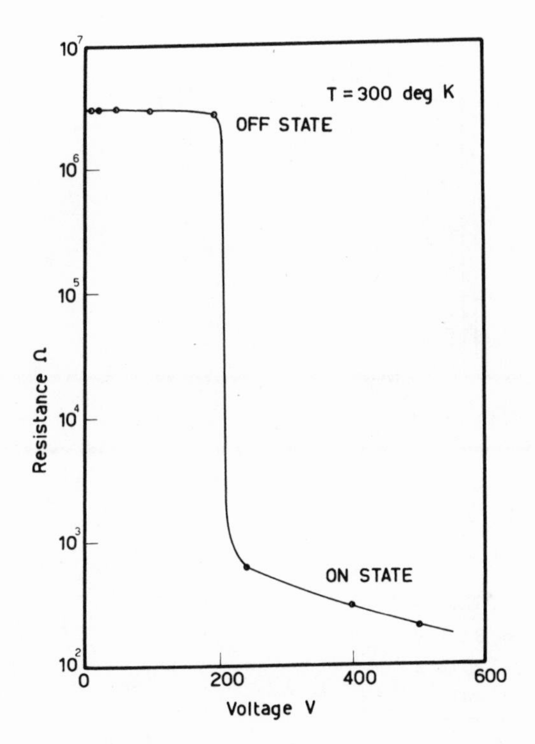

Fig. 4: Variation of resistance of an evaporated sample (10 mole% Fe_2O_3) with applied voltage before and after switching.

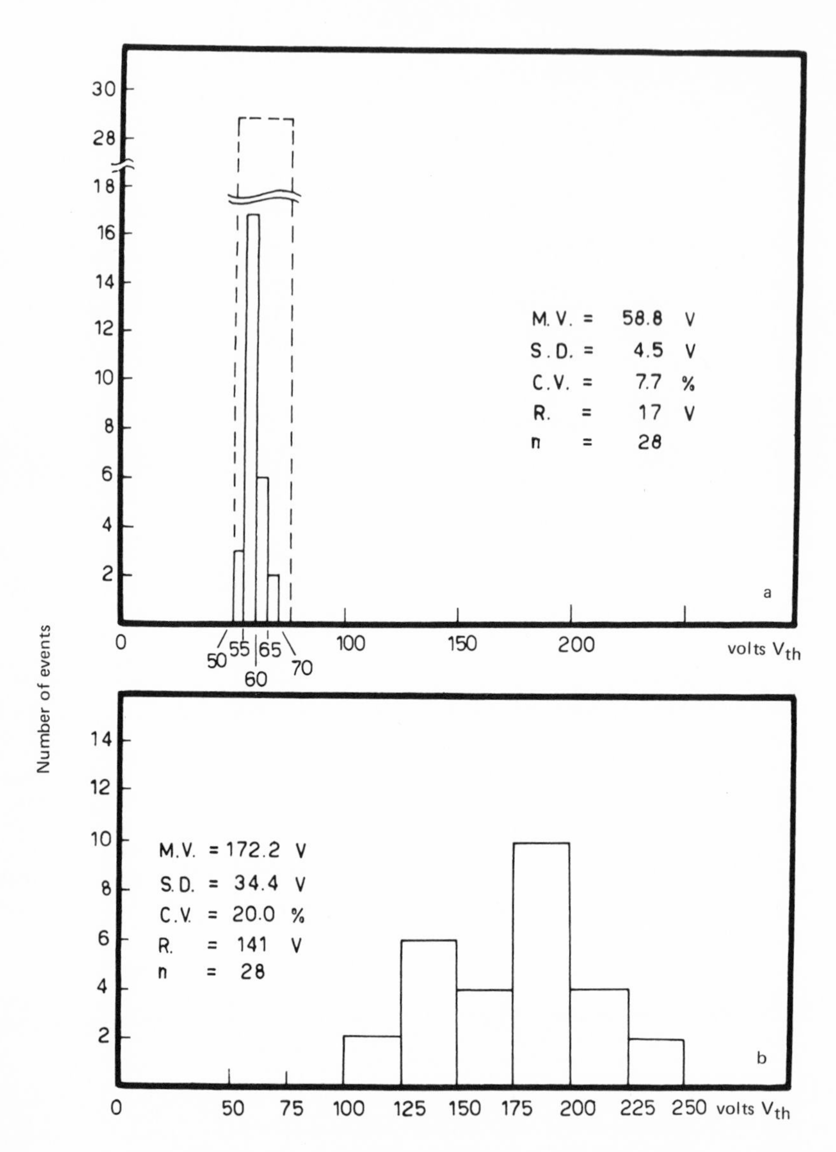

Fig. 5: Threshold voltage for (a) evaporated and (b) melted film samples of phosphate glass containing 30 mole% Fe_2O_3.

1. Device Temperature. In general, on increasing the device temperature, the threshold voltage was found to decrease. This was also observed in the borate system,[5] and the same effect was observed for chalcogenide glasses.[4]

The dependence of the threshold voltage on the temperature for a 20 mole % Fe_2O_3 evaporated device is shown in Fig. 6.

2. Effect of Heat Treatment. It was found that on heat treating the samples, the threshold voltage decreased with increasing the temperature and/or the time of heat treatment. Data for evaporated samples from a glass containing 10 mole % Fe_2O_3 annealed for 15, and 30 minutes at 450°C are shown in Fig. 7.

3. Sample Composition. The effect of increasing the iron oxide content in all samples appeared to reduce the threshold voltage; however, the general switching characteristics of the device remained unchanged. The effect of varying the iron content on the threshold voltage is shown in Fig. 8.

4. Electrode Configuration. Attempts were made to use different electrode configurations and different electrode materials, as mentioned above. The results are shown in Fig. 9. In fact, no significant changes in the threshold voltage were observed when different electrode materials of approximately the same area were used. However, the number of times that the device with evaporated and painted electrodes could be switched reversibly before breaking down was less than when a tungsten electrode was used.

5. Sample Thickness. Experiment indicated that as the thickness of the device was decreased, the voltage required to switch the device to the on-state decreased. Samples of different thicknesses were prepared and the threshold voltages for each thickness were determined. The distribution of the threshold voltages for three samples of different thickness $t_3 > t_2 > t_1$, are shown in Fig. 10. Two points were chosen for the thickness t_3 because the threshold voltage was quite irregular. Microscopic examination of the film surface indicated the sample's high granularity and rough topography.

DISCUSSION

There has been considerable debate concerning the possible mechanisms for the fast reversible switching observed in semiconducting glasses and several theories have been advanced to account for this phenomenon. There are two major lines of thought reading the origin of the electrical switching phenomenon. Some workers (Hindley[8], Haderland and Stiegler[9], Lucas[11], Fagen and Ovshinsky[11], Austin[12].

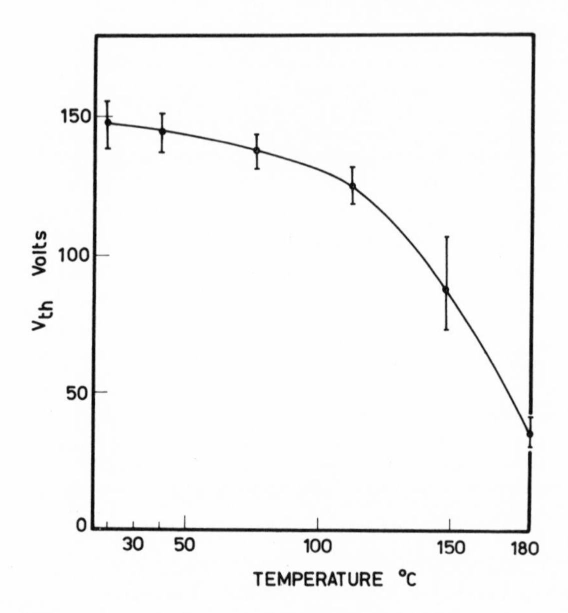

Fig. 6: Dependence of threshold voltage on sample temperature of an evaporated sample containing 20 mole% Fe_2O_3.

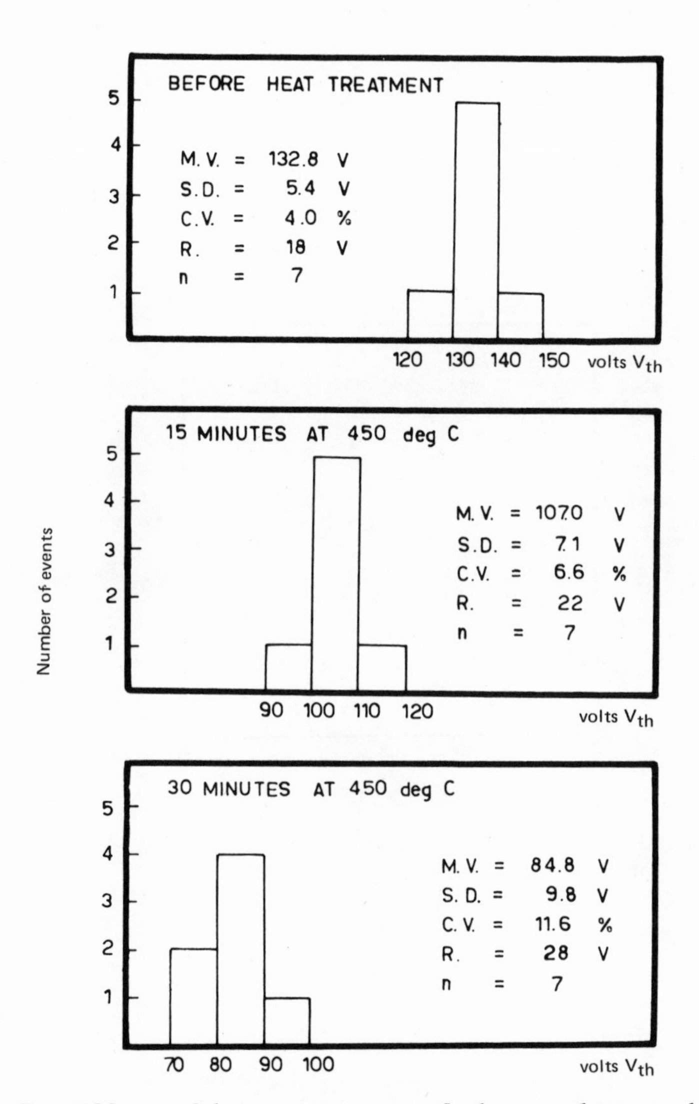

Fig. 7: Effect of heat treatment of the samples on threshold voltage. Evaporated films containing 10 mole% Fe_2O_3.

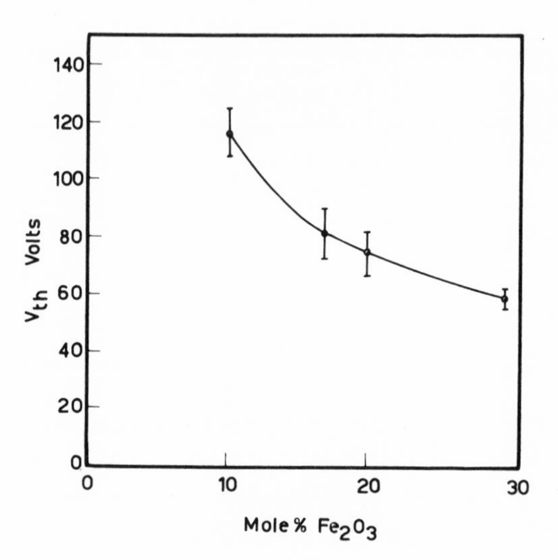

Fig. 8: Dependence of threshold voltage on iron-oxide concentration in the samples.

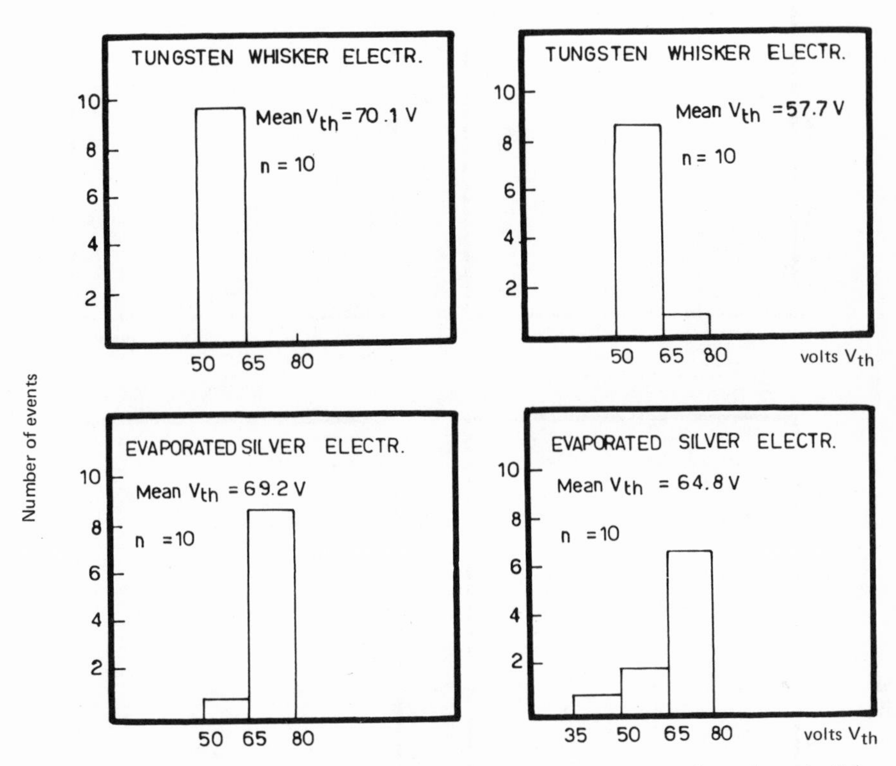

Fig. 9: Effect of electrode configuration on the threshold voltage needed for switching.

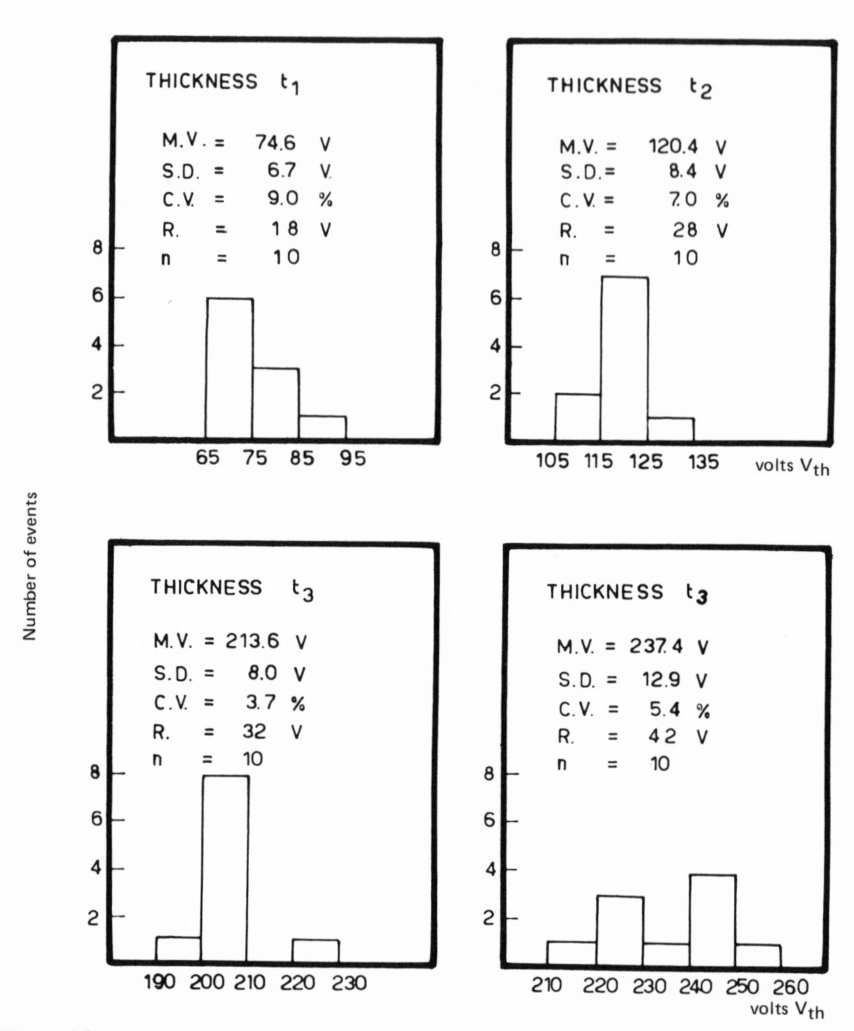

Fig. 10: Dependence of threshold voltage on sample thickness ($t_3 > t_2 > t_1$).

Scanlan and Engel[4] etc.) favour the idea that switching in semiconducting glasses is mainly due to electronic processes.

On the other hand, many workers believe that switching is due to thermal processes. The thermal mechanism for switching has been supported by Stocker, Barlow and Weiranch[13], Croitoru and Popescu[14], Kaplan and Adler[15], Warren[16], Thomas and Male[17], Pearson[18], Eaton[19], Omar[5], and others.

The memory switching characteristics that we have invariably observed in our samples (prepared from calcium phosphate glass containing iron) suggest a thermal mechanism producing a phase change which maintains the more conducting state, (on-state). Our glasses have a low thermal conductivity, and a negative temperature coefficient of resistivity[21]. The two major conditions for Joule self-heating of the glass are therefore satisfied. Joule heating in such glasses would lead to the formation of a highly conducting path (or filament) extending between the electrodes and maintain the On-memory state which is stable under zero bias. The formation of such filaments have been reported by many workers.[20,21,22] In the study on switching in borate glasses containing iron[5], the formation of crystalline paths extending between the electrodes was observed.

I-V characteristics traced at different high temperatures supported the idea that a thermally produced filament is responsible for switching. The passage of a large current pulse in the memory state would result in the fusion of hot spots along this filament, and the sharp fall of this pulse would quench these fused regions back to their original glassy state, thereby reestablishing the original high resistance (off-state).

In the case of melted films, a thin layer was formed on top of the samples. This layer is of much higher electrical resistance than the bulk of the sample and therefore a high forming voltage is needed to switch the melted device for the first time. In this way the tungsten whisker penetrates through this insulating layer to make contact with the semi-conducting glass sample. Thereafter the device would behave in the normal fashion. This is supported by the fact that no forming voltage was needed in the case of evaporated samples. Our observation that the threshold voltage decreases with increasing temperature of the glass as shown in Fig. 6, supports the proposed mechanism of Joule heating. This is also consistent with the fact that as the temperature is raised, molecular re-arrangement and relaxation become easier in the glass, thereby facilitating the formation of the crystalline filaments.

In addition the reduction of the threshold voltage by heat treatment of the samples indicates that the samples were devtrified and hence the threshold voltage was lowered. Devitirification of calcium phosphate glasses containing iron by heat treatment has also been shown through Mossbauer and x-ray diffraction techniques.[6] This lends further support to the proposed mechanism for switching due to Joule heating in devices.

The lower threshold voltages that were required to switch the samples of higher iron content, shown in Fig. 8 indicate the importance of the presence of iron in the glasses. The role of iron as it appears, is to promote the formation of the conducting crystalline phase at lower threshold voltages.

The decrease of the threshold voltage with decreasing thickness as shown in Fig. 10, is also consistent with the ease of filament formation across the smaller electrode distances.

REFERENCES

1. S.R. Ovshinsky, Electronics 32 76 (1959); Control Engineering 6 121 (1959).

2. A. Pearson, W.R. Nothover, J.F. Dewald, W.F. Peck Jr., in Advances in Glass Technology, Plenum, New York (1962) 357.

3. B.T. Kolomiets, E.A. Lebedev, Radio Eng. Electron. USSR 8 1941 (1963).

4. C.F. Drake, I.F. Scanlan, A. Engel, Phys. Stat. Sol. (1969) 193.

5. M.H. Omar, M. El Hamamsy, A. Bishay, IXth Intern. Congress on Glass, Versailles (1971) pt. 1, 521.

6. R.S. Motran, A.M. Bishay, D.J. Johnson, Second Cairo Solid State Conference (1973), this volume.

7. C. Feldman, K. Moorjani, J. Non-Cryst. Solids 2 82 (1970).

8. N.K. Hindley, J. Non-Cryst. Solids 8-10 557 (1972).

9. D.R. Haberland, H. Stiegler, J. Non-Cryst. Solids 8-10 408 (1972)

10. I. Lucas, J. Non-Cryst. Solids 6 145 (1971).

11. H.K. Heneish, E.A. Fagen, S.R. Ovshinsky, J. Non-Cryst. Solids 4 538 (1970).

12. I.C. Austin, J. Non-Cryst. Solids 2 474 (1970).

13. H.J. Stocker, C.A. Barlow Jr., D.F. Weirauch, J. Non-Cryst. Solids 4 523 (1970).

14. N. Croitoru, C. Popescu, Phys. Stat. Sol. 3 1047 (1970).

15. T. Kaplan, D. Adler, J. Non-Cryst. Solids 8-10 538 (1972).

16. A.C. Warren, J. Non-Cryst. Solids 4 613 (1970).

17. D.L. Thomas, J.C. Male, J. Non-Cryst. Solids 8-10 522 (1972).

18. A.D. Pearson, J. Non-Cryst. Solids 2 1 (1970).

19. D.L. Eaton, J. Amer. Ceram. Soc. 47 554 (1964).

20. P.R. Eusner, L.R. Durden, L.H. Slack, J. Amer. Ceram. Soc. 55 43 (1972).

21. C.H. Sie, J. Non-Cryst. Solids 4 548 (1970).

22. C.H. Sie, M.P. Dugan, S.C. Moss, J. Non-Cryst. Solids 8-10 877 (1972).

ROLE OF IRON AND ELECTRICAL CONDUCTIVITY IN IRON CALCIUM PHOSPHATE GLASSES*

R. S. Motran, A. M. Bishay, and D. P. Johnson

Solid State and Materials Research Center

The American University in Cairo, Egypt

Mossbauer spectroscopy, X-ray diffraction and magnetic susceptibility techniques were used to study the role of iron in calcium phosphate glasses before and after heat treatment. Factors affecting the electrical conductivity of these glasses included the total iron concentration, ferrous-ferric ratio, devitrification by heat treatment, and the type of local order in the glass. An exponential decrease in log ρ *was observed with increasing total iron concentration from about 8 to 17 wt.%* Fe_2O_3. *The range at which there is a change in slope of the exponential (about 11-13 wt.%* Fe_2O_3*) is characterized by the presence of a new compound (*$FePO_4$*) in the heat treated samples, as confirmed by X-ray diffraction studies. These samples, heated for 88 hours at* 750^oC, *showed a minimum resistivity. The foregoing observations support the postulate that local order in the glasses of this series is similar to that in corresponding crystals separated on devitrification of these glasses by heat treatment. The appearance or disappearance of these crystals (eg.* $FePO_4$*) at certain compositions was found to coincide with changes in physical properties such as conductivity.*

---

* Based on a thesis submitted by R.S. Motran for M.Sc. degree in Solid State Science at The American University in Cairo.

This research has been sponsored in part by the Air Force Office of Scientific Research through the European Office of Aerospace Research (OAR), United States Air Force, under Grant AFOSR 71-2099.

The Mossbauer isomer shift results showed that glasses melted under normal conditions gave a Fe^{2+}/Fe_t *ratio of 0.26 before heat treatment and 0.22 after heat treatment in air at 750°C for 88 hours. This ratio was practically the same for the series of glasses melted under the same conditions, irrespective of the total iron concentration. This was confirmed by magnetic susceptibility results. Isomer shift results also suggest that the majority sites for* Fe^{2+} *are octahedral, though a small concentration may be present in tetrahedral coordination. Ferric ions are mainly present in tetrahedral sites, with some in octahedral coordination. Quadrupole splitting indicates significant distortion in the ferrous and ferric sites.*

INTRODUCTION

Although iron is not generally a major constituent of inorganic glasses a knowledge of its behavior is important because of its common occurrence as a troublesome impurity. In addition, since very high concentrations of iron oxide could be introduced in different glasses, new types of glasses with certain characteristic properties are now being produced. One example is a glass semiconductor containing a high concentration of Fe^{3+} and Fe^{2+} ions.

The presence of iron in the form of Fe^{+2} or Fe^{+3} depends on the method of preparation. Samples prepared under normal atmospheric conditions are expected to have low Fe^{+2} content, while samples prepared under reducing conditions have a high Fe^{+2} content, depending on the severity of the reduced conditions. Kurkjian and Buchanan[1] found that the Fe^{+2}/Fe_{total} = 0.2 for phosphate glasses prepared under normal atmospheric conditions.

There has been considerable discussion of the possible coordination number of both di- and trivalent ions in glass.[2,3] Research workers considered the behavior of the U.V. cut-off and visible color of the glass as an indication of the coordination. Recently, ligand field theory has been applied to Fe^{3+} and Fe^{2+}. However, disagreement still exists because of the similarity of the spectra to be expected from the $3d^5$ ions in octahedral or tetrahedral sites. Because of ligand field stabilization in $3d^5$ ions, Fe^{+2} would be expected to be octahedrally rather than tetrahedrally coordinated. However, since Ni^{+2} occurs in tetrahedral sites in some glasses, it is possible that Fe^{+2} may behave similarly, as shown by Bishay and Kinawi.[4] Recently, Bishay and Makar,[5] from electron spin resonance and optical studies on calcium phosphate glasses, suggested the presence of both 4 and 6 coordination for Fe^{+2} as well as Fe^{+3} ions.

The electrical properties of phosphate glasses containing iron have been investigated by many workers. Hassan[6] has reported the d.c. conductivity behavior of a 55 FeO - 45 P_2O_5 glass with varying Fe^{3+}/Fe_{total} ratios. Hansen et al[7] have also reported dielectric dispersion in three iron phosphate glasses with different Fe^{3+}/Fe_{total} ratios. Their X-ray and transmission electron microscope investigations of these glasses revealed no evidence of crystalline phases or liquid-liquid separation after annealing at 500°C. Kinser[8] observed a.c. and d.c. electrical properties of a 55 FeO - 45 P_2O_5 glass heat treated for 1 hour at varying temperatures. Devitrification during treatment at high temperatures was observed by transmission electron microscopy, and appearance of dispersions in a.c. data was correlated with the onset of devitrification. Later Dozier et al[9] correlated changes in electrical properties to the ratio Fe^{3+}/Fe_{total} and to devitrification induced by heat treatment.

The aim of this work is to study the electrical a.c. resistivity of a series of iron calcium phosphate glasses before and after different conditions of heat treatment, and to try to relate the results to glass structure, to the ratio Fe^{3+}/Fe_t as well as to bond strength. Mossbauer spectroscopy, X-ray diffraction and magnetic susceptibility techniques were the major tools used in the study. X-ray fluorescence technique was used to determine the total iron content in the different glass samples.

All samples were of the molar composition 1.0 CaO - 1.0 P_2O_5 with increasing Fe_2O_3 up to 18.45 weight %.

EXPERIMENTAL

Sample Preparation. Calcium oxide and phosphorus pentoxide were introduced as calcium dihydrogen phosphate $Ca(H_2PO_4)_2H_2O$, while iron was introduced as ferric oxide "Fe_2O_3". The batch was melted in air at about 1200°C for four hours in Zircon crucibles, in a silicon carbide heated furnace. The glass was then poured in a brass mould and annealed at a temperature of about 600°C.

Heat treatment of the different samples was conducted in another electric furnace. The glass samples were put in this furnace at room temperature, then heated to the required temperature. The heating time is defined as the time between reaching the required temperature and turning off the furnace. The heat treated samples were taken out from the furnace after it cooled to room temperature.

A.C. Resistivity. Samples were cut to the form of a parallelepiped of 1x1x0.1 cm. The two square surfaces placed in contact with the electrodes were painted with a thin layer of silver to produce a layer of high electrical conductivity.

The resistance was measured using a bridge (internal frequency 1 Kc/s), the accuracy of which is about ± 5%. All measurements were taken under vacuum conditions ($\sim 10^{-3}$ mm. Hg) to avoid the presence of water vapor which proved to decrease the sample resistance due to surface conductivity.

X-Ray Diffraction. X-ray diffraction patterns were obtained for the sample powder using a "General Electric XRD-6" diffractometer with a Co target tube. An iron filter was used to cut down the K_β radiation and the rest of the continuous X-rays. A.S.T.M. cards were used to detect the compounds present in the devitrified glasses.

Mossbauer Spectroscopy. Since the maximum velocity required is of the order of few centimeters a second, the device used to produce this velocity in this laboratory is a modified loudspeaker. An identical loudspeaker is mechanically coupled with the first speaker in order to produce a voltage proportional to velocity. This analog voltage is converted into digits which are used as the address for a "PDP-8" computer. This computer is used to store the number of counts instead of a multi-channel analyser. The number of channels used was 1024. The source used is a 5 m.c. Co^{57} in Cu matrix. The detector is a gas (a mixture of 20% Kr - 10% N) filled proportional counter. The pre-amplifier and the single channel analyser are "Camberra" Models, while the amplifier was "home made".

Magnetic Susceptibility. The Faraday method was used for measuring the magnetic susceptibility of some of samples before and after heat treatment using a four-inch "V-2300A" Varian electromagnet and a semimicro Mettler balance. The samples used were about 80 mg each. The standard material used was $(NH_4)_2SO_4 \cdot FeSO_4 \cdot 6H_2O$.

RESULTS AND DISCUSSION

Electrical Conductivity. Figure 1 shows the a.c. resistivity for different samples containing from 4.75 to 18.45 wt.% Fe_2O_3 before heat treatment. The a.c. resistivity decreases appreciably as the amount of iron oxide increases. The a.c. resistivity at a temperature of 690°K ($10^4/T = 14.5$) is 2.24 MΩ-mt (2.24×10^8 Ω.cm) for the sample containing 4.75 wt.% Fe_2O_3 and 1.78 KΩ-mt (1.78×10^5 Ω.cm) for the sample containing 18.45 wt.% Fe_2O_3, i.e., a change by a factor of 10^3.

The change in ΔH with iron oxide concentration for the same series of glasses is shown in Fig. 2. In the high temperature range, a change of ΔH from 25.8 Kcal/mole to 17.5 Kcal/mole is observed as the Fe_2O_3 concentration increases from 4.75 to 18.45 wt.%.

Figure 3 shows the a.c. resistivity for the same group of samples after heat treatment at 750°C for 88 hours. A general decrease in resistivity was noted after different conditions of heat treatment. However, most of the samples showed an initial decrease in resistivity with increasing heat treatment time followed

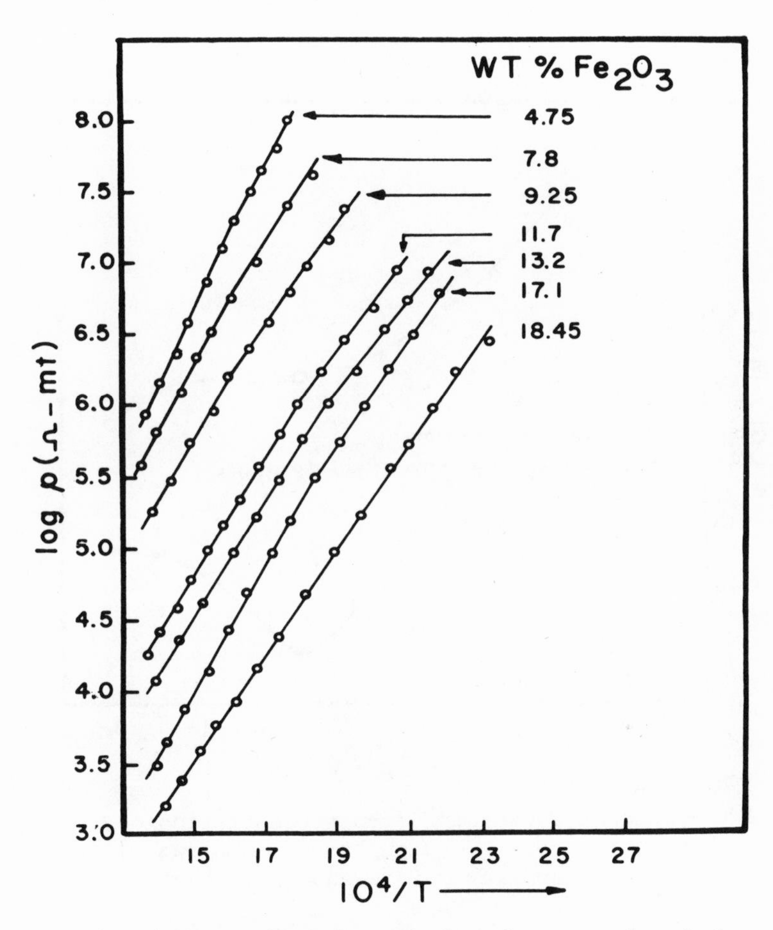

Fig. 1: A.C. resistivity of the glass samples before heat treatment.

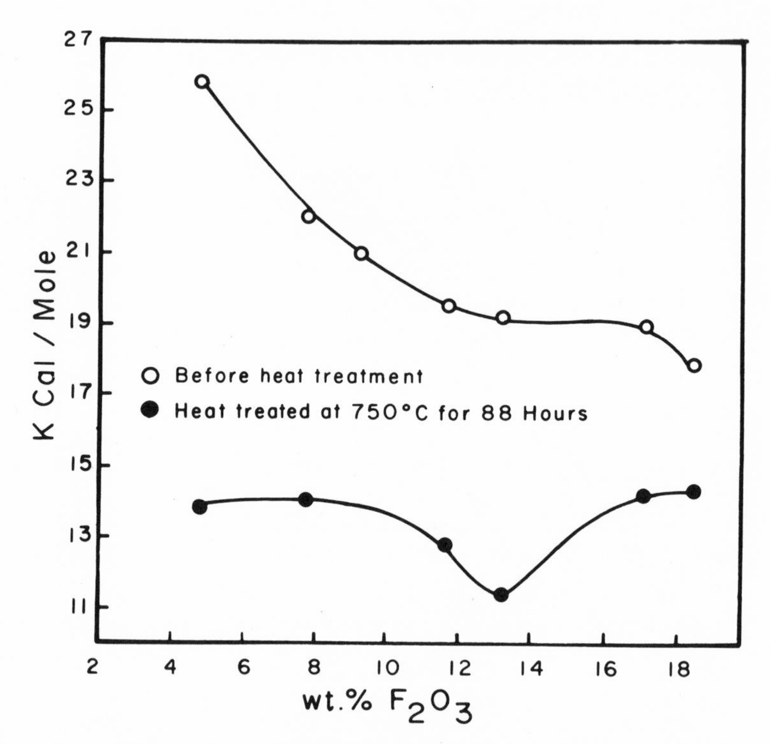

Fig. 2: Effect of increasing iron concentration on the activation energy (high temperature range) for the series of samples before and after heat treatment at 750°C for 88 hours.

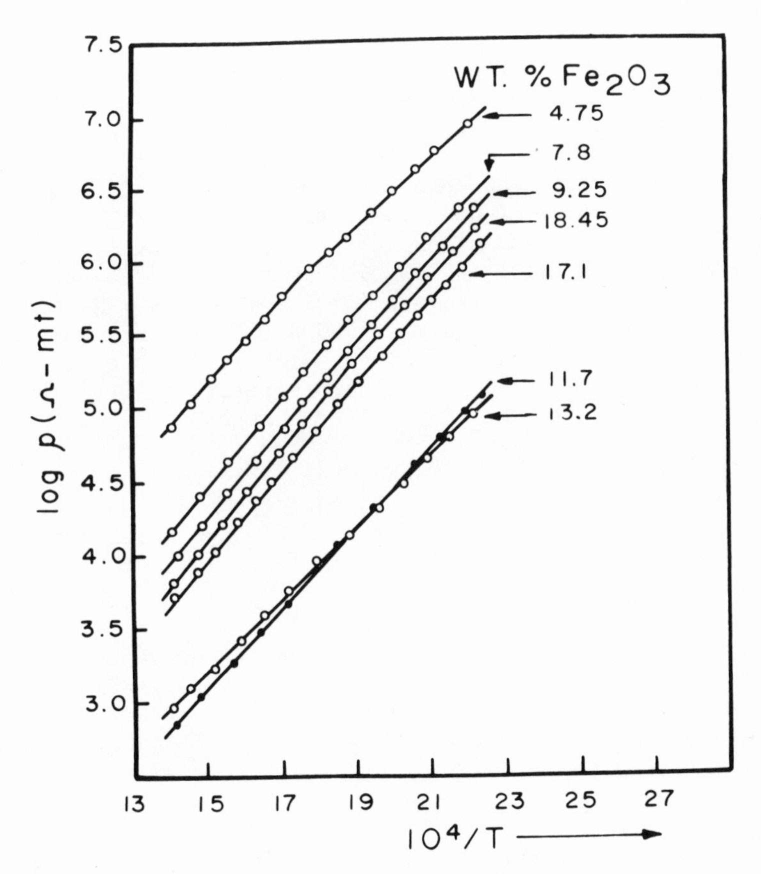

Fig. 3: A.C. resistivity of the samples after heat treatment at 750°C for 88 hours.

by some increase in resistivity after longer heating time. It is believed that this behavior is dependent upon the types of crystals formed on devitrification. This may explain the opposite behavior observed by Dozier[9] for a 55 mole % FeO-45 mole % P_2O_5 glass in which an initial rise in resistivity was observed, followed by a decrease in resistivity.

The following observations can also be made:

1. In general, the decrease in resistivity as a result of heat treatment is more pronounced in samples containing low iron concentration than in those containing high iron concentration.

2. The rate at which the change in resistivity and activation energy occurs as a result of heat treatment differs according to the concentration of iron oxide. Thus, glasses containing low iron concentrations (9.25 wt.% Fe_2O_3 or lower) show a pronounced decrease in resistivity from 3.55 MΩ-mt (3.55×10^8 Ω.cm) to 79.5 KΩ-mt (7.95×10^6 Ω.cm) at a temperature of 590°K ($10^4/T = 17$) as a result of heat treatment for 24 hours at 700°C. Increasing the heat treatment time (from 24 to 75 or 88 hours) or temperature (from 700 to 750°C) does not seem to cause any further changes.

On the other hand, samples containing high iron concentrations (17.1 wt.% Fe_2O_3 or higher) show small variations in resistivity after different conditions of heat treatment. However, in this case a clear change in activation energy (from 19 to 14.2 Kcal/mole) is observed as a result of heat treatment.

Samples containing intermediate concentrations of iron (11.7 and 13.2 wt.% Fe_2O_3) showed a gradual decrease in resistivity with increasing time or temperature of heat treatment. These samples when heat treated for 88 hours at 750°C gave the least resistivity observed in our studies of a series of phosphate glasses containing iron (Figs. 3 & 4).

The minimum resistivity obtained for this group is 2.32 KΩ-mt (2.32×10^5 Ω.cm) at 590°K for the glass containing 11.7 wt.% Fe_2O_3 and heat treated for 88 hours at 750°C.

Figures 4 and 2 show the effect of increasing the total iron concentration on the resistivity (log ρ) and activation energy of the different glasses before and after heat treatment at 750°C for 88 hours. A decrease in the resistivity and activation energy of the series of glasses is observed before heat treatment. However, a change in slope is clearly shown in the range from about 11 to 14 wt.% Fe_2O_3. On the other hand, a minimum is observed in the range of compositions (11-14 wt.% Fe_2O_3) for both the resistivity and activation energy of the heat treated samples.

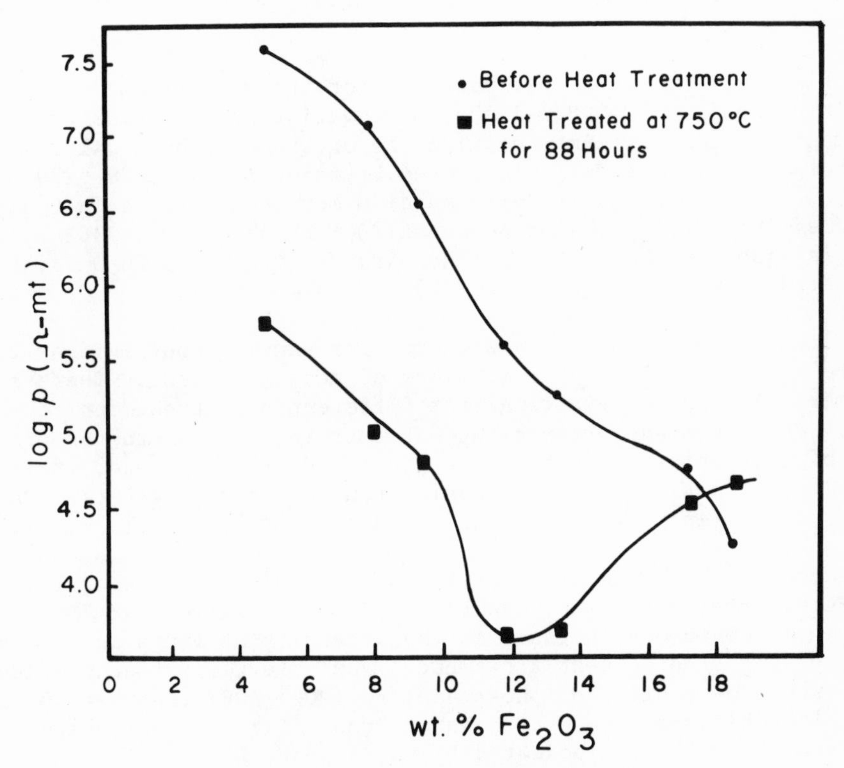

Fig. 4: Log ρ at 590°K for the series of samples before and after heat treatment at 750°C for 88 hours.

The above results show that heat treatment generally causes a decrease in resistivity and activation energy. Furthermore, the increase in total iron concentration up to about 11 wt.% Fe_2O_3 is accompanied by a pronounced decrease in resistivity and activation energy. However, the changes observed in both resistivity and activation energy curves (before and after heat treatment) suggest that a new factor is responsible for the observed results at high iron concentrations. Changes in the glass structure, specially in the range between 11-14 wt.% Fe_2O_3 may be responsible for these observations. This will be discussed in more detail when considering X-ray diffraction results.

X-Ray Diffraction. The X-ray patterns for the heat treated samples containing no iron ($CaO.P_2O_5$) showed no variation with time of heat treatment. The major lines confirm the presence of beta calcium metaphosphate, β-$(CaP_2O_6)_x$ (indicated by major peaks at $2\Theta = 29.5^o$, 33.2^o, 27.7^o) and possibly delta calcium metaphosphate, δ-$Ca(PO_3)_2$ crystals (indicated by major peaks at $2\Theta = 29.5^o$, 36.9^o, 33.2^o). There is also a possibility of minor traces of $CaH\ PO_4.2H_2O$ crystals (indicated by major peaks at $2\Theta = 13.6^o$, 24.4^o, 34°).

Figure 5 shows the X-ray patterns for samples containing 9.25 wt.% Fe_2O_3 after different conditions of heat treatment. There is no remarkable change resulting from different heat treatment conditions. The observed patterns suggest that in addition to the presence of beta calcium metaphosphate, β-$(Ca\ P_2O_6)_x$ crystals, it is possible to identify traces of basic iron phosphate $Fe_6(PO_4)_4.(OH)_6.7H_2O$ crystals (indicated by major peaks at $2\Theta = 26.5^o$, 31.7^o, 34.5^o).

Figure 6 shows the X-ray patterns of samples containing 13.2 wt.% Fe_2O_3 after different conditions of heat treatment at 700°C. There is a pronounced increase in the intensity of lines as a result of increasing time of heat treatment. The observed patterns suggest the presence of basic iron phosphate, $Fe_6(PO_4)_4(OH)_6.7H_2O$, as well as the possible presence of traces of iron (III) orthophosphate dihydrate, $FePO_4.2H_2O$ (indicated by major peaks at $2\Theta = 34.6^o$, 28.3^o, 24.5^o) and beta calcium metaphosphate, β-$(CaP_2O_6)_x$.

The X-ray patterns of samples containing 17.1 wt.% Fe_2O_3 were also obtained after different conditions of heat treatment. No remarkable changes were observed as a result of increasing time of heat treatment. The observed patterns suggested the presence of basic iron phosphate, $Fe_6(PO_4)_4(OH)_6.7H_2O$; traces of delta calcium metaphosphate δ-$Ca(PO_3)_2$; iron (III) orthophosphate dihydrate, $FePO_4.2H_2O$ and traces of calcium ferrite $Ca_2Fe_2O_5$ (indicated by major peaks at $2\Theta = 38.5^o$, 37.4^o, 55°).

The X-ray patterns for samples containing from 0.0 to 18.4 wt.% Fe_2O_3 were obtained after heat treatment at 750°C for 88 hours. The

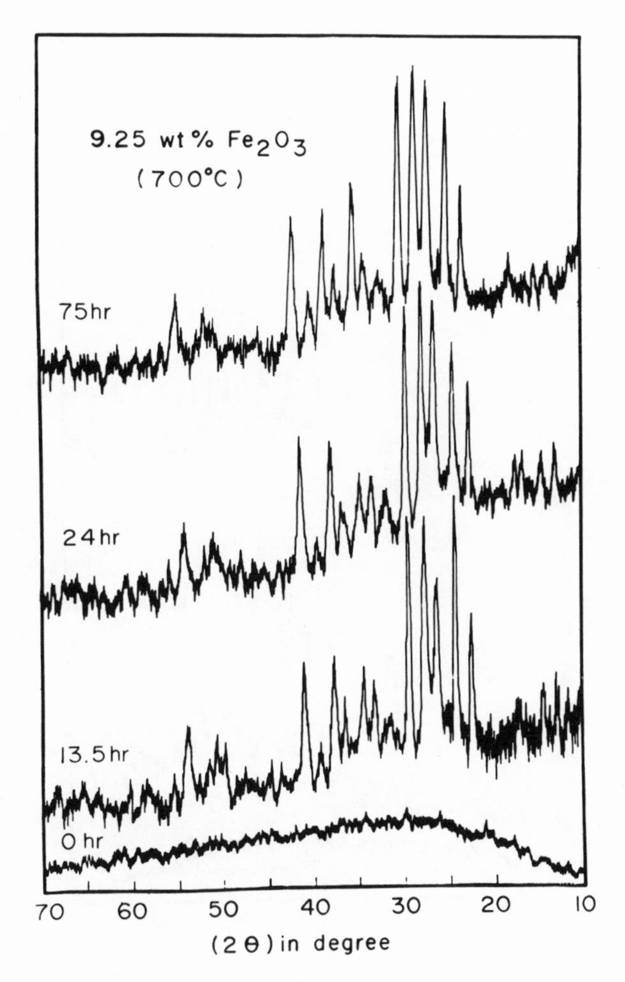

Fig. 5: X-ray diffraction patterns for samples containing 9.25 wt.% Fe_2O_3 before and after different conditions of heat treatment.

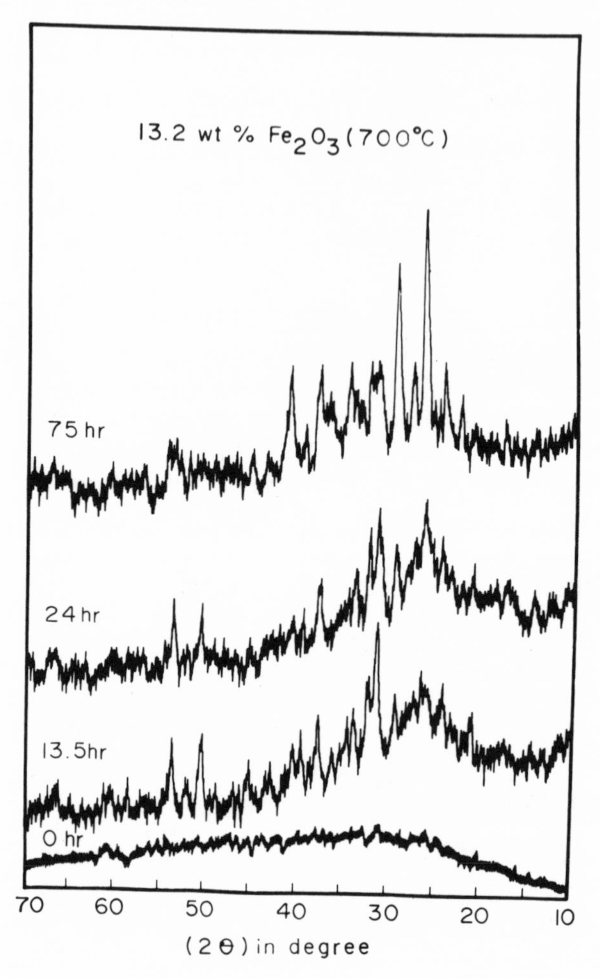

Fig. 6: X-ray diffraction patterns for samples containing 13.2 wt.% Fe_2O_3 before and after different conditions of heat treatment.

patterns for four representative samples are shown in Fig. 7. It was possible to divide the X-ray patterns obtained into three groups:

1. Samples containing up to about 9 wt.% Fe_2O_3. The X-ray patterns of these samples are very similar to samples containing no iron, with the major compound identified being beta calcium metaphosphate, β-$(CaP_2O_6)_x$.

2. Samples containing from about 11 to 13 wt.% Fe_2O_3. The devitrified compound characteristic for these heat treated glasses is iron (III) orthophosphate ($FePO_4$) (indicated by major peaks at $2\Theta = 34.5^o, 29.8^o, 23.5^o$). It should be noted that this compound was not identified in these glasses when heat treated at 700°C only. Traces of beta calcium metaphosphate, β-$(CaP_2O_6)_x$ and basic iron phosphate, $Fe_6(PO_4)_4(OH)_6 \cdot 7H_2O$ can also be detected.

3. Samples containing high iron oxide concentrations, from 17 to 18.5 wt.% Fe_2O_3. The X-ray patterns show that the characteristic compound for these samples is basic iron phosphate, $Fe_6(PO_4)_4(OH)_6 \cdot 7H_2O$.

The results obtained in this study clearly indicate a difference in the structure of the devitrified samples as a result of increasing iron concentration. It may be possible to conclude that each of the above devitrified groups has a specific compound characteristic of that group:

1. β-$(CaP_2O_6)_x$ for the range containing up to about 9 wt.% Fe_2O_3.
2. $FePO_4$ for the range containing from 11 to 13 wt.% Fe_2O_3.
3. $Fe_6(PO_4)_4(OH)_6 \cdot 7H_2O$ for the range containing from 17 to 18.5 wt.% Fe_2O_3.

It should be noted, however, that each range of compositions contains traces of compounds characteristic of other ranges. Thus, the range from 11 to 13 wt.% Fe_2O_3 contains traces of β-$(CaP_2O_6)_x$ and $Fe_6(PO_4)_4(OH)_6 \cdot 7H_2O$ in addition to its characteristic compound ($FePO_4$). This compound ($FePO_4$) may be responsible for the minimum resistivity and activation energy observed for this range of compositions (Figs. 2 and 4).

Mossbauer Spectroscopy. Figure 8 shows a typical Mossbauer spectrum for the calcium phosphate glasses containing iron. The spectrum is resolved (dotted lines) into Fe^{3+} and Fe^{2+} doublets. A similar spectrum was observed by Kurkjian[10] for an $Fe(PO_3)_3$ glass.

Figure 9 shows the Mossbauer spectra for samples containing 9.25 wt.% Fe_2O_3 before and after different conditions of heat treatment. A gradual decrease in width due to heat treatment is observed

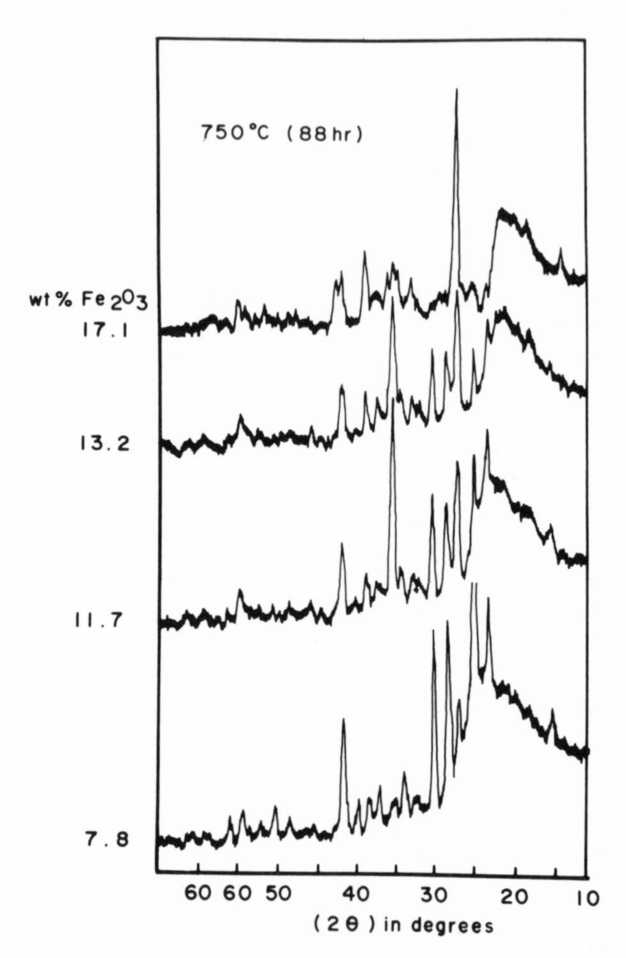

Fig. 7: X-ray diffraction patterns for four representative samples after heat treatment at 750°C for 88 hours.

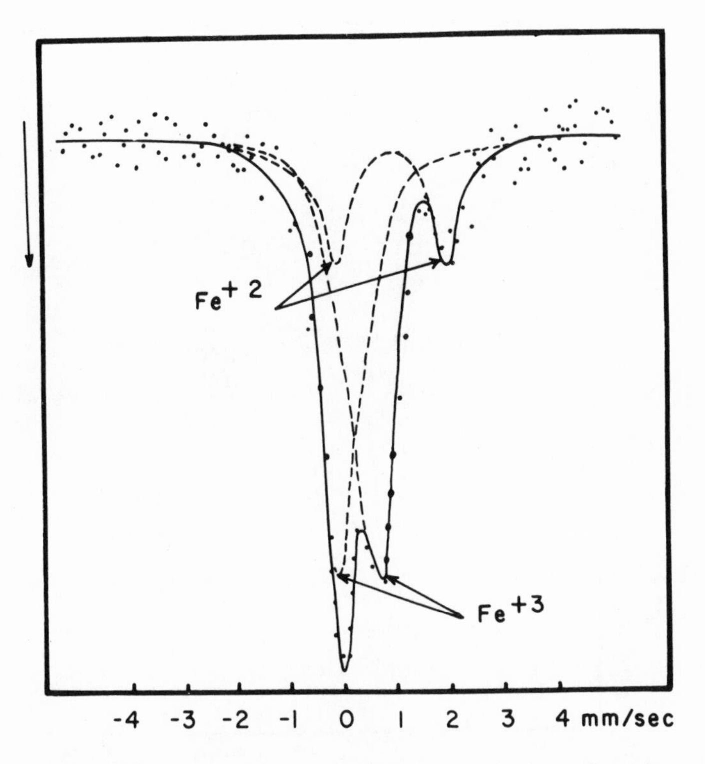

Fig. 8: Typical Mossbauer spectrum for calcium phosphate glasses containing iron. The spectrum is resolved qualitatively to Fe^{3+} and Fe^{2+} doublets.

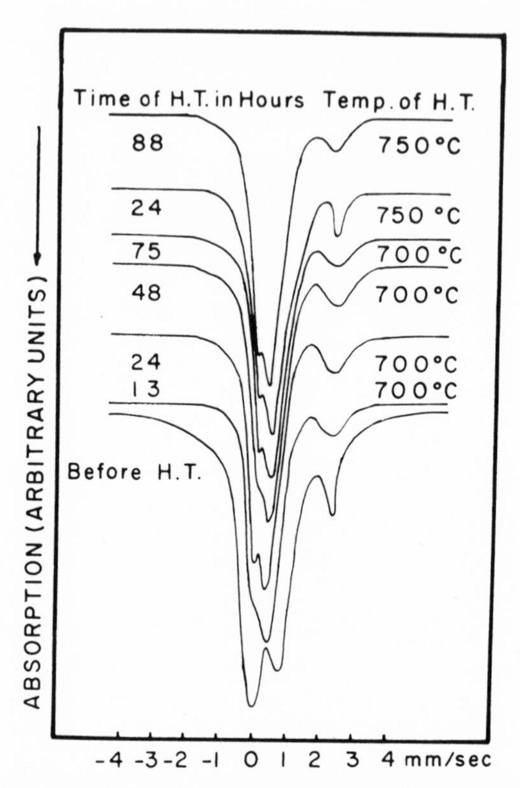

Fig. 9: Mossbauer spectra for samples containing 9.25 wt.% Fe_2O_3 before and after different conditions of heat treatment.

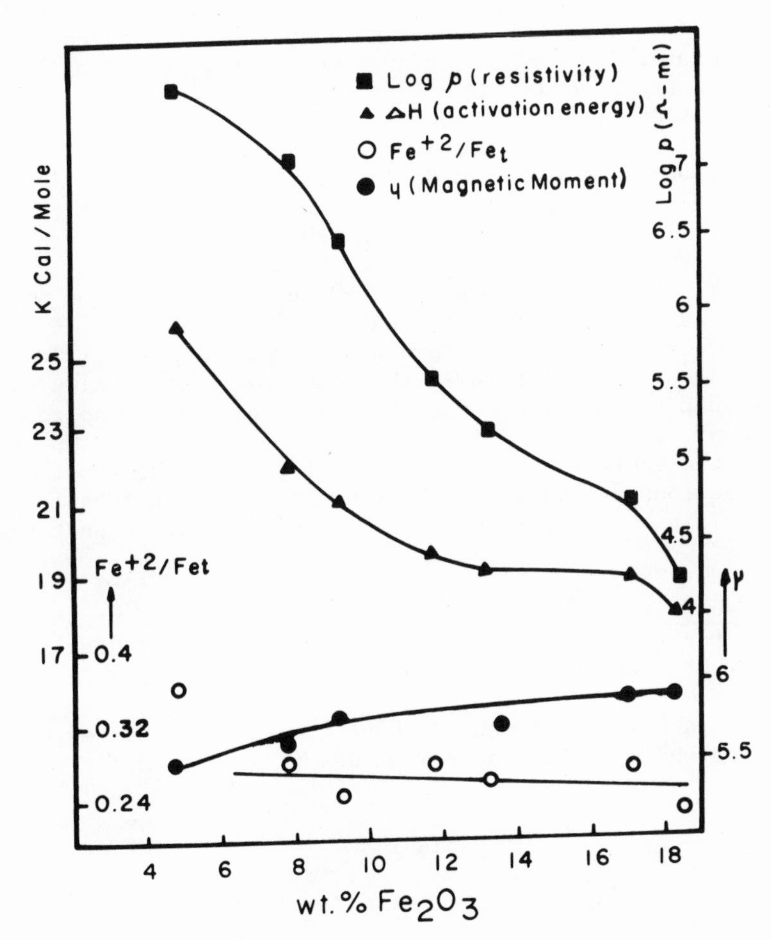

Fig. 10: A.C. resistivity, activation energy, magnetic moment and Fe^{2+}/Fe_{total} for the series of samples before heat treatment.

(from 0.075 to 0.05 cm/sec). This is in line with the observation made by Kurkjian and Buchanan[1] (from 0.064 to 0.053 cm/sec) for an $Fe(PO_3)_3$ glass before and after heat treatment. These results imply a decrease in the number of non-identical sites due to heat treatment. There is also a decrease in quadrupole splitting for both Fe^{2+} and Fe^{3+} (from 0.228 to 0.211 cm/sec for Fe^{2+} and from 0.073 to 0.036 cm/sec for Fe^{3+}). This decrease in quadrupole splitting could be due to less distortion in the cubic symmetry as a result of heat treatment. In other words there is more long range order, as confirmed by X-ray diffraction. It is also noted that as a result of heat treatment, the isomer shift for Fe^{2+} increases from 0.113 to 0.142 cm/sec; while in the case of Fe^{3+} the I.S. decreases from 0.037 to 0.018 cm/sec (all values of isomer shift are with respect to iron). Similar observations were noted for other iron concentrations in this series and for other phosphate glasses reported in the literature.[10] In general the ratio Fe^{2+}/Fe_t decreases from about 0.26 to 0.22 as a result of heat treatment. The decrease in isomer shift for Fe^{3+} suggests that before heat treatment the ferric ions have a net valence somewhat lower than the theoretical value of "3" and as a result of heat treatment the electronic charge approaches the valence of value of "3", i.e. the valence increases. On the other hand, the increase of isomer shift for Fe^{2+} suggests that the valence of ferrous decreases slightly due to heat treatment. This may also explain the slight decrease in Fe^{2+}/Fe_t due to heat treatment.

Magnetic Susceptibility. Table 1 shows the results of magnetic susceptibility for the different samples before and after heat treatment at 750°C for 88 hours.

There is a slight increase in μ as a result of increasing iron concentration which could possibly be due to a decrease in the Fe^{2+}/Fe_t.

Table I

Samples Before Heat Treatment

wt.% Fe_2O_3	4.75	7.8	9.25	13.6	17.1	18.45
$\lambda \times 10^6$	7.58	12.9	16.25	23.2	31	33.7
μ (Bohr Magn.)	5.53	5.62	5.8	5.72	5.88	5.91

(Table 1, cont.)
Samples Heat Treated at 750°C for 88 Hours

wt.% Fe_2O_3	4.75	7.8	9.25	13.6	17.1	18.45
$\lambda \times 10^6$	7.65	12.95	17	23.3	31.9	34.3
μ (Bohr Magn.)	5.55	5.63	5.92	5.73	5.97	5.96

However, there seems to be no notable effect on the magnetic susceptibility as a result of heat treatment. The average value of μ is about 5.76 Bohr magnetons indicating that most of the iron in the samples melted under normal conditions is in the ferric state, in agreement with the Mossbauer results.

DISCUSSION AND CONCLUSIONS

The different techniques described in this paper, X-ray diffraction, magnetic susceptibility and Mossbauer spectroscopy were used to study the role of iron in calcium phosphate glasses before and after heat treatment. Factors affecting the electrical conductivity of glasses included the total iron concentration, ferrous-ferric ratio, devitrification by heat treatment, and the type of local order in the glass.

Total Iron Concentration. The total iron concentration in the different glasses was determined by X-ray fluorescence. Figure 10 shows the effect of increasing total iron concentration on the conductivity of the series of glasses studied. It is possible to consider that there is an exponential decrease in log ρ (resistivity) with increasing total iron concentration from about 8 to 17 wt.% Fe_2O_3. The range at which there is a change in slope of the exponential glasses containing about 11-13% Fe_2O_3 corresponds to the presence of a new compound, $FePO_4$, in the heat treated samples, as confirmed by X-ray diffraction studies. The compound was not detected in any of the other heat treated samples. It is interesting that the heat treated samples (750°C for 88 hours) showed a minimum resistivity in this range of compositions (11-13% Fe_2O_3).

The glass containing 4.75 wt.% Fe_2O_3 showed a slightly lower resistivity than the expected value if the exponential was extrapolated. Figure 10 shows that on the basis of Mossbauer spectroscopy, supported by magnetic susceptibility results, the Fe^{2+}/Fe_{total} ratio is practically constant at about 0.26 in this series of glasses.

This is in line with the findings of Kurkjian and Buchanan[1] who estimated that about 20% of the total iron is present in the divalent form for $Fe(PO_3)_3$ glass. The value of 0.378 given for the Fe^{2+}/Fe_t ratio in the glass containing 4.75 wt% Fe_2O_3 is nearer to the theoretical optimum value of 0.5 (i.e. $Fe^{2+}/Fe^{3+}= 1.0$) which may explain the higher conductivity (lower resistivity) for this glass as compared to the expected value.

Ferric/Ferrous Ratio. The study of the Mossbauer isomer shift for samples prepared under normal conditions indicates that both Fe^{+3} and Fe^{+2} ions are present with the major concentration of iron in the Fe^{+3} state. Glasses with a total iron content from 8 - 18 wt.% Fe_2O_3 gave an average value for Fe^{2+}/Fe_{total} ratio of about 0.26 before heat treatment and 0.22 after heat treatment at 750°C for 88 hours. These results indicate that heating in normal atmosphere (air) oxidizes part of the ferrous ions in the glass.

It is interesting to note that this ratio was practically the same for the series of glasses melted under the same conditions, irrespective of the total iron concentration. Only one exception was observed, namely the glass containing 4.75 wt.% Fe_2O_3 which was the least iron concentration studied. It is reasonable to expect that glasses containing lower total iron concentration may have different ratios of Fe^{2+}/Fe_{total}. However, in our range of compositions (> 7 wt.%) this ratio may be considered constant. This is also confirmed by the magnetic susceptibility results.

Since the Fe^{2+}/Fe_{total} ratio in the majority of these glasses did not show much change, it is reasonable to conclude that the observed changes in conductivity are mainly due to increasing total iron concentration. It is necessary to investigate the effect of changing the Fe^{2+}/Fe_{total} ratio for a series of glasses having the same total iron concentration. When this is accomplished, we may be in a better position to clarify the effect of the Fe^{2+}/Fe_{total} on electrical conductivity.

Devitrification by Heat Treatment. The X-ray diffraction results showed that the glasses devitrified as a result of different heat treatments. The identified compounds separated after heat treatment $\beta-(CaP_2O_6)_x$, $Fe_6(PO_4)_4(OH)_6.7H_2O$ and $FePO_4$. The appearance of the latter compound was accompanied by a significant decrease in the resistivity of the heat treated glasses (11 - 13 wt.% Fe_2O_3). It should be also noted that the rate of devitrification of glasses in this range (11 - 13 wt.% Fe_2O_3) was much slower than in the rest of the series (as shown by X-ray diffraction results). Similarly, a gradual change in properties (conductivity and Mossbauer) was observed with increasing time or temperature of heat treatment for the glasses in this medium range of compositions (11 - 13 wt.% Fe_2O_3) as compared to other glasses in the series which showed an abrupt change in properties directly after the first heat treatment.

The above observations support the postulate that crystallinity has a major effect on the conductivity of these glasses. Thus, when devitrification occurred abruptly after the first heat treatment (9.25 wt.%), a sudden decrease in resistivity was observed. Further heat treatment did not cause any additional pronounced devitrification and accordingly, practically no changes in conductivity. On the other hand, glasses containing 11 - 13 wt.% Fe_2O_3 showed gradual devitrification as a result of different heat treatments and correspondingly gradual decreases in resistivity.

It is interesting to note here that as a result of Mossbauer spectroscopy and magnetic susceptibility studies, the Fe^{2+}/Fe_t ratio seems to decrease after heat treatment. This may be partly responsible for the slight increase in resistivity of the glass containing about 18 wt.% Fe_2O_3 as a result of heat treatment. It is possible that there is some ordering in the structure of this glass even before heat treatment, though not to the extent that it could be seen by X-ray diffraction technique. Accordingly, the conductivity did not increase as a result of heat treatment, as was the case in other glasses. On the contrary, there was some decrease in conductivity (increase in ρ) which could be attributed to the decrease in Fe^{2+}/Fe_{total} as a result of heat treatment in air (Fig. 4).

It is worth noting here that the change in the slope of the conductivity curve (for glasses containing 11 - 13 wt.% Fe_2O_3 before heat treatment) corresponded to a minimum resistivity in the heat treated glasses and the appearance of a new compound in these devitrified glasses ($FePO_4$). On the other hand, $Fe_3(PO_4)_2$ crystals were tentatively identified by Kinser[8] in the heat treated 55 feO - 45 P_2O_5 glass.

Structural Configuration. On the basis of the results of isomer shift it is suggested that the majority sites for Fe^{2+} ions are octahedral, though a small concentration may be present in tetrahedral coordination. On the other hand, the Fe^{3+} ions are mainly present in tetrahedral sites with some in octahedral coordination. Kurkjian and Buchanan[1] suggested an octahedral co-ordination for both Fe^{2+} and Fe^{3+} in the $Fe(PO_3)_3$ glass. Kurkjian[10] explained that due to ligand field stabilization in $3d^6$ ions, Fe^{2+} would be expected to be octahedrally rather than tetrahedrally coordinated. However, since Ni^{2+} occurs in tetrahedral sites in some glasses, it is possible that Fe^{2+} may behave similarly. In crystalline $Fe(PO_4)$ the iron is known to be in tetrahedral coordination. The presence of quadrupole splitting, however, implies that both coordinations suffer significant distortion.

As a result of heat treatment, a slight increase in isomer shift for Fe^{2+} and a slight decrease in isomer shift for Fe^{3+} was observed. These results suggest that there is a decrease in covalency for ferrous ions and an increase in covalency for ferric ions. This is in line with the suggestion that Fe^{2+} is predominantly octahedral (network modifier) and Fe^{3+} is predominantly tetrahedral (network former) in this series of glasses.

It is worth noting here that the change in the slope of the conductivity curve (for glasses containing 11 - 13 wt.% Fe_2O_3 before heat treatment) corresponded to a minimum resistivity in the heat treated glasses and the appearance of a new compound in these devitrified glasses ($FePO_4$).

The above observations support the postulate that local order in the glasses of this series is similar to that in corresponding crystals[11] separated on devitrification of these glasses by heat treatment. The appearance or disappearance of crystals (eg: $FePO_4$) at certain compositions was found to coincide with changes in physical properties such as conductivity.

REFERENCES

1. C.R. Kurkjian, D.N.E. Buchanan, Phys. Chem. Glasses 5 63 (1964).
2. A.M. Bishay, J. Am. Ceram. Soc. 42 403-7 (1959).
3. A.M. Bishay, J. Am. Ceram. Soc. 44 16-21 (1961).
4. A.M. Bishay, A. Kinawi, Proc. of the Int. Conf. Delft, July 1964.
5. A.M. Bishay, L. Makar, J. Am. Ceram. Soc. 52 605-9 (1969).
6. K.W. Hansen, J. Electrochem. Soc. 112 994 (1965).
7. K.W. Hansen, M.T. Splann, J. Electrochem. Soc. 113 895 (1966).
8. D.L. Kinser, J. Am. Ceram. Soc. 117 546 (1970).
9. A.W. Dozier, L.K. Wilson, E.J. Friebelle, D.L. Kinser, J. Am. Ceram. Soc. 55 373 (1972).
10. C.R. Kurkjian, J. Non Cryst. Sol. 3 157 (1970).
11. A.M. Bishay, M. Maklad, I. Gomaa, S. Arafa, in Interaction of Radiation with Solids, Proc. of the First Cairo Solid State Conference 1966, Plenum Press (1967).

PROPERTIES OF SIMULTANEOUSLY SPUTTERED THIN METAL FILMS

F. Ismail and L.M. El Nadi

Physics Department, Faculty of Science

Cairo University, Giza, Egypt

The physical properties of thin single-metal, bimetal, and multimetal films were studied. The thin films were deposited on glass substrates by the cathodic sputtering method. Bimetal or multimetal films were sputtered simultaneously, using a specially designed multilayer cathode. The effect of sputtered carbon on the properties of the single or multimetal films was tested using an additional cathode. Techniques included visual examination of color changes, electron microscopy and diffraction, and electrical resistivity. Some of the bimetal films showed no alloying at room temperature, exhibiting the structure of the metal with higher sputtering rate. Most of the multimetal films exhibited properties differing from those of the constituting metals. Carbon sputtered with single and multimetals (although having a low sputtering rate) showed changes in structure and in crystal shape and size in the sputtered films. Although the substrate temperature increased during sputtering its temperature was lower than the melting point of the sputtered metals. Studies of the alloying of multi-metal films indicated that alloying occured during preparation and at temperatures much lower than the recrystallization temperature of any of the metals involved.

INTRODUCTION

The crystal structures of some thin metal films are known to be different from those of the bulk material.[1,2,3] Chopra et al.[4,5] reported the observation of stable f.c.c. structure in sputtered films of some metals whose crystal behaviour in thick films are known to be of the b.c.c. type.

The general behaviour patterns of some twenty elements in bimetal films were studied by R. Belser[6] over a considerable temperature range (25° - 600^oC). The results indicated that thin bimetal films, evaporated successively, alloy when exposed to temperatures near the recrystallization temperature of the metal of higher melting point.

The aim of the present work is to study the structure and properties of thin nickel films when singly sputtered or when simultaneously sputtered with iron, copper and brass on amorphous pyrex glass substrates. These elements are known to be crystallized in cubic structures and to have nearly crystal parameters. The possibility of alloying during simultaneous sputtering under the effect of heat during preparation, and the effects of sputtered carbon on the properties of the single or multimetal films were investigated.

EXPERIMENTAL AND MATERIALS

Thin Film Preparation. Single or multimetal thin films were deposited on ultrasonically cleaned pyrex glass slides by the cathodic sputtering method. A standard sputtering unit (1A, 2A Edwards High Vacuum) was used with nitrogen gas of pressure 10^{-2}mm Hg as the sputtering ions. For single metal films, metal cathodes of high purity and highly polished square surfaces were used. Bimetals and multimetal thin films were sputtered on a specially designed multimetal cathode. The carbon (or graphite) cathodes had larger smooth areas than those of the metal cathodes in order to compensate for the low sputtering rate of carbon.

The vacuum chamber contained the cathode at the top, connected to the negative high voltage terminal. The substrates receiving the sputtered material were placed centrally below the cathode on a small aluminium table which functioned as the anode. The cathode - substrate distance was 6 cm. The power supply of the sputtering unit could be rated up to 2 KV and 200 mA. Film thicknesses for singly sputtered metal films were approximately 75- 100 A^o; for bimetals it did not exceed 150 A^o and for trimetals it was 250-270Å.

The substrate temperature was found to reach 60- 80^oC during preparation of all the films. Before removing the substrates with the sputtered films, they were kept under vacuum for 30 minutes after switching off the voltage in order to minimise the probability of oxide layer formation when the films were subjected to air.

Electron-microscopic Studies. Structural studies were carried out by transmission electron diffraction and microscopy, using a Carl

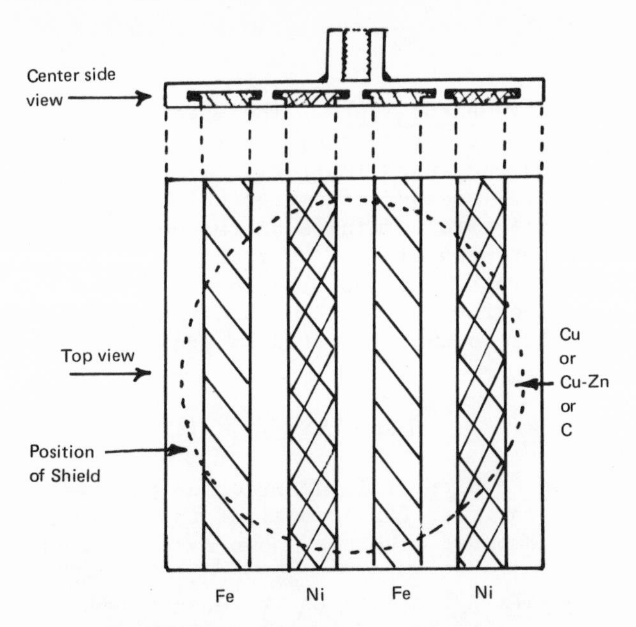

Fig. 1: Side and plane views of the multicathode used in the sputtering experiments.

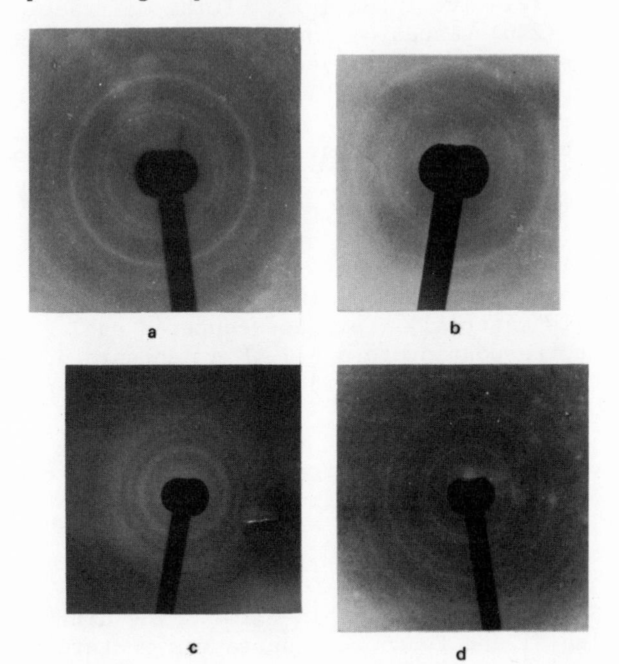

Fig. 2: The electron diffraction patterns for
(a) Nickel (b) Iron
(c) Copper (d) Brass

Zeiss-Jena EF5 electron microscope (electromagnetic type operated at 50 KV).

Electrical Resistivity Measurements. The Wheatstone's bridge method was used to measure the electric resistivity of the films. The current passing through a known length (1 cm) of the film was measured by a sensitive spot galvanometer calibrated by standard resistances. The thin film terminals on the nonconducting glass slide were connected to the circuit by means of spot mercury electrodes of known cross-section. The mercury spots were pressed to guarantee good electrical contact without spoiling the thin film. This method is similar to that reported by Kruidhof and Moret.[7]

RESULTS

Electron Diffraction Results. The diffraction patterns obtained in transmission were identified for several of the thin films prepared. These results were compared with the known data of the bulk material.

Singly Sputtered Metals. The results are summarized in Table I. The nickel thin film (Fig. 2a) diffraction lines spacings were indexed on the basis of the face-centered cubic lattice, $a_o = 3.524$ A^o. The relative intensities of the diffraction pattern lines suggest the (220) planes of the cubic lattice to be parallel to the substrate surface. No changes in the diffraction pattern were noticed when carbon was simultaneously sputtered with nickel.

In the case of thin iron films, the faint diffraction pattern, (Fig. 2b), presumably due to small crystallite size and population, suggested an f.c.c. structure with planes (111) parallel to the substrate surface.

Copper was found to be formed in the same crystalline system as that of the bulk material, i.e. face-centered cubic, $a_o = 3.615$ A^o. But a weak line corresponding to d = 1.48 A^o appears on the diffraction pattern (Fig. 2c). Oxide formation would not explain the presence of such a line.

With respect to brass the diffraction pattern agrees with the f.c.c. system for (Cu- Zn) crystalline structure having $a_o = 3.66$ A^o (Fig. 2d). Sputtered carbon and graphite showed amorphous structure.

Simultaneously Sputtered Bimetal Thin Films. The diffraction pattern of the combination Ni and Fe showed the same structure as that for singly sputtered nickle. When graphite was simultaneously sputtered with these bimetals, the diffraction pattern (Fig. 3b) indicated d-spacings equal to 2.516(m), 2.119(VS), 1.777(S), 1.367(m),

TABLE I

Ni	Standard d	2.034	1.765	1.246	1.0624	1.0172	0.881	0.808
	I/I_1	100	42	21	20	7	4	14
	hkl	111	200	220	311	222	400	331
	Exp. d	2.042 S	1.750 S	1.248 VS	1.063 VW	1.015 W	0.875 VW	0.814 W
-Fe	Standard d	2.091	1.810	1.282	1.090	1.042	0.905	0.830
	I/I_1	100	46	27	17	4	3	9
	hkl	111	200	220	311	222	400	331
	Exp. d	2.109 VS	1.785 S	1.262 S	1.050 M			
Cu	Standard d	2.088	1.808	1.278	1.090	1.0436	0.903	0.829
	I/I_1	100	46	20	17	5	3	9
	hkl	111	200	220	311	222	400	331
	Exp. d	2.010 VS	1.827 S	1.283 S	1.140 M	1.036 W	(extraline	1.45) M
Cu-Zn	Standard d	2.113	1.830	1.30	1.22	1.10	1.06	
	I/I_1	100	60	80	30	100	60	
	hkl	111	200	220	211	311	222	
	Exp.	2.100 VS	1.818 S	1.280 S	1.230 W	1.081 VS		

Standard d : Spacing in angstrom from A.S.T.M Tables.

Exp.: experimental results with relative intensities (W, VW, S, and VS).

1.250(VS) and 0.917(VM) A°. The letters QA° indicate the relative intensities of the diffraction lines: VS, very strong; S, strong; m, medium strength; W, weak and VW, very weak. This diffraction pattern differs from that of the involved elements. Although some of the lines could be related to iron oxide and others to nickel oxide, the whole pattern could not be due to any of the two, or to both together.

Bimetal thin films of nickel and brass showed the same diffraction pattern as that of singly sputtered nickel. No effect of simultaneous sputtering of either carbon or graphite was observed on their structure.

Thin films of both nickel and copper showed deviation from that of the individual constituents. The d-spacings and their relative intensity (Fig. 3a) were 2.064(VS), 1.674(VW), 1.260(S), 1.033(m) and 0.824(VW), respectively. Oxide formation would not explain their presence. When carbon or graphite were simultaneously sputtered with this metal combination, a faint diffraction pattern was obtained which was very difficult to identify.

Sputtered Multimetal Thin Films. The structure of nickel, iron and brass simultaneously sputtered to form a thin film showed the same diffraction pattern as that of singly sputtered nickel. Simultaneously sputtered carbon or graphite produced no effects on the diffraction pattern.

The combination of nickel, iron and copper resulted in an amorphous structure, but when carbon was introduced, a sharp new diffraction pattern was detected (Fig. 3c). The d-spacings and relative intensities were 2.144(VS), 1.924(S), 1.591(VS), 1.347(S), 1.252(m) and 1.171(m) A° respectively. These lines differ from those of any of the involved elements or their oxides.

Analysis of Micrographs. Nearly all the prepared thin films showed the existance of polycrystalline structure. Micrographs for singly sputtered metals and for combination of metals which showed deviations from the singly sputtered ones are shown in Fig. 4. The results of the studies and analyses of all prepared films for particle shape, average grain size estimated by Martin's method,[10] and particle population are given in Table II.

Electrical Resistivity. The results of the electrical resistivity measurements are given in Table II. The results for singly sputtered thin films are found to be higher than those of the bulk materials known to have electric resistivity 6.9, 8.8, 1.7 and 7.0 micro ohm-cm for nickel, iron, copper and brass respectively. It was also found that when anomalous structure was detected the electrical resistivity increased.

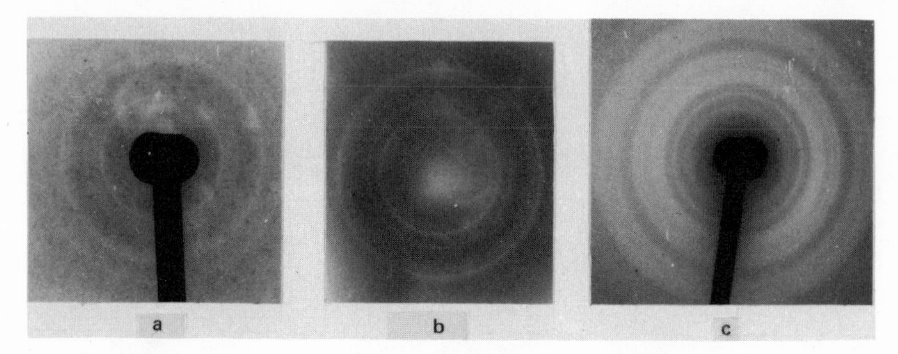

Fig. 3: The electron diffraction patterns for

(a) Ni+Cu (b) Ni+Fe+g
(c) Ni+Fe+Cu+C

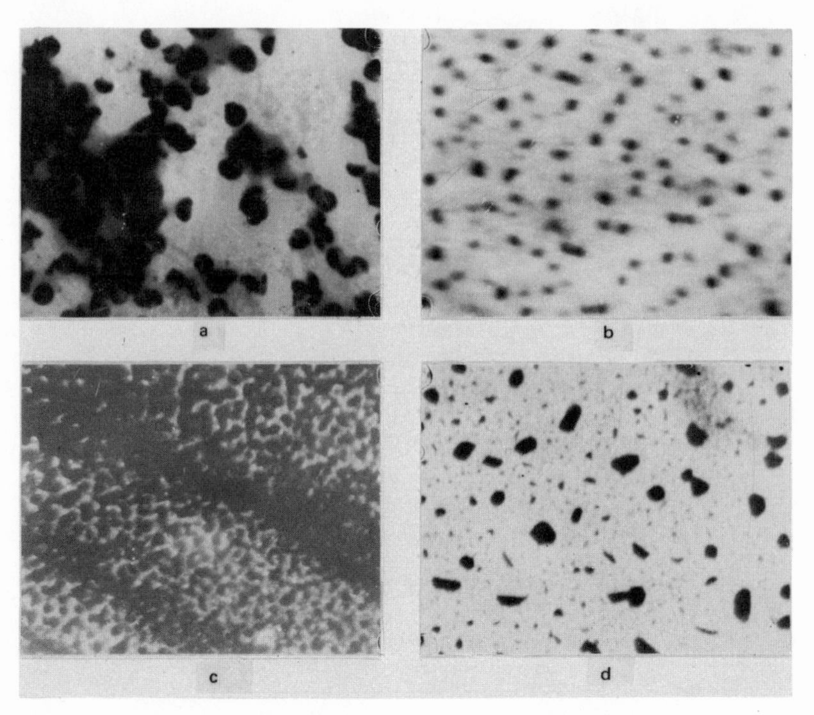

Fig. 4: Electron micrographs of sputtered thin films of

(a) Nickel (b) Iron
(c) Copper (d) Brass

TABLE II

Element	Shape	Av. Grain Size μ	Population grain/μ^2	ρ/ρ_{Ni}
Ni	Hemispheres & spheres	0.419	6	1
Ni+C	nonuniform	0.627	13	1.78
Ni+g	spheres	0.282	9	5.33
Fe	triangles	0.164	4	0.22
Cu	triangle & nonuniform	0.178	10	1.42
Brass	tetrahedron	0.150	3	1.05
Ni+Fe	triangles	0.354	15	2.25
Ni+Fe+C	nonuniform	0.043	95	9.00
Ni+Fe+g	spheres	0.346	14	8.42
Ni+Cu	nonuniform	0.316	72	12.20
Ni+Cu+C	hexagon & spheres	0.779	10	6.00
Ni+Brass	spheres	0.243	18	1.92
Ni+Fe+Cu	nonuniform	0.221	140	9.50
Ni+Fe+Cu+C	nonuniform	0.763	10	8.20

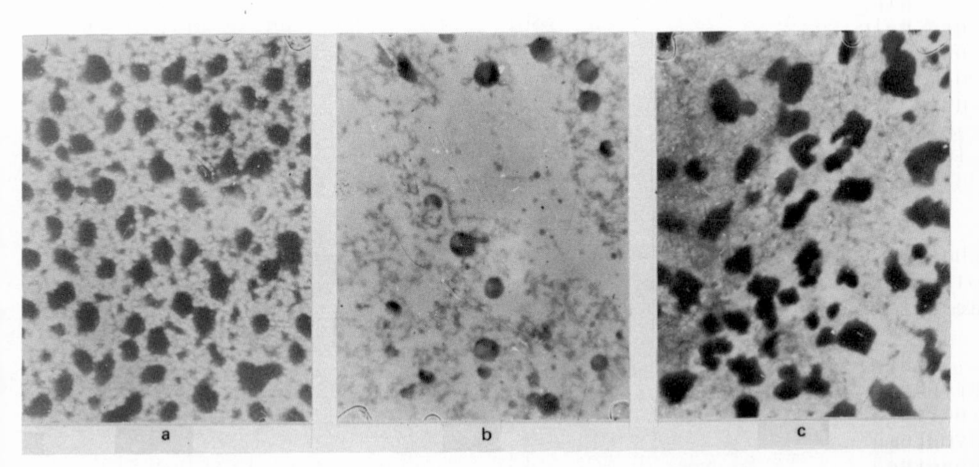

Fig. 5: Electron micrographs of sputtered multimetal thin films of

(a) Ni+Cu (b) Ni+Fe+g

(c) Ni+Fe+Cu+C

CONCLUSION

The data reported here suggest the following:

Singly sputtered thin metal films, grown at substrate temperature below 80°C appear to crystallise immediately on the substrate. The crystalline structure was found to be similar to that of the bulk material except for that of iron which showed a face-oriented cubic system. The factors affecting the (220) orientation for the f.c.c. structure of nickel could not be defined, since no information is available on the behaviour of the vapour phase of nickel during sputtering.

For most of the sputtered multimetals, the structure resembled that of nickel. This might be attributed to the fact that nickel, with its higher sputtering rate, predominated in a non-alloying state at this stage of substrate temperature.

Systems of interstitial or alloy structures were formed during preparation of the thin films at temperatures well below the recrystallisation temperature of any of the involved materials. Symptoms of the existence of such structures could be clearly observed from the changes in all the film properties of nickel+-copper, nickel+iron+graphite, and nickel+iron+copper+carbon.

The effect of carbon sputtering with single or multimetals was clearly demonstrated on grain size and grain shape, and on the values of electrical resistivity, even for cases which showed no structure changes in their diffraction patterns.

The effects of temperature annealing on the properties of all these films are still under investigation.

ACKNOWLEDGMENT

The authors express their thanks to M. El-Nadi, Dean of the Faculty of Science, Professor R. Kamel, and to the staff of the electron microscope service department, National Research Center

REFERENCES

1. C.J. Calbic, N. Schwartz, 9th Natl. Vac. Symp. 81 (1962).

2. B.G. Lazarov et al., Zh. Eksp. Teor. Fiz. 37 1461 (1959).

3. W.L. Bond et al., Phys. Rev. Lett. 15 210 (1965).

4. K.L. Chopra, J. Appl. Phys. 37 2249 (1966).

5. K.L. Chopra et al., Phil. Mag. 16 261 (1967).

6. R. Belser, J. Appl. Phys. 31 562 (1960).

7. E.W. Kruihof, H. Moret, J. Sci. Inst. 39 132 (1962).

8. A.I. Bublik, B.I.A. Pines, Dokl. Akad. Nauk. SSSR 87 215 (1952).

9. C.W.B. Grigson, D.B. Dove, J. Vac. Sci. Tech. 3 120 (1966).

10. R.D. Cadle, Particle Size, Reinhold Pub., New York (1965) 4.

INDIRECT EXCITON WITH DEGENERATE VALENCE BAND IN THE TWO-PHOTON ABSORPTION IN SEMICONDUCTORS

A. R. Hassan

Ain Shams University, Cairo, Egypt

We have developed from third-order time-dependent perturbation theory the theory of indirect transitions to exciton states with two-photon absorption when the valence band is degenerate in semiconductors. A group theoretical analysis is performed for the case of GaP (or III - V compounds) in order to determine which transition is allowed in the process, the symmetry of the excitons, and the symmetry of the phonons which contriubte in the transitions. A numerical application has been performed on GaP compound and compared with the literature. The GaP compound seems a good candidate for an indirect-type semiconductor.

INTRODUCTION

In the development of laser technique over the last ten years, multiphoton processes have been extensively studied. Direct two-photon interband transitions theory has been developed at all critical points.[1] The theory of the direct two-photon exciton transitions has been treated theoretically by Loudon,[2] using a four-band model, and by Mahan,[3] using a two-band model. Loudon's model predicts the appearance of only s-exciton states, while in Mahan's model p-exciton states are allowed. Many experimental observations of the two-photon exciton transitions have been made in semiconductors[4] and alkali halides.[5]

One-photon transitions to indirect exciton states have been studied and classified,[6] taking into account the degeneracy of the valence band and the anisotropy of the electron masses of the conduction band at different points in the first Brillouin zone, where a complete analysis of the exciton spectra with non-zero wave vector has been performed. Experimental observations of the one-photon indirect excitons in different solids have been studied.[7]

The theory of indirect two-photon interband transitions has been developed from third-order time-dependent perturbation theory.[8] The frequency and temperature dependence of the non-linear absorption coefficient were shown to be very similar to those of the one-photon indirect transitions but selection rules were different. Preliminary evidence for the relevance of two-photon phonon-assisted transitions has been given by Ashkinadze et al.[9] Some observation peaks in the spectrum have been atrributed to the electron-hole interactions.

We have developed in a previous work[10] the theory of phonon-assisted two-photon-exciton transitions in semiconductors for a simple valence and conduction bands where the effective masses are isotropic. Explicit expressions of the absorption coefficient through three different band models is given, in terms of six different matrix elements which couple different band and exciton states, depending on the band model adopted in the calculations. Numerical estimates for the case of GaP and a determination of the symmetry of the phonons which contribute to the transitions is performed. The results show that the three-band model gives the largest contribution to the absorption coefficient for this type of problem.

FORMULATION OF THE PROBLEM

Diamond and zincblende structures of all semiconductors have a degenerate valence-band maximum at $\underline{k} = 0$. The absolute minimum of the conduction band is, generally, at the same point of the first Brillouin zone. In some cases, however, such a minimum is at $\underline{k} \neq 0$ and its location depends on the particular substance under consideration. For Ge, the minimum is at the point $L = \frac{\pi}{a}(1,1,1)$, for Si, it is along the axis $\Delta = (\vec{K}, 0,0)$. Zincblende or III-V compounds have indirect gap at the point $X = \frac{2\pi}{a}(1,0,0)$. The conduction band has a non-degenerate minimum and the valence band has a three-fold degenerate maximum (neglecting spin). When spin is included, the valence band becomes sixfold degenerate and is split by spin-orbit interaction into an upper fourfold and a lower twofold degenerate band separated by a spin-orbit splitting Δ. It is the purpose of this paper to develop the theory of indirect two-photon exciton transitions where the values of the exciton binding energy have been computed, taking into account the details of the valence band and the anisotropy of the conduction band.

For this kind of structure, the Hamiltonian for the relative electron-hole[11] is (neglecting the electronic spin):

$$H_{ex} = H_e(\vec{p}) - H_h(\vec{p}) - \frac{e^2}{\varepsilon r} \quad (1)$$

where r and $\vec{p}$ are the relative electron-hole coordinate and momentum respectively; ε is the static dielectric constant; e, h refer to the electron and hole, respectively; H_e is the kinetic energy of the electron near the conduction-band minimum; and H_h is the well known[12] 6 x 6 matrix which describes the hole kinetic energy near $\vec{k} = 0$.

The explicit expression for the operator H_e depends on the position of the conduction-band minima. Since these minima generally occur on high symmetry directions for which the electron has cylindrical symmetry, we can write:

$$H_e = \frac{p_1^2 + p_2^2}{2\, m_{e\perp}} + \frac{p_3^2}{2\, m_{e\parallel}} \quad (2)$$

where $m_{e\perp}$ and $m_{e\parallel}$ are the transverse and longitudinal electron masses, respectively. The operator (2) is written with respect to the electron ellipsoidal axes 1, 2 and 3 which are, in general, different from the crystal cubic axes x, y and z used for the hole.

It has been noted that under the operations of the rotation group, the operators included in the matrix of the exciton Hamiltonian have different symmetry properties. Thus one can divide H_{ex} into the following parts:

$$H_{ex} = H_s + H_p + H_d \quad (3)$$

where H_s, H_p and H_d are 6 x 6 matrices which contain only s-like, p-like, and d-like operators, respectively. The first term represents an exciton which results from the Coulomb interaction between the electron and the isotropic part of the hole, H_d describes the effect of the anisotropy in the valence band, and H_p the effects of the inversion symmetry (only in the case of zincblende). H_p and H_d are small with respect to H_s. So one can consider these terms as a perturbation. It is easily seen that[6] for the 1s state, the first non-vanishing contribution comes from second-order perturbation theory, and the binding energy takes the form:

$$E_b = R_o \left[1+0.16\left(\frac{\mu_o}{\mu_{2h}}\right)^2 + 0.64\left(\frac{\mu_o}{\mu_{1h}}\right)^2 + 0.64\left(\frac{\mu_o}{\mu'}\right)^2\right] \quad (4)$$

where $R_o = \frac{\mu_o e^4}{2\hbar^2\varepsilon^2}$ is the Rydberg relative to an effective mass μ_o and to a dielectric constant ε.

$$\frac{1}{\mu_o} = \frac{1}{\mu_{oe}} + \frac{1}{\mu_{oh}}, \quad \frac{1}{\mu_{oe}} = \frac{1}{3}\left(\frac{2}{m_{e\perp}} + \frac{1}{m_{e\parallel}}\right), \text{ and}$$

$$\mu' = \begin{cases} \mu_{1\mp} & \text{where the conduction-band minima lie on (0,0,1) direction} \\ \mu_{1\pm} & \text{where the conduction-band minima lie on (1,1,1) direction} \end{cases}$$

$$\frac{1}{\mu_{1\pm}} = \frac{1}{\mu_{1e}} + \frac{1}{\mu_{1h}}, \quad \frac{1}{\mu_{1e}} = \frac{1}{3}\left(\frac{1}{m_{e\perp}} - \frac{1}{m_{e\parallel}}\right), \text{ and}$$

$$\frac{1}{\mu_{oh}} = -\frac{2}{h^2} A, \quad \frac{1}{\mu_{1h}} = -\frac{1}{h_2} B$$

$$\frac{1}{\mu_{2h}} = \frac{2}{h^2}\left(C^2 + 3B^2\right)^{1/2}$$

where A, B and C are defined in reference (11).

From third-order time-dependent perturbation theory[13], we find the following expression for the absorption coefficient $K_{ex}^{(3)}(\omega_1)$ of the photon $\hbar\omega_1$ in a process in which two photons of frequency ω_1 and ω_2 are absorbed and simultaneous absorption (-) or emission (+) of a phonon $(\Omega_{\vec{q}})$ takes place to form final exciton states of non-zero wave vector $\vec{k}_{ex} = \vec{k} - \vec{k}'$.

$$K_{ex}^{(3)}(\omega_1) = \frac{4\pi\hbar^4 N_2}{n_1 n_2^2 m_o^4 C\alpha_{ex}^{3/2}\omega_1\omega_2}(1+\mathcal{P}_{12}) \times$$

$$\frac{\vec{R}_{f\ell}\ (\text{or } \vec{S}_{f\ell}) \cdot \vec{P}^{(1)}_{\ell\ell'}\ (\text{or } \vec{C}^{(1)}_{\ell\ell'}) \cdot \vec{D}^{(2)}_{\ell' i}}{(E_\ell - E_i - \hbar\omega_1 - \hbar\omega_2)^2\ (E_{\ell'} - E_i - \hbar\omega_1)^2} \times$$

$$\left\{ \frac{\eta(\hbar\omega_1+\hbar\omega_2-\hbar\,\Omega_{\vec{q}}-E_g+E_b)}{1-e^{-\hbar\,\Omega_{\vec{q}}/kT}} (\hbar\omega_1+\hbar\omega_2-\hbar\,\Omega_{\vec{q}}-E_g+E_b)^{1/2} + \right.$$

$$\left. \frac{\eta(\hbar\omega_1+\hbar\omega_2+\hbar\,\Omega_{\vec{q}}-E_g+E_b)}{e^{\hbar\,\Omega_{\vec{q}}/kT}-1} (\hbar\omega_1+\hbar\omega_2+\hbar\,\Omega_{\vec{q}}-E_g+E_b)^{1/2} \right\} \quad (5)$$

where n_1 and n_2 are the refractive indices at the frequency ω_1 and ω_2, respectively, N_2 is the density of the photon $\hbar\omega_2$, α_{ex} is the inverse effective mass of the exciton which includes all the details of the valence and the conduction bands, m_o being the free-electron mass, E_ℓ and $E_{\ell'}$ are the energies of the intermediate states ℓ and ℓ', which can be band or exciton states depending on the transition adopted.

The step function is: $\eta(x) = \begin{cases} 1, & x > 0 \\ 0, & x < 0 \end{cases}$

We included in the above expression the temperature dependence through the Bose-Einstein statistics term:

$$n_q = 1/(e^{\hbar\,\Omega_{\vec{q}}/kT} - 1)$$

We introduced the following expression for the matrix elements which appear in the above expression:

$$\vec{R}_{g,ex} = \langle \text{ground}|\text{phonon}|\text{exciton}\rangle,$$

$$\vec{S}_{ex,ex} = \langle \text{exciton}|\text{phonon}|\text{exciton}\rangle,$$

$$\vec{P}_{cv} = \langle \text{band}|\text{photon}|\text{band}\rangle,$$

$$\vec{C}_{g,ex} = \langle \text{ground}|\text{photon}|\text{exciton}\rangle,$$

$$\vec{D}_{ex,ex} = \langle \text{exciton}|\text{photon}|\text{exciton}\rangle.$$

In expression (5) we have adopted the three-band model (see Fig. 1) which gives the largest contribution[10]. This is due to the type of matrix elements included in this model and to the small

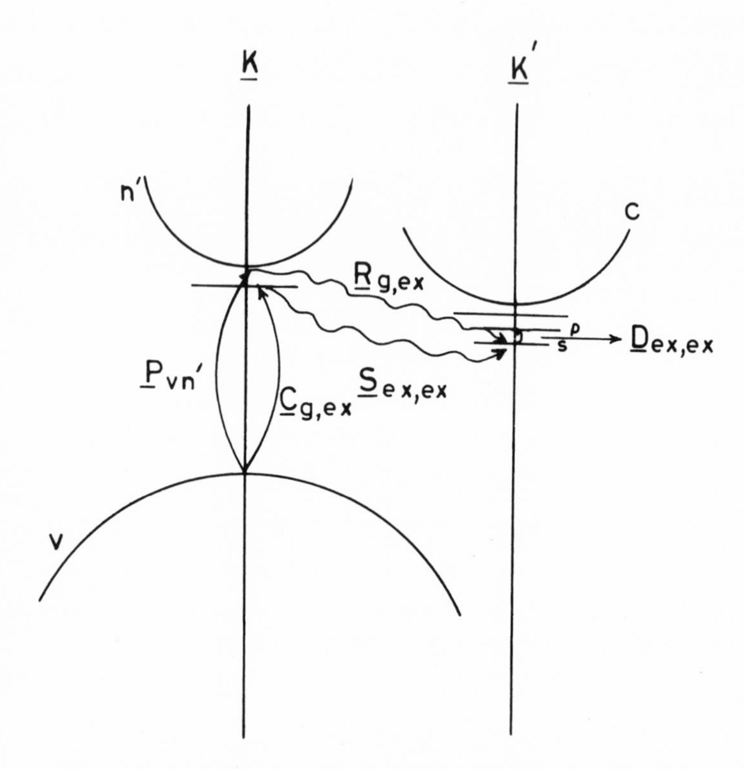

Fig. 1: Three band model scheme for third-order indirect exciton transitions with participation of two photons and one phonon. The different matrix elements are indicated.

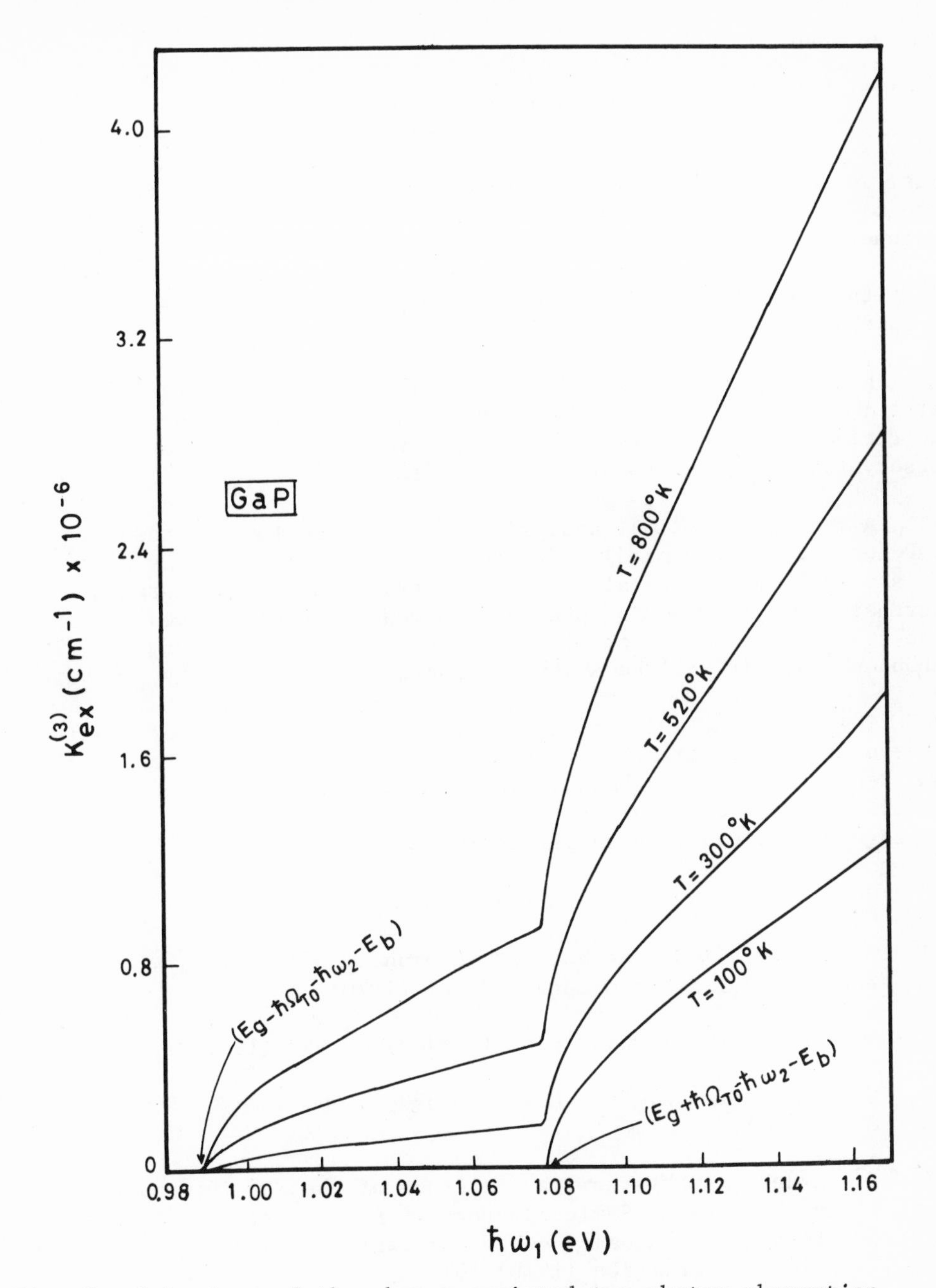

Fig. 2: Behaviour of the phonon-assisted two-photon absorption coefficient $K_{ex}^{(3)}(\omega_1)$ in the exciton region, as a function of $\hbar\omega_1$ at different temperatures for the case of GaP, where the valence band is degenerate.

energy denominators.

RESULTS AND CONCLUSIONS

A numerical application has been performed for the absorption coefficient $K_{ex}^{(3)}(\omega_1)$ for the case of GaP. We have substituted for the binding energy the expression (4) which takes into account the degeneracy of the valence band and the two masses of the conduction band. The frequency dependence in Fig. 2 for GaP shows that the magnitude of $K_{ex}^{(3)}(\omega_1)$ is of the same order as that found[10] for simple valence and conduction bands.

Experiments by Ashkinadze et al[9] for GaP, and by Stafford et al[14] for KI are good evidence for this type of process. Sufficient agreement has been found between the experimental observations and the theoretical results.

A group theoretical analysis has been done for the case of GaP (or III - V compound). The results show that excitons with X_1 and X_5 symmetry are allowed in the transitions for s-states. Furthermore, all types of phonons, in general, are allowed, so one can consider any type of phonon symmetry without worrying about whether it will be allowed or forbidden by symmetry.

The main conclusion of this work is that the order of magnitude of the absorption coefficient $K_{ex}^{(3)}(\omega_1)$ is the same for simple and degenerate valence bands. Thus the previously used simple model, in which the degenerate valence bands are replaced by an average simple band, is satisfactory.

REFERENCES

1. F. Bassani, A.R. Hassan, Opt. Commun. 1 371 (1970); A.R. Hassan, Nuovo Cimento 70B 21 (1970).

2. R. Loudon, Proc. Phys. Soc. (London) 80 952 (1962).

3. G.D. Mahan, Phys. Rev. Lett. 20 332 (1968); Phys. Rev. 170 825 (1968).

4. F. Pradere, A. Mysyrowicz, in Proc. of the Xth Intern. Conf. on the Physics of Semiconductors, S.P. Keller, J.C. Hensel, F. Stern, eds., Atomic Energy Commission, Divison of Technical Information (1970) 101.

5. D. Frolich, *ibid.*, p. 59.

6. N.O. Lipari, B. Baldereschi, Phys. Rev. 3 2497 (1971).

7. T.P. McLean, Progr. Semicond. 5 55 (1960); D.G. Thomas, J.J. Hopfield, M. Power, Phys. Rev. 119 570 (1960); M. Gershenzon, D.G. Thomas, R.E. Dietz, Proc. of the Intern. Conf. on the Physics of Semiconductors (Exeter), A.C. Strickland, ed., The Institute of Physics and Physical Soc., London (1962) 752.

8. F. Bassani, A.R. Hassan, Nuovo Cimento 7B 313 (1972).

9. B.M. Ashkinadze, A.I. Bobryshiva, E.V. Vitiu, V.A. Kovarsky, A.V. Lelyakov, S.A. Moscalenko, S.L. Pyshkin, S.I. Radautsan, Proc. of the IXth Intern. Conf. on Physics of Semiconductors, Nauka, Moscow (1968) 189.

10. A.R. Hassan, Intern. Center for Theoretical Physics, Trieste, Italy (preprint); to be published in Solid State Commun. 1973; A.R. Hassan, Intern. Center for Theoretical Physics, Trieste, Italy (preprint) and Nuovo Cimento 13B 19 (1973).

11. G. Dresselhaus, A.F. Kip, C. Kittel, Phys. Rev. 98 368 (1955); M. Luttinger, Phys. Rev. 102 1030 (1956).

12. M. Luttinger, W. Kohn, Phys. Rev. 97 869 (1955).

13. L.D. Landau, E.M. Lifshitz, Quantum Mechanics, Vol. 3, Course of Theoretical Physics, Oxford (1962); G. Baym, Lectures on Quantum Mechanics, New York (1969).

14. R.G. Stafford, K. Park, Phys. Rev. Lett. 25 1652 (1970); Phys. Rev. 4 2006 (1971).

STRUCTURE OF CHALCOGENIDE GLASSES

Norbert J. Kreidl*
Department of Ceramic Engineering, and

Werner Ratzenboeck†
Material Research Center, University of Missouri at Rolla
Missouri, U.S.A.

When group V and IV elements are added to group VI elements, glasses are formed in which structural groups V $VI_{3/2}$ and IV $VI_{4/2}$ modify and replace their chain and ring structures. In the resulting binary and polynary systems atomic structure and microstructure can be correlated with properties of increasing technological significance. Threshold and memory switching, diode, photocopy, acousto-optical and infrared optical devices are based on these properties. The increase in softening temperature attainable by the incorporation of V-VI structures and microstructures is particularly promising for the design of temperature-stable infrared-transmitting materials.

1. INTRODUCTION

The chalcogens S and Se form glasses by themselves with ring and chain structures, while O occurs in O_2 molecules not quenchable to glass. Nor does Te easily form a glass by itself. However, Te as well as S and Se can participate in binary and polynary glasses in large proportion and many variations. The position in the periodic system and the electronic character of elements participating in complex chalcogenide glasses has been successfully analyzed by Krebs (1953-1970) Klemm (1963) and in the excellent survey by Rawson

---

*1972/73 Fulbright Professor Institut f. Phys. Chemie, University of Vienna, Austria.

†Present Address: Bunzl u. Biach AG, Vliesstoff, A-2762 Ortmann, Austria.

(p. 250) (1967). Melts tend to increase in covalency and polymerization much above the liquidus and can be quenched to stable glasses from those high temperatures. Bonds in the resulting glasses are often much more covalent than in crystals of related composition (Betts et al 1972). Glass forming ability decreases, and the Tg of glasses formed also decreases generally with increasing atomic weight of elements added, excepting the first period (Hilton, 1966). In polynary system, ranges of glass formation are often more extended in, and towards, fields of stability of binary compounds in correspondence to classical postulates of compatibility of constituent elements by Dietzel (1942). Chain structures with intrachain covalent bonds are frequent (Vaipolin 1966), but high Tg chalcogenide glasses will contain three-dimensional networks as well and resemble more conventional oxide glasses.

Since large, low field-strength constituent elements give rise to low energy vibrations (Tanaka et al 1965), chalcogenide glasses are candidate materials for ir optics.

The investigation and development of high Tg or "harder" ir lens materials was based on the successful introduction of Ge or Si into chalcogenide glasses by Hilton (since 1963).

Structures of glasses and crystals of identical composition may be very similar (As_2S_3) or very different (As_2Te_3). As in oxide glasses, and for the same thermodynamic reasons, non-crystalline phase separation is frequent and significant. Phase separation above the liquidus is shown in many phase diagrams and phase separation below the liquidus can be anticipated from other phase diagrams. Thus Myers et al (1972) expect, above liquidus separation in glasses, S-Sb, Sn, In, Pb, Tl; Se-Sn, Bi, Pb, Tl, Te-As, Ga, and below liquidus separation in glasses, S-Te, Bi, Se-Sb, Ge etc. Experimentally, glass-glass phase separation has been reported in systems As-Se (also with Sb, Bi), As-Te (with I), As-Se-Te (with Tl), Ge-Se (also with Pb, and with Na,Li), Ge-Te (with As, Ga, and Si). Dramatic electronic changes may accompany large or small structural changes and are conducive to threshold and memory switching in amorphous semiconductors (Ovshinsky 1966, 1970). Current controlled rapid (10^{-10}s) decrease of resistance by orders of magnitude (to 10^6) after a somewhat less rapid (10^{-6} - 10^{-9}s) delay has been the subject of an imposing array of investigations stimulated by the design of switching devices. A candid and pertinent summary of these efforts was given by Fritzsche (1972). A typical device needs a critical field of 10^5 V/cm. In threshold switching the process is reversible, in memory switching the "on" (low resistance) state remains locked. The "on" stage is associated with a crystallized filament. Crystallization in some cases may be preordained by a separated non-crystalline phase. The process is electronic but, at least in thick devices and those above the threshold voltage, Joule heating may be involved.

The potential application of amorphous materials to switching devices has also caused a large amount of research to be concerned with the theory of amorphous semiconductors. A comprehensive and extensive summary of progress in this direction is that by Adler (1971). In transferring the successful energy band model from ordered to disordered solids (Anderson, Mott) sharp band edges are removed, tails extend into the gap and localized states appear, separated from delocalized states by the mobility gap. Fritzsche has proposed a model of potential fluctuations in amorphous semiconductors in which predominantly p and n type regions alternate, explaining contradictory Hall data and coexistence of ohmic and blocking contacts in lead electrodes.

Many chalcogenide glasses are photoconductive. The photoconductivity of non-crystalline elemental Se, small but vastly larger than its dark conductivity, is the foundation of a dry photocopy process.

Chalcogenide glasses with low acoustic loss, low sound velocity, high acoustooptical effect and high refraction, a critical parameter, are acoustooptical candidate materials (Krause et al 1970).

2. Sulfur. Sulfur crystallizes in several allotropic forms characterized by different stacking of S_8 rings (Prins 1960). It melts at 115°C to a liquid consisting of monomeric rings which at 160°C stiffens (10^3P) to one consisting of increasing concentrations of polymeric chains (Myers et al 1967) of 10^5 to 10^6 atoms. This viscous liquid can be quenched to a non-crystalline solid or glass with a Tg of -27°C in which the chains persist. Above 300°C oriented fibers can be drawn from the liquid which consist of winding chains with rings tucked in. A summary of these modifications, their structures, and the comparative structures of Se and Te is presented in Table 1. (For details consult Rawson 1967).

S-chains can be broken by small concentrations of monovalent ions, e.g. I, or crosslinked by, e.g. P, (Rawson 1967).

Table 1

Comparison of Structures in S, Se and Te

Rings S_8	Se_8
(1) Stable crystalline forms $\alpha\beta\gamma$	(1) metastable crystalline forms
(2) Metastable crystalline ε	(2) part of liquid
(3) Extract from fibers (Φ):ω	(3) part of glass
(4) Liquid < 160°C Λ	
(5) part of liquid > 160°C-(μ)	
(6) " " glass (κ)	
(7) " " fibers (Φ)-ψ	

Chains S_∞	Se_∞	Te_∞
(1) Extracted fibers $(\Phi)\psi$	(1) Stable crystalline form	(1) Stable crystalline form
(2) part of liquid >160°C (μ)		
(3) part of glass (κ)	(2) part of liquid	(2) liquid
(4) " " fibers $(\Phi)\psi$	(3) part of glass	(3) film
longest	medium	shortest

3. Selenium. Se_8 rings are not stable, but do appear in metastable monoclinic Se and in the melt. In stable trigonal metallic Se atoms are arranged in spiral chains (Marsh et al 1953). In the melt near the m.p. (220°C) the average length of the chains is 10^4 atoms (much smaller than that of S). Se is quenched to a glass more easily, doubtless due to the relative instability of the more easily crystallizable ring structure.

The d.c. conductivity of (oxygen-free) Se is as low as 10^{-17} $\Omega^{-1}cm^{-1}$. Delocalized conduction is through the valence band, mostly by free holes. The mobility gap is 2 eV, about that of the gap of states, probably because spirals and rings do not differ too much in short range order. (Adler 1971). The small photoconductivity is so much larger (0.6 eV edge difference) than the dark conductivity that it is of well known technological significance for photocopy. Lucovsky (1969) based his interpretation on the assumption that Frenkel excitons in rings do not contribute to conductivity.

4. Tellurium. The only crystalline form of Te is isomorphous with trigonal Se, i.e., Te atoms are arranged in spiral chains. (Bragg p. 31). It melts at 453°C to a very viscous liquid. Above 465°C chains begin to break and the melt behaves like that of a metal. Glass formation is very difficult (Keller et al 1965, Rawson 1967), presumably because bonding is less discrete and directional, or more metallic, electrons being less tightly held. Te shows hole conductivity and is weakly photoconductive (Keller et al 1965).

5. Binary Systems IV/VI and V/VI. Among the many possible combinations of chalcogenides those with group V elements As and Sb, and with group IV elements Si and Ge are most important. Crosslinked chains and networks involving As(Sb)X3/2 and Si(Ge)X4/2 are characterized by stronger covalent bonds, resulting in higher Tg in large areas of glass formation. The fact that unlike O, the elements S, Se, Te are condensed phases at the temperatures of synthesis and use is significant for the wide ranges of existence of glasses deviating from formal stoichiometry (Rawson 1967). Maximum Tg is reached in Ge-S, Ge-Se, Si-Te systems; Si-S and Si-Se bonds are chemically unstable. In contrast, monovalent elements such as Na, K, Tl, Cl, Br, I act as chain terminators lowering Tg (Kolomiets, Nazarova et al

1962, Flaschen et al 1960).

6. As-VI Systems.

6.1 Crystalline Compounds: In the binary As-S there are two compounds: AS_2S_3, orpiment, monoclinic, with each As bonded to three S, each S to two As (Fig. 1), AS_4S_4, realgar (Fig. 2), also monoclinic, in which each As is covalently bound to another As and two S and each S to two As (Bragg).

In the binary As-Se two analogous compounds occur: As_2Se_3 crystallizing in the orpiment (AS_2S_3) structure (Vaipolin 1966, Zallen et al 1971). As_4Se_4 corresponds with realgar (As_4S_4) but melts incongruently decomposing into As_2Se_3+As (Myers et al 1967). The structure of As_2Te_3, monoclinic, is completely different from that of As_2S_3 and As_2Se_3. The As has coordinations 6 and 3 in octahedral and trigonal groupings with Te, which in turn has coordinations 3 and 2 (Carron 1963), Fig. 3.

6.2 As-VI Glasses: AS_2S_3 and As_2Se_3 form glasses easily. Their structures are closely related to the orpiment structure (As_2S_3: Petz 1961, Hopkins et al; As_2Se_3: Vaipolin 1960, Myers & Felty 1967). Interatomic distances are slightly larger. Amorphous As_2Te_3 is difficult to obtain, (Hruby et al 1971) usually only as a film (Fitzpatrick et al 1971) and/or by doping. The structure of non-crystalline As_2Te_3, however, has no relation to that of the corresponding crystalline As_2Te_3. It most likely possesses the orpiment structure of the As_2S_3 and As_2Se_3 glasses (Fitzpatrick et al 1971). The structural relations of crystalline compounds and glasses As-VI are summarized in Table II. In the binary As-S, the range of glass formation is large (Flaschen 1959), extending to either side of As_2S_3, at which composition a sharp maximum of Tg is observed (200°C) (Myers & Felty 1967, Tsuchihashi et al 1971, Maruno et al 1972, Soklakov et al 1963). The increase of Tg between S and As_2S_3 is nearly linear. Apparently the addition of S to As_2S_3 leads to the monotonous decay of $AsS_{3/2}$ bonds, with formation of S_8 rings (monomers). DTA data show polymerization temperatures in this area (Myers & Felty 1967). At very low As content Tg is very low (20°C) and crystallization (depolymerization) occurs (Myers & Felty 1967, Pearson 1964, Soklakov and Zhanov 1963).

Between S and As_2S_3, As-S glasses are mixtures (copolymers) of As_2S_3, S_8 and S_∞ groupings. Between $As_2S_{2.6}$ and $As_2S_{2.5}$ constant Tg (176°C) suggests immiscibility (Maruno et al 1972). For As:S larger than 2:3, As_4S_4 groups may break the As_2S_3 network (Myers and Felty 1967). As-Se glasses consist of groupings of S_∞, Se_8, $AsSe_{3/2}$ and and As-richer arrangements. (Myers & Felty 1972). Subliquidus immiscibility is observed and extends into the ternary As_2Se_3-As_2Te_3 (Kinser et al 1972).

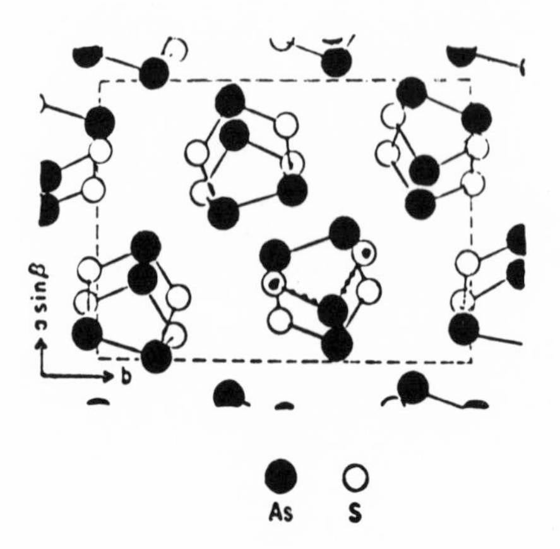

Fig. 1: Atomic arrangement in crystalline As_2S_3 (orpiment).

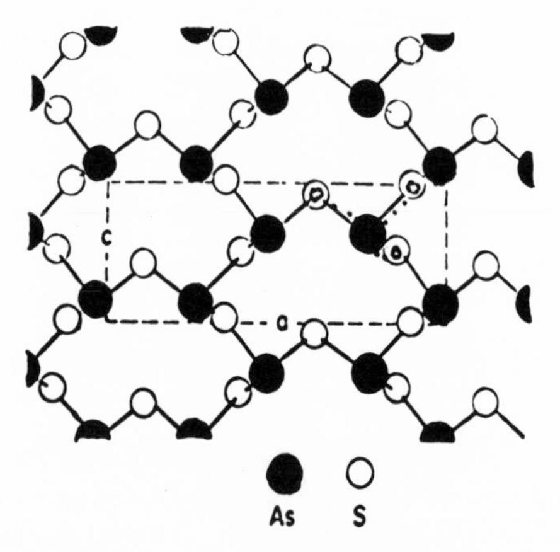

Fig. 2: Atomic arrangement in crystalline As_4S_4 (realgar).

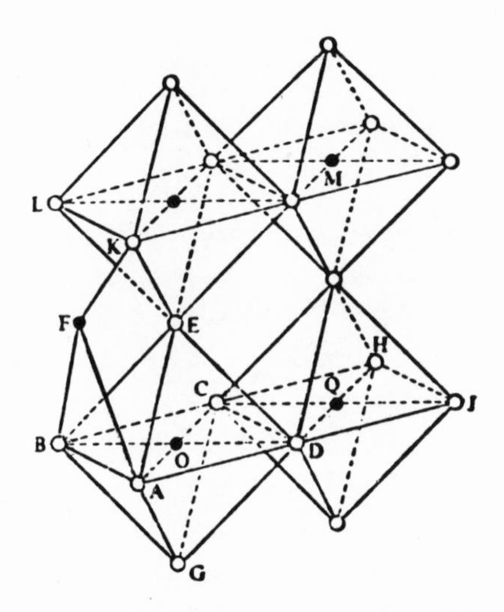

Fig. 3: Atomic arrangement in crystalline As_2Te_3.

"As_2S_3 glass", discovered by Schultz-Sellak (1870), was rediscovered by Frerichs (1950) when the application of chalcogenide glasses to ir optics was indicated. It is prepared by distillation in nitrogen at 735°C, S being added in excess, and cast in aluminum-covered molds (Fraser & Jerger 1953, Glaze et al 1957).

As_2Se_3 glass exhibits memory effects (Thornburg 1972).

Table 2

Structural Relations in As--X Glasses

X = S, Se, Te, cr = crystalline, gl = glassy,
O = orpiment (As_2S_3) structure, r = realgar (As_4S_4) structure,
m = monoclinic As_2Te_3 structure, imm = immiscibility, () = unstable,
Tm =melt point

X =		S		Se		Te
X	cr	X_8		$X_\infty(X_8)$		X_∞
	gl	X_8	+	X_∞		–
X to As_2X_3gl		X_8	+	X_∞	+	O
As_2X_3gl				O		
	cr	O				m
As_2X_3+As gl		r+O imm.		O cross-linked		
As_4X_4	cr	r		Tm r → As^+ +O		

7. Chalcogenide Glasses Containing Ge and Si. Bonds between Ge or Si and S, Se or Te are strongest among those combinations still capable of wide ranges of glass formation. Many chalcogenide glasses containing Ge, Si or both have properties of technological significance, particularly where high Tg is desired.

7.1 Binary Compounds: Compounds between Si or Ge and S, Se or Te are GeS, GeSe, $GeSe_2$, $GeSe_2$, GeTe, GeTe, SiS, SiSe, SiS_2, $SiSe_2$, SiTe, $SiTe_2$ (doubtful) and Si_2Te_3.

7.2 Ge--S and Ge-Se Glasses: In binary Ge-S glasses (Nielsen et al

1962, Hilton 1964, Bailey 1964, Myuller et al 1966, Imaoka 1967, Kiwamoto et al 1971) two regions may be distinguished: (a) GeS_9 to GeS_2 with crosslinked S chains, (b) $GeS_{1.5}$-$GeS_{1.3}$ with GeS_4 tetrahedra, similar to crystalline GeS_2; and GeS_6 octahedra, similar to crystalline GeS. In region (a) S rings form between GeS_4 and GeS_5, as indicated by linear property changes and CS_2 solubility. For more S than in GeS_4 immiscibility appears to occur.

Ge-As-S glasses (Nielsen et al 1962, Imaoka 1967) have been considered for acoustooptical uses (Krause et al 1970) because the critical refractive index which enters the merit factor in the 6th power is as high as 2.5-3.0, while sound velocity and loss are reasonably low.

Binary Ge-Se glasses containing up to 30% Ge are easily obtained from mixtures melted in sealed SiO_2 glass tubing at 1-10 torr and quenched (Savage et al 1965). As Ge is added to pure Se, the structure of Se-glass (its chains and rings), becomes modified first by single, then netted tetrahedral $GeSe_{4/2}$ units (Kolomiets et al[c,d,e] 1962, 1963, 1964, Nemilov et al 1963, 1964, Avetkyan et al 1969). A second lower viscosity glass region exists between $GeSe_2$ and GeSe. Structures are based on Ge-Ge bonds in $Ge_2Se_{6/2}$ groups or on the appearance of Ge^{2+} (Feltz et al 1973).

The extension of glass formation in ternary systems with group V elements is associated with the existence of binary compounds. The area is very large in the system Ge-As-Se (to about 40% Ge and 50% As), where the compounds GeAs and $GeAs_2$ exist, and extends towards the field of stability of these compounds. It is more limited in the ternary Ge-Sb-Se in which no binary compounds Ge-Sb exist. The areas in polynary Se systems are generally smaller than in corresponding S-, and larger than in corresponding Te-systems. The region of highest softening temperatures as well as small crystallization rates in the Ge-As system lies around $Ge_{25-35}\ Se_{50-60}\ As_{10-20}$.

In these ternary glasses $GeSe_{4/2}$ as well as $AsSe_{3/2}$ groups contribute to cross linking manifested in relatively high softening ranges (Hilton[cd] 1965/66, Krebs 1970). An ir optical glass ($Ge_{33}As_{12}Se_{55}$) of technical significance has a softening temperature (as standardized by Jones et al 1968) as high as 450°C. High softening range also is related to high radiation stability (Hench et al 1972).

In ternary Ge chalcogenide glasses conduction is generally by holes, yet Hall mobility is generally n-type (Adler 1971), explicable by microstructure. The optical gap generally exceeds the conduction activation energy. The glasses are not photoconductive, perhaps because of the absence of the transitions characteristic of chalcogenide rings (Se). Ge-As-Se glasses are also applicable to acoustooptics (Kurkjian et al 1971) and to diodes in which they are placed between Al and In (or Sn, Ag) (Panus et al 1967-1969).

Switching and memory effects may be found in regions of substantially lower Ge and higher chalcogenide content; however, the Te systems are more important in this application (Sec. 7.4). But IGO, et al (1973) formulated erasable light-based memories (photochromics) in the area $Ge_{30}As_{25}Se_{45}$, in which Ge-As bonds must appear.

7.3 Ge-Te Systems: There is only one compound, GeTe, in the binary system (Schubert et al 1951, Panson 1964, Goldak et al 1971, Efimov et al 1971) with both atoms having 6-coordination. The telluride glasses are more difficult to obtain than the corresponding S and Te glasses (Kokorina 1971, Krebs 1970) and where glasses form the structure differs entirely from that of GeTe cr. In the glasses, bonds are much more covalent (Betts et al 1972). When 5% Ge are added to pure Te the beginning of glass formation is observed, and the tendency increases as one approaches the eutectic at 20% Ge. In these glass $GeTe_{4/2}$ units probably form, the coordinations for Ge and Te most likely being 4 and 2, but certainly not Ge=2 (Bienenstock et al 1970, Betts et al 1970, 1972, De Neufville 1972). The Tg rises with increasing formation of these units. Phase separation leads a phase consisting almost entirely of Te (Takamori 1970).

Non-crystalline GeTe (Adler 1971) is a p-semiconductor with hole conduction, with a carrier concentration of 10^{18} cm^{-3} (Howard et al 1970). Crystalline GeTe with a carrier concentration of above 10^{20} cm^{-3} has about 10 x higher conductivity. (Kolomiets et al 1964). The optical gap of 0.7 eV corresponds with the conductivity. The memory effect - irreversible transition at a threshold V_{th} to a lower resistance state - in GeTe has been described by Evans et al 1970. The crystallization of Te in the Te-rich separated phase may be the significant process in the formation of the low resistance filament. (Takamori et al 1970, 1971). Some low As Ge-As-Te glasses are superconductive (Sample et al 1972).

Table 3

Glass	T at viscosity 10^{13}	$10^{9.8}$
$Si_{35}As_{25}Te_{40}$	370	442
$Si_{35}As_{15}Ag_{10}Te_{20}Se_{20}$	500	560

There is evidence for the black P structure for GeS, GeSe, and GeTe. Phase separation in Ge-Te and Ge-Se may be present even in areas where it was not observed by transmission electron microscopy (Bienenstock 1973). The coordination 4 for Ge, 2 for Te was confirmed by n-scattering (Nicotera et al 1973).

7.4 Chalcogenide Glasses Containing Silicon: The ternary Si-As-Te contains one of the largest fields of glass formation that is encountered among chalcogenide systems. The large field of glass formation (Fig. 4) has been associated with the occurrence of the large number of binary compounds larger (as indicated in the figure) than in e.g. the Ge-Sb-Se or Ge-As-Se systems. This system contains stable chalcogenide glasses with the highest softening temperatures, mainly because the field extends to low Te and high Si contents (Feltz et al 1972). The highest softening temperatures occur near the pseudobinary join SiTe-As, e.g. in $Si_{35}As_{25}Te_{40}$ glass (Anthonis 1972, 1973). In this region, Si-As bonds may occur. The fact that SiTe is isoelectronic with As may be significant (Deneufville 1972, Krebs 1953, 1958, Suchet 1971). There seems to be some difference from the Ge-As-Te glasses in which the major network is based on $GeTe_2$ units.

In spite of their high transmission in the far ir and their high softening temperature their application to ir optics is handicapped by their susceptibility to oxygen impurity which gives rise to damaging absorption bands. This susceptibility is based on the unfavorable thermodynamics governing the reaction of SiO_2 in the conventional SiO_2 containers used in their preparation, but vycor containers may extract O from the glass (Moynihan et al 1972). Since Si-Se glasses generally decompose (Sholnikov 1965) it was not possible to obtain glass of higher softening temperatures by introducing Se into Si-As glasses. Monovalent ions appear to permit the introduction of Se in an As-rich phase of Si-As-Te glasses which separates on their addition (Anthonis et al 1972, 1973). Thus viscosities in a glass containing Ag and Se were increased by more than 100°C in these biphasic glasses (Table 3).

Ge-Si-Te glasses were investigated by Feltz et al 1972.

8. Other Chalcogenide Glasses. Tl, Cl, Br, I modify various chalcogenide glasses generally breaking chains and networks. (Flaschen 1960, Mel'Nichenko et al 1971, Kolomiets et al[bgh] 1958, 1963, 1965, Dembrovski 1971, Eaton 1964, Quinn et al 1972, Dorr 1971, Uphoff et al 1961, Feltz et al 1972). In Tl glasses photoconductivity occurs frequently and was used by Kolomiets[h] to establish localized levels in line with postulates of Gubanov 1960, 1962, 1965.

Ag, Cu, Au, Fe were introduced into As(Sb)-Se(S,Te)-Si glasses

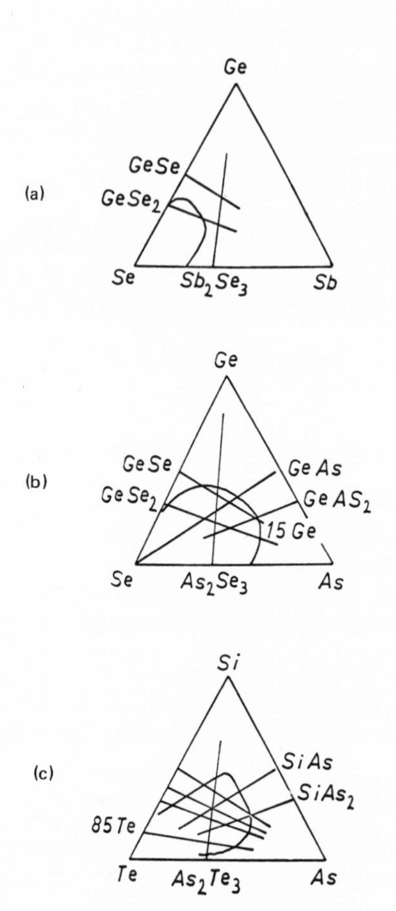

Fig. 4: Glass formation ranges in Systems (a) Ge-Sb-Se, (b) Ge-As-Se, (c) Si-As-Te.

(Three M Patent 1968, Kolomiets et al[k] 1961, Borisova 1970, Anthonis et al 1972, 1973). Alkali sulfur, selenium and tellurium glasses have large areas of phase separation (Plumat 1968). Au accelerates memory crystallization in CuAsSe glasses (Asahara et al 1973).

B-Se glasses, also some containing Te were studied by Dembrovski et al 1971. Binary Bi-Se glasses can only be obtained by vapor deposition and show phase separation, the Bi_2Se_3 rich liquid crystallizing rapidly (Myers & Berkes 1972). Bi-Se glasses might be of interest for special photocopy problems (Schottmiller et al 1969). In complex glasses Bi was introduced by Kislitskaya et al 1971. It increases alkali resistance (Borisova 1971).

BIBLIOGRAPHY

H. ANTHONIS, N.J. KREIDL, *(a)* Non-Cr. Sol. 11 257 High softening point Si-As-Te-Ag gl (1972). *(b)* ibid. to be publ. High Tg polynary Si-As-Te-Se gl. (1973).

D. ADLER, Amoprhous Seminconductors. Crit. Views in Sol. State Sci. Oct. (1971) Comprehensive survey of the entire field.

Y. ASAHARA, T. IZUMITANI, J. Non-Cr. Sol. 11 407 (1972/73).

G.B. AVETKYAN, LA.A. BAIDAKOV, L.P. STRAKHOV, Inorg. Mat. 5 1411 Ge-Se gl (1969).

L.G. BAILEY, M. BRAU, A.R. HILTON, US Pat. 3 154 242 Oct. (1964) High softening point ir transmitting gl.

F. BETTS, A. BIENENSTOCK *(a)* and OVSHINSKY, J. Non-Cr. Sol. 4 554 Ge-Te gl (1970). *(b)*D.T. KEATING and J.P. DE NEUFVILLE, ibid. 7 417 dto. (1972). *(c)* ibid. 8-10 56 dto. (1972). *(d)*and C.W. BATES ibid. 8-10 364 dto. (1972).

A BIENENSTOCK, F. BETTS, D. KEATING, J. DE NEUFVILLE *(a)* Bull. Am. Phys. Soc. 15 1616 $Ge_{17}Te_{83}$ gl (1970). *(b)* J. Non-Cr. Sol. 11 447 (1972/73).

Z. BORISOVA, *(a)*Stekloobraznoe Sostoyannie Cu in Chalcog. gl 5 89 (1970). *(b)*A.V. PAZIN and M.A. GARBUSOVA, Vestn. Leningr. U. Fiz. Khim 3 99 Bi-G-Se gl (1971).

G. CARRON, Acta Cryst. 16 338 $CrAs_2Te_3$ (1963).

S.A. DEMBROVSKII, V.V. KIRILENKO, YU. A. BUSLAEV, Neorg. Mat. 1 510 Tl-B-Se gl, I_2-Si-S gl (1971).

J.P. DE NEUFVILLE, J. Non-Cr. Sol. 8-10 85 Ge-Te gl. (1972).

A. DIETZEL, Z. Elektrochem. 48 9 Field strength of cations in relation to compound formation (1942).

R.C. DOOR, C.R. KANNEWURF, J. Non-Cr. Sol. 6 113 Chalcog. gl. containing Tl (1971).

D.L. EATON, J. Am. Cer. Soc. 47 554 As-Te-halide gl. (1964).

A.M. EFIMOV, B.F. KOKORINA, Stekloobraznoe Sostoyannie (Proc. V All-Union Congr. 1969) Nauka (1971) 92; "GeTe".

E.A. EGOROVA, V.F. KOKORINA, *(a)* Structure of Glass Vol. 2 (1959) trans. Cons. Bur. 1960, p. 430 High softening point chalcog. gl. *(b)* Zh. Prikl. Khim. Leningrad 3 440 (1968), Chem Abs 1968 69 5740, 61250 c Ge-Se-Se gl.

E.J. EVANS, J.H. HELBERS, J. OVSHINSKY, J. Non-Cr. Sol. 2 333 Memory effect in Ge-Te gl (1970).

A. FELTZ, M.J. BÜTTNER, F.J. LIPPMANN, W. MAUL, J. Non-Cr. Sol. 8-10 64 Phase sep. in GeSiTe, GeTeI etc. gl (1972). *(b)* and F.J. LIPPMANN, to be publ. 25 anorg. Chem. 331 (1973).

H. KELLER, J. STOKE, J. Phys. Sta. Sol. 8 831 Similarities of amorphous Te, Se, Ge (1965).

D.L. KINSER, L.K. WILSON, H.R. SANDERS, D.J. HILL, J. Non-Cr. Sol. 8-10 823 Immiscibility in As_2Se_3-(As_2Te_3) gl, and elec. prop. (1972).

E.A. KISLITSKAYA, V.F. KOKORINA, Zh. Prikl. Khim. 44 646 (1971) Sn for Ge in Sb-Se glasses.

W. KLEMM, NIERMANN, Angew. Chem. Intern. Ed. 2 523 (1963) Glass formation in chalcog. systems.

V.F. KOKORINA, Stekloobraznoe Sost. (Proc. V All-Union Congr. 1969) Nauka (1971) 95 Decreasing gl. formation S Se Te.

B.T. KOLOMIETS *(a)* and N.A. GORYUNOVA, Ac. Sci. USSR Phys. Ser. 20 1372 (1956) As-Se-S gl. *(b)* N.A.G. and V.P. SHILO, Sov. Phys. Techn. Phys. 3 912 (1958) As-Se-Tl gl. II Stekloobr. Sost. 456 (1960) Pb, Sn in chalcog. gl. *(c)* T.F. NAZAROVA and V. P. SHILO, Int. Conf. Phys. Semicond. Exeter (1962) 259 Ge introduces 3 dim. nw into As-Se gl. *(d)* L.G. AIO and V.F. KOKORINA, Optiko-Mekh. Promyshl dto (1963). *(e)* Phys. Stats. Sol. 7 362, 364, 713 Glass formation in system Ge-As-S, Se. *(f)* E. YA. LEV and L.M. SYSOEVA, Phys. Stat. Sol. 5 2101, 6 551 Semicond. GeTe. *(g)* T.N. MAMONTOVA and G.I. STEPANOV, Phys. Stat. Sol. 7 1320, 9 (1967, 1927 [?] Photocond. As-Se-Te-Tl gl. *(h)* M. BEN'YAMINOVICH, B.T.K. VEKSLER, T.N. MAMONTOVA and G.I. STEPANOV, Phys. Sol. Stat. 7 13235 (1965) dto. *(i)* ibid. 9 19.27 (1967). *(j)* YU. V. ROKHLYADEV, Phys. Sol. 8 2201 389 (1967) Ge-As-Se(Sn) gl. *(k)* YU. V. ROKHLYADEV and V.P. SHILO, J. Non-Cr. Sol. 5 389 (1971) As-Se-Ag(Cu). *(l)* ibid. p. 402 As-Se-In-Ga-Tl gl. *(m)* and E.A. LEBEDEV, Radio Tekn. i Elektron. 8 2097 (1963) $As_7Tl_7Se_2Te_{12}$ gl.

J.T. KRAUSE, C.R. KURKJIAN, D.A. PINNOW, E.A. SIGETY, Appl. Phys. Lett. 17 367 (1970) Low acoustic loss chalcog. gl.

H. KREBS, *(a)* Appendix to HILTON(c). *(b)* Angew. Chemie 65 293 (1953). *(c)* ibid. 70 615 (1958); Z. anorg. allg. Chem. 265 156 (1961). *(d)* and F. FISCHER, Faraday Conf. Vitr. State 1970, Paper 4, Ge-As-Te gl.

C.R. KURKJIAN, J.T. KRAUSE, E.A. SIGETY, IX Intern. Glass Congr. Vol. 1 (1971) 503 Acoustooptical effects in Ge-As-Se gl.

G. LUCOVSKY, *(a)* Mat. Res. Bull. 4 (1969). *(b)* Physics of Se and Te, C. COOPER ed., Pergamon, Oxford (1969) 255 Structure of Se,As-Se glasses. Ir data.

R.E. MARSH, L. PAULING, J. MCCULLOUGH, Acta Crist. 6 71 (1953) Structure of cr. Se.

S. MARUNO, M. NODA, J. Non-Cr. Sol. 7 (1972) As_2S_x gl.

T.N. MEL'NICHENKO, I.P. MIKHAL;KO, D.G. SEMAK, I.D. TURYANITSA, D.V. CHEPUR, Neorg. Mat. 7 1065 (1971) AsSe(S)I(Br) gl.

N.F. MOTT (and W. TWOSE) Adv. Phys. 10 107 (1961), 16 49 (1967), and many pertinent texts and papers.

C.T. MOYNIHAN, P.B. MACEDO *(a)* M. MAKLAD, R. MOHR, Fall Mtg, Am. Cer. Soc. 1972, to be publ. *(b)* I.D. AGGARIVAL, U.E. SCHNAUS, J. Non-Cr. Sol. 6 322 (1971) Phase sep. in Ge-As-Se-Pb gl.

M.B. MYERS, *(a)* E. FELTY, Mat. Res. Bull. 2 715 (1967) Structure of S,Se,As-S,As-Se, As-S-Se glasses. *(b)* and J.S. BERKES, J. Non-Cr. Sol. 8-10 804 (1972) Phase sep. in chalcog. glasses (see also B. and M.J. Electrochem. Soc. 118 1485 (1971).

R.L. MYULLER, B.M. ORLOVA, V.N. TIMOFEVNA, G.I. TERNOVA, Solid State Chem. (Transl. Consult. Bur. NY) 1966, 232 (from Tverd Tela 1965 12 LGV, GeS, AsSeSi gl., etc.).

S.V. NEMILOV, and PETROVSKII *(a)* Zh. P. Kh. 36 977 (1963). *(b)* 28 1783 (1964) Cross linking of As-Se gl. by additives.

E. NICOTERA, M. CORCHIA, G. DE GIORGI, F. VILLA, M. ANTHONINI, J. Non-Cr. Sol. 11 417 (1972/73).

S. NIELSEN, *(a)* Infrared Phys. 2 117 (1962) GeS gl. *(b)* J.A. SAVAGE, S. Nielsen, Glastech Ber. VII Congress (1965) paper III.2,105 AsGeS gl.

A.D. PEARSON, Modern Aspects of Vitr. State (J.D. MACKENZIE ed.) Butterworths, Washington (1964) 29 S,Se,Te glasses, review.

S.R. OVSHINSKY, *(a)* US Pat 3 271 591, 6 Sept. 1966, "Ovonics." *(b)* Phys. Lett. 21 1450 (1968) dto. *(c)* J. Non-Cr. Sol. 2 99 (1970) Survey of switching and memory devices.

A.J. PANSON, Inorg. Chem. 3 940 (1964) crst. GeTe.

V.R. PANUS, *(a)* and I. BOROSOVA, J. Appl. Chem. 40(S)964(1967). *(b)* YA. M. KSENDSOV, Neorg. Mat. 4 885 (1968) Ge-As-Te gl. *(c)* Vest. Leningr. Gos. Un. Phys. Chem. 22 135-138 (1969) Phase sep. in GeAsTe gl.

J. PETZ, R. KRUH, G. AMSTUTZ, J. Chem. Phys. 34 576 (1961) Structure of As_2S_3 gl.

E. PLUMAT, J. Am. Cer. Soc. 51 499 (1968) Glass formation and phase sep. in alkali containing chalcog. gl.

J.A. PRINS, Non-Crystalline Solids (V.D. FRECHETTE ed.) Wiley, New York (1960) 322.

R.K. QUINN, R.T. JOHNSON, J. Non-Cr. Sol. 7 53 (1972) As-Te-I gl.

H. RAWSON, Inorganic Glass Forming Systems, Academic Press, New York (1967) Glass textbook containing excellent and complete survey of chalcogenide glasses.

H.H. SAMPLE, L.J. NEURINGER, J.A. GERBER, J.P. DE NEUFVILLE, J. Non-Cr. Sol. 8-10 50 (1972) Superconductivity in system Ge-As-Te.

J.A. SAVAGE, S. NIELSEN, *(a)* Phys. Chem. Gl. 5 82 (1964). *(b)* ibid. 6 90 (1965). *(c)* 7 56 (1966). *(d)* Infrared Phys. 5 195 (1965) Ge-Chalcog. gl.

J.C. SCHOTTMILLER, T.W. TAYLOR, F.W. RYAN, Appl. Opt. Suppl. 3 55 (1969) BiSe.

K. SCHUBERT, H. FRICKE, Z. Naturforsch. 6a 781 (1951) Crst. GeTe.

C. SCHULTZ-SELLAK, Ann. Phys. 139 162 (1870) Transmission of As_2S_3 gl.

E.V. SHOL'NIKOV *(a)* Vestnik Leningr. U. 4 115, 120 (1965) Instability of Si-Se. *(b)* Stekloobr. Sost. 5 94 (1970) GeAsSe(Sn,Si,Pb) gl.

A.I. SOKLAKOV, G.S. ZHANOV, Sov. Phys. Cryst. 7 447 (1963) As-S gl.

J-P. SUCHET, Mat. Res. Bull. 6 491 (1971) Chain group systematics in Ge-As(Sb) gl.

T. TAKAMORI, R. ROY, McCARTHY, *(a)* Mat. Res. Bull. 5 529 (1970) Ge-As-Te gl. *(b)* J. Appl. Phys. 42 2577 (1971) Phase sep. and memory in GeTe gl.

M. TANAKA, T. MINAMI, Jap. J. Appl. Phys. 4 1023 (1965) Vibrations in glasses containing large cations.

D. THORNBURG, J. Non-Cr. Sol. 11 113 (1972) Memory effects in As-Se gl. *(b)* 3M corp. Brit. I 117, 19 June (1968) Switching gl. Sb(Cu,Ag,Au)S(Te)II.

S. TSUCHIHASHI, Y. KAWAMOTO, *(a)* J. Non-Cr. Sol. 5 286 (1971) As-S gl.

G.Z. VINOGRADOVA, S.A. DEMBOVSKII, S.A. LUZHNAYA, N.P. LUZHNAYA, Zh. Neorg. Khim. 13, Russ. J. Inorg. Chem. 13 758 GeAsSe. [?]

H.L. UPHOFF, J.H. HEALY, Contr. NONR 2965 (00) A-1-(NR 017446) 11 July 1961.

A. VAIPOLIN, *(a)* Sov. Phys. Cryst. 10 509 (1966) Crst. chalcog. structures. Orpiment structure of As_2Se_3. Ge-Cd-S(Se) gl. *(b)* and E.A. PORAI-KOSHITS, Sov. Phys. Sol. State 5 178, 186 497 (1963) Orpiment structure of As_2Se_3, As_2Se_3. Structure of glasses.

R. ZALLEN, G. LUCOVSKY, Selenium (W.C. Cooper, R.A. ZINGRAD eds.) Van Nostrand-Reinhold, New York (1971). *(b)* M.I. SLADE, A.T. WARD, Phys. Rev. B3 4257 (1971) Crst. As_2Se_3.

EFFECT OF HEAT TREATMENT ON ELECTRICAL PROPERTIES AND STRUCTURE OF As-Te-Ge CHALCOGENIDE THIN FILMS

F.H. Hammad
Atomic Energy Establishment, Cairo

and
A.A. Ammar, M.M. Hafiz, and A. El-Nadi
Assiut University, Assiut, Egypt

The electrical properties and structure of Te_{53}-As_{36}-Ge_{11} chalcogenide glass were investigated with various conditions of heat treatment and preparation. This material was found to exhibit negative resistance with a memory and can be converted from high to low resistance state thermally by heating to a transition temperature or electrically by passing a transition current. The activation energy for conduction was found to be 0.36 and 0.3 eV in the high and low resistance states respectively.

Annealing in the range 403-477°K showed that the time required to attain the conducting state decreased upon increasing the temperature. Similarly, increasing substrate temperature during evaporation caused a continuous decrease of room temperature resistivity until a conducting state was reached.

Transmission electron microscopy showed that electron beam heating caused the formation of a dendritic crystalline phase in the original amorphous film. X-ray diffraction also showed that the high resistivity film was amorphous and the low resistivity film was crystalline. The decrease in electrical resistivity during annealing was attributed to progressive nucleation of a second crystalline phase.

INTRODUCTION

Certain amorphous chalcogenide glasses have received growing attention because they exhibit interesting memory and threshold switching properties. These glasses usually contain Se and Te alloyed with one or more elements such as Ge, I, As, Sb, Ga or Tl. Amorphous semiconductors can be made from binary, ternary or quaternary systems. The properties of various compositions of the ternary system As-Te-Ge have been studied.[1,2]

The switching properties of chalcogenide glasses have been related to regenerative structural changes such as phase separation[3,4] or thermally induced crystallisation.[5] Correlation between electrical properties and structure of a chalcogenide glass Te-As-Ge (53,36,11 % by weight) is reported here; this composition was not investigated previously.

EXPERIMENTAL

The alloy was prepared by melting the required amounts of As, Te and Ge in an evacuated silica tube. The melt was kept at 1450°K for one hour and was vigorously shaken to ensure homogenization, then air cooled. Thin films were prepared from bulk specimens by vacuum deposition (10^{-6} mm Hg) on a glass substrate. A planer electrode configuration was used and gold film electrodes were deposited before the specimen. The thickness of the film was 3000 A° and the inter electrode seperation 1 mm. unless otherwise specified.

RESULTS

I-V characteristics. The I-V behaviour of a bulk specimen shown in Fig. 1 is typical of a negative resistance device with memory.[6] At low applied voltage (up to Vo) the device is ohmic and has a high resistance. At V > Vo the behaviour becomes non-ohmic and ends up at the threshold voltage V_{th}, beyond which a region of current controlled negative resistance (CCNR) sets in. These characteristics are reproducible and reversible provided the current does not exceed the transitional current I_t. The conductive, ohmic (or memory) state is established when a current exceeding I_t is passed. The device remains in this ON state when the current is decreased gradually. The high resistance or OFF state can be reestablished by passing a pulse of high current density through a circuit containing 1 K. series resistance. Thin film specimens were also found to display the same I-V characterestics.

The I-V behaviour of thin film specimens (having an inter-electrode separation of 135 μm) is strongly influenced by temperature as shown in Fig. 2. As the temperature is increased, the threshold voltage decreases and the threshold current increases as shown in Fig. 3.

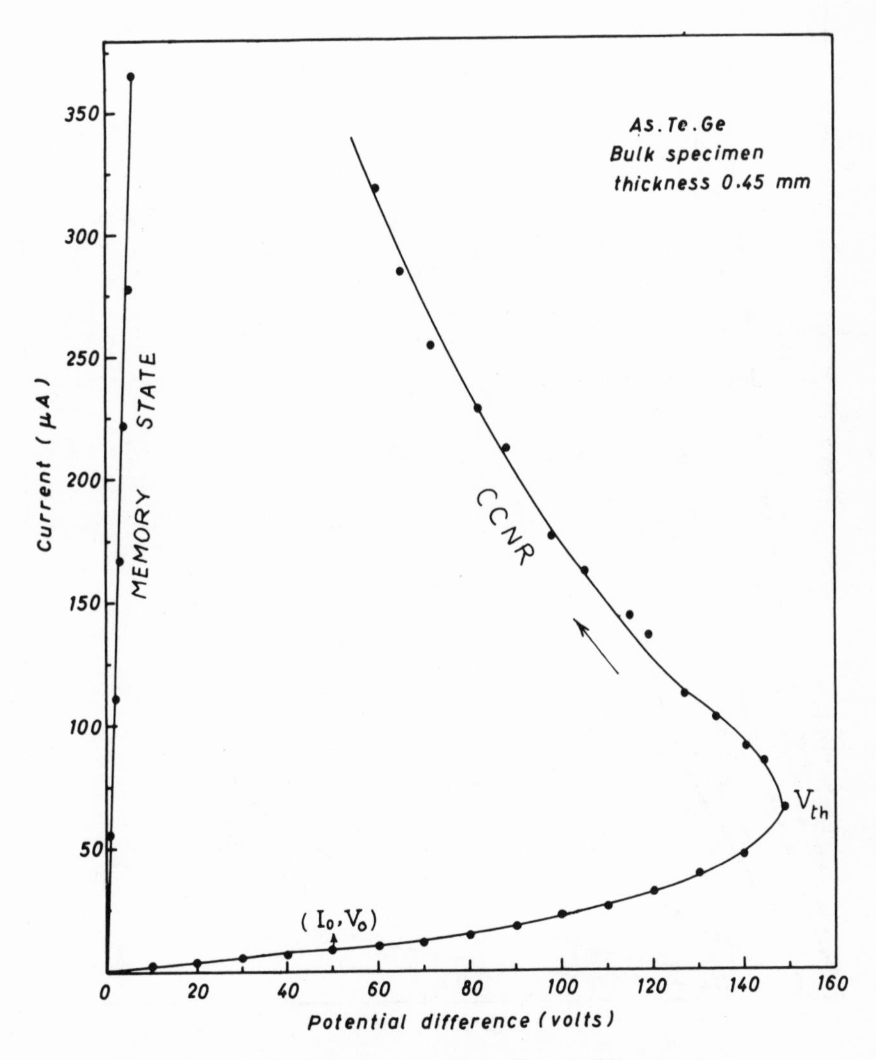

Fig. 1: I-V characteristics for bulk specimen.

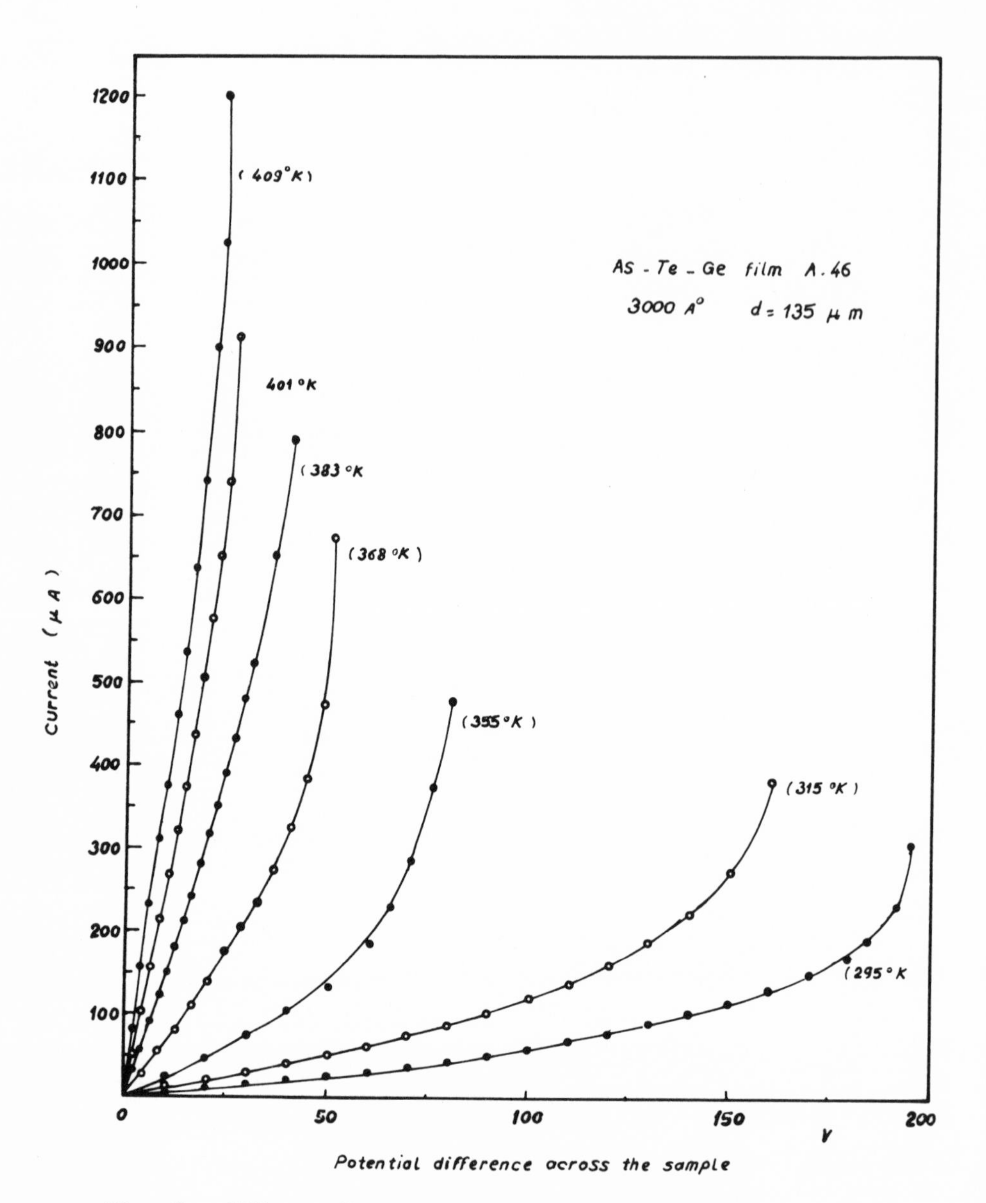

Fig. 2: Effect of temperature on the I-V characteristics of thin film specimen.

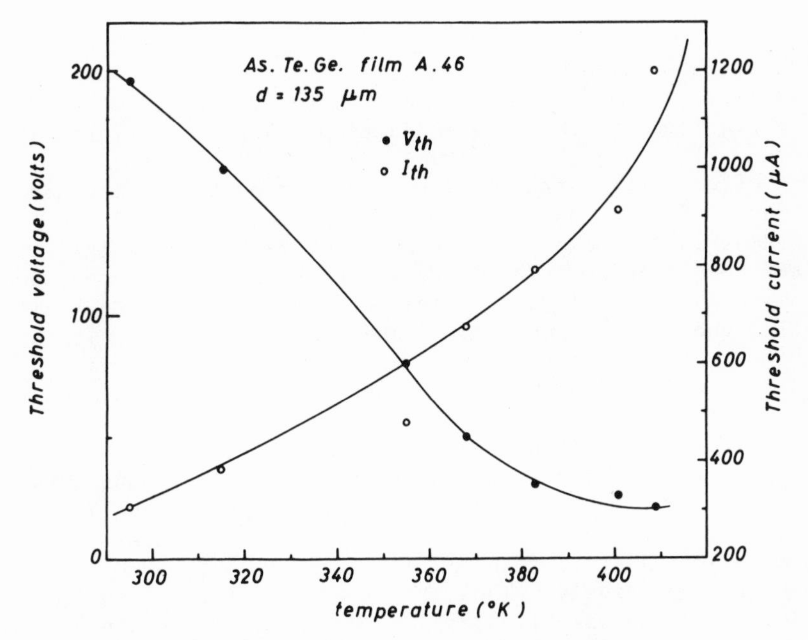

Fig. 3: Effect of temperature on threshold voltage and threshold current.

Effect of heat treatment on electrical resistance. The decrease in resistance upon heating (at a rate of 0.8°K/min.) is shown in Fig. 4. A linear relationship between log R and 1/T°K in the range from room temperature to 370°K indicates an intrinsic behaviour. The activation energy ΔE was found to be 0.36 eV. This is in close agreement with the values reported by Pinto[7] and Fagen and Fritzsche.[8]

A sharp drop in resistance is noted above 37°K which is not characterized by a single activation energy. A conducting state is reached at the transition temperature 471°K which persists during cooling (at a rate of 2°K/min). The activation energy of this state was found to be 0.03 eV in agreement with the findings of Pinto[7] and Adler.[3]

The effect of isothermal heat treatment at various temperature shown in Figs. 5 and 6 indicates that as the temperature is increased the time required to reach the conducting state is decreased.

The effect of substrate temperature during deposition has a significant influence on resistivity, as shown in Fig. 7. A film deposited at 297°K had a resistivity of 1.27×10^5 Ω cm which decreased upon heating. The decrease is sharp in the range 330-340°K. Above 340°K the resistivity remained constant at 3.5×10^{-2} Ω cm indicating the establishment of a conducting state.

X-ray diffraction and electron microscopy. X-ray diffraction (using Cu K_α, λ = 1.5405 A°) patterns for 10,000A° thick films which received various heat treatments are shown in Fig. 8. The virgin specimens gave broad peaks characteristic of an amorphous structure. Marked crystallinity was observed after 1 hr. at 160 and 230°C. The resistivity of the virgin specimen was 2×10^5 Ω cm while that of specimens annealed at 160°C and 230°C dropped to 2.35×10^4 and 5×10^{-3} Ω cm respectively. Thus, the decrease of resistivity with temperature and the establishment of the conducting state are associated with crystallization.

The structure of 800Å thick films was examined by transmission electron microscopy (JEM-6C Electron Microscopy operated at 80 K.V) As shown in Fig. 9 the as-deposited film is structureless and the diffraction pattern (insert) shows halos typical of conventional amorphous structure. Electron beam heating caused a sudden and very rapid (the speed could not be measured) local transformation of the structure into a crystalline phase after an incubation period; the new phase had a dendritic structure, as shown in Fig. 10.

Heating the specimens outside the microscope at 100 and 140°C for 15 minutes did not result in phase transformation. Increasing the annealing temperature or time, e.g. heating at 150°C for 30 minutes, led to observation of the crystalline second phase

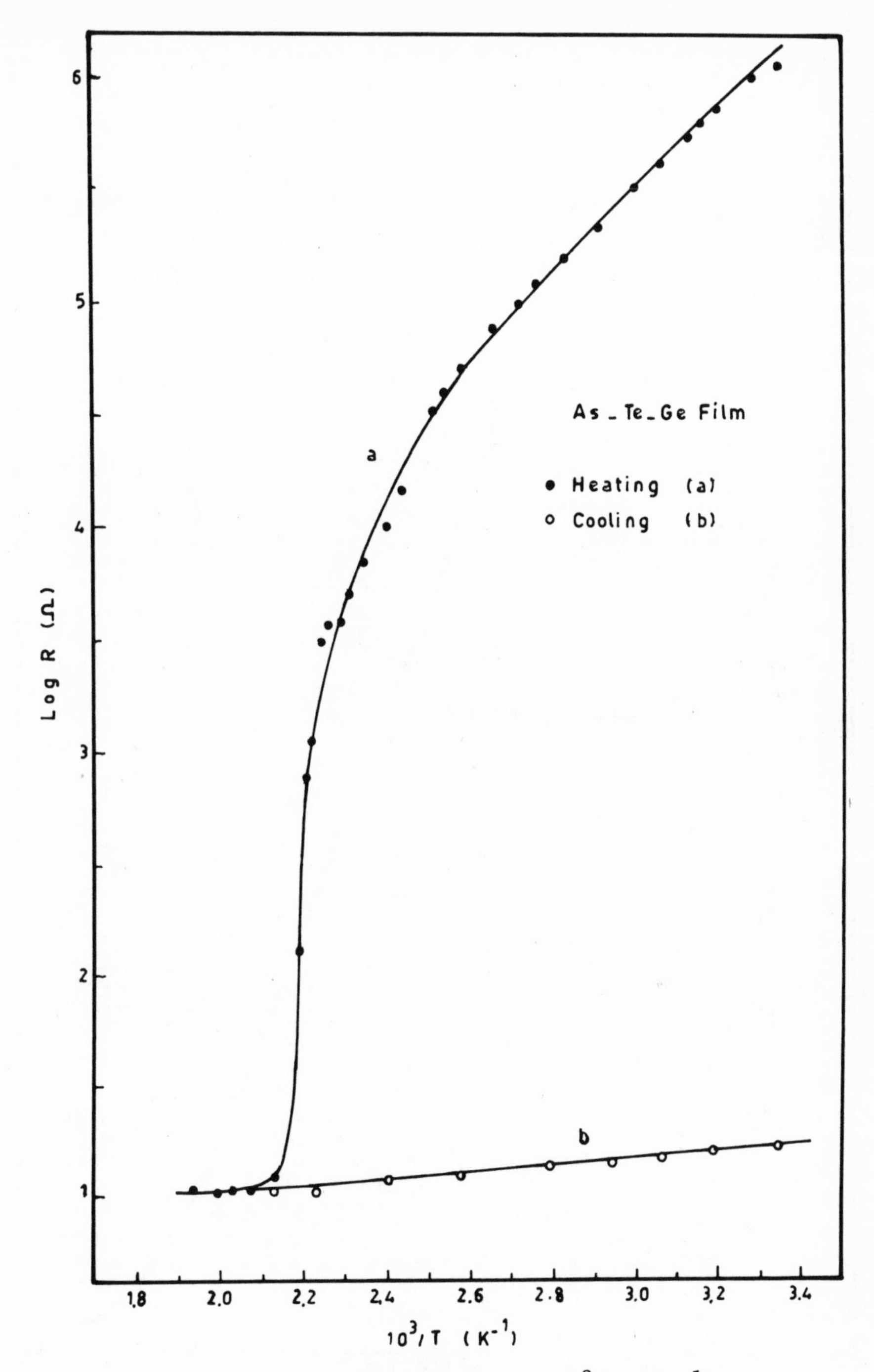

Fig. 4: Log resistance vs. $10^3/T$ ($^oK^{-1}$).

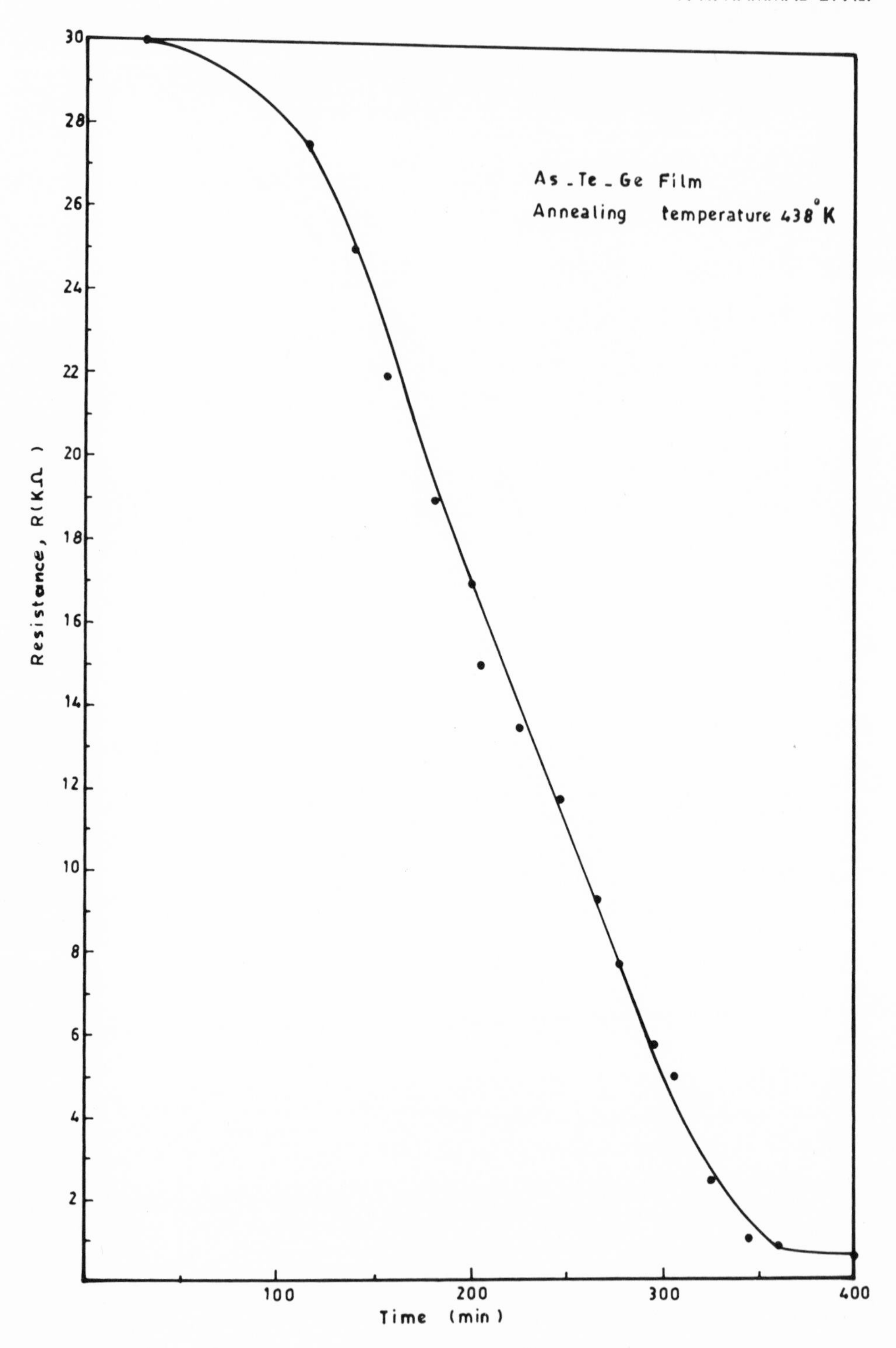

Fig. 5: Effect of annealing at 438°K on film resistance.

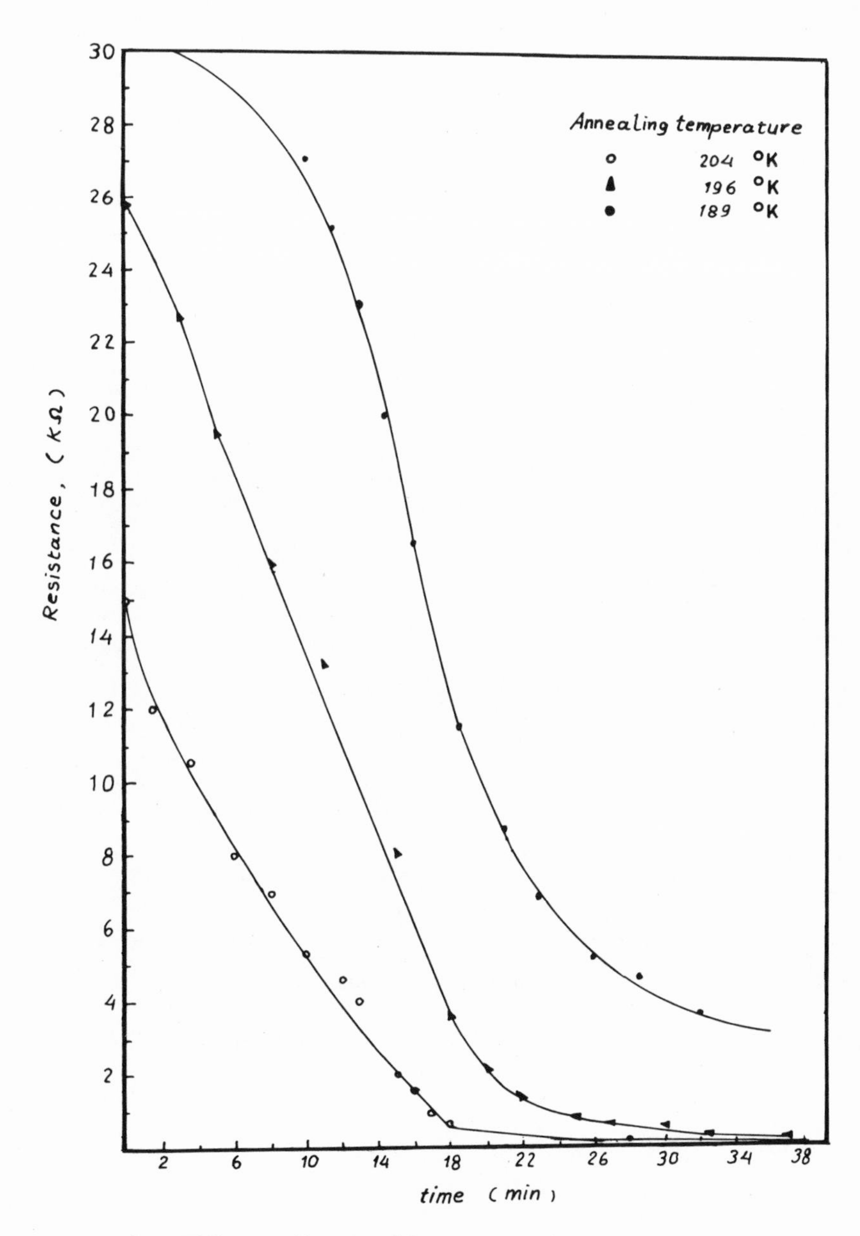

Fig. 6: Effect of annealing at various temperatures on film resistance.

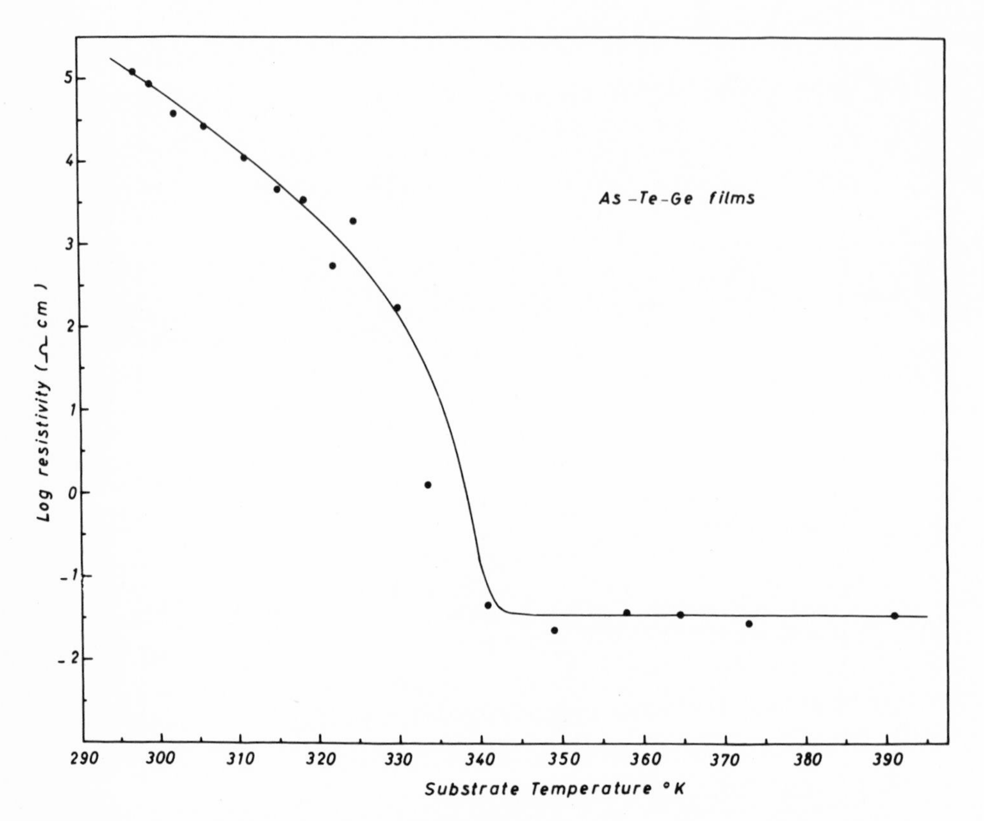

Fig. 7: Effect of substrate temperature during evaporation on the film resistivity.

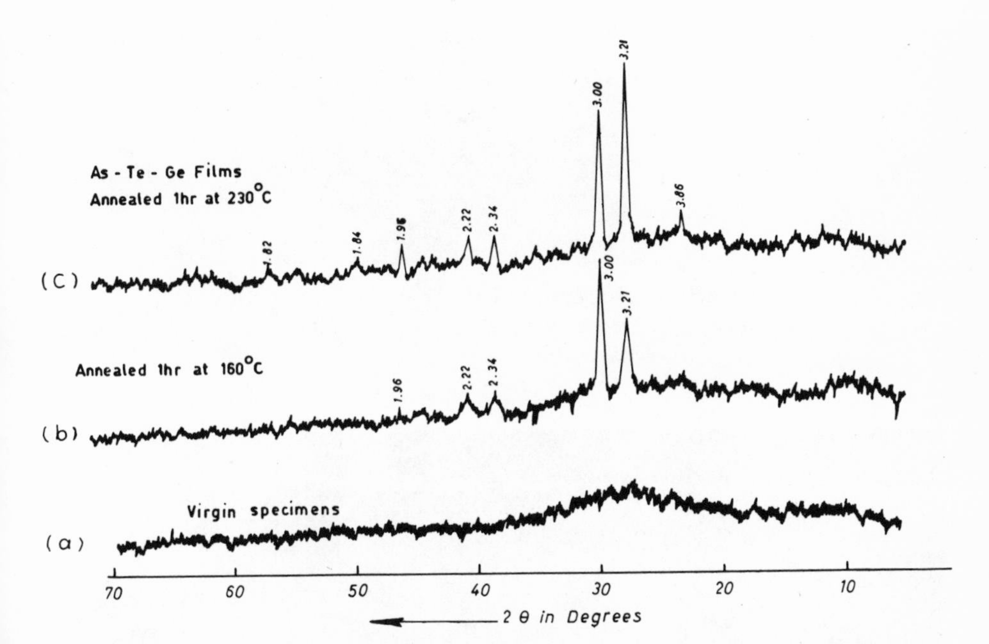

Fig. 8: X-ray diffraction of virgin and annealed thin film specimens.

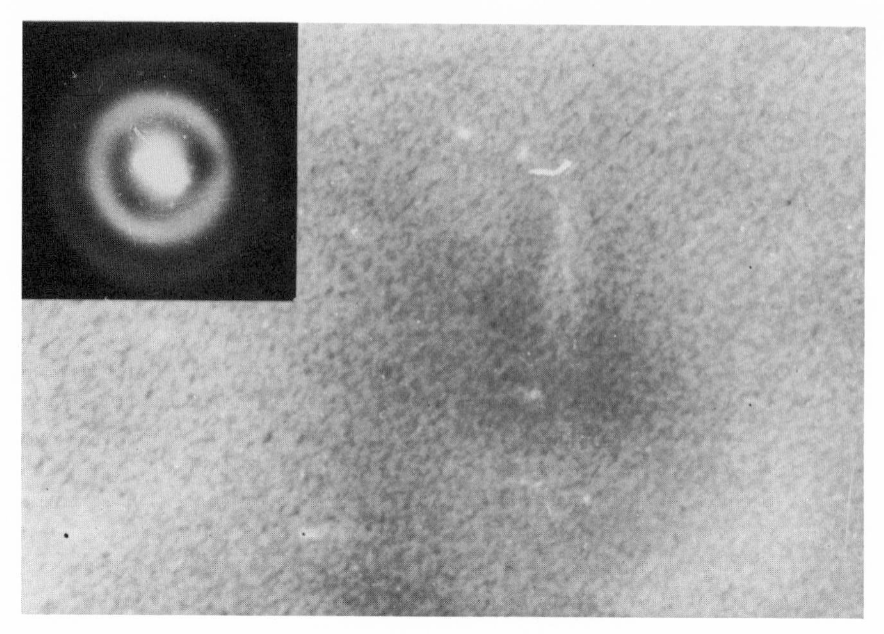

Fig. 9: Transmission electron micrograph of as-deposited film. X12,500.

Fig. 10: Transmission electron micrograph of thin film after electron beam heating. X14,500.

immediately after inserting the specimen in the microscope and before beam heating effects became considerable.

The structures of As-Te-Ge alloys of different compositions were examined by various techniques.[2,3,7,9-11] Annealed and slowly cooled specimens showed the presence of crystallites while non-heat-treated specimens showed no feature of crystallinity.

Isothermal annealing. The effect of heating at 451°K on the room temperature electrical resistivity of identically prepared specimens is shown in Fig. 11. The heating and cooling rates were slow (1°K/min). After an incubation period, a decrease in resistivity was noted after 45 minutes and became sharp after 60 minutes. The conducting state was attained after 105 minutes.

The change of electrical resistivity with temperature was measured during cooling after the end of the heat treatment and the intrinsic activation energy for conduction was determined for all specimens. Figure 12 shows that ΔE of the high resistivity state initially decreases to that of a conducting state during the heat treatment in a manner similar to that of resistivity.

Metallographic examination showed a dendritic second phase dispersed in a homogeneous matrix after 45 minutes of annealing, as shown in Fig. 13. This coincides with the time at which the decrease in resistivity and ΔE was noted. As the time was increased the amount of the second phase increased until it covered the whole surface. The conducting state was reached (after 105 minutes) when the dendritic branches touched each other (Fig. 14).

Relation between structure and electrical properties. The results indicate that Te-As-Ge (53%-36%-11%) can exist in two stable states: an amorphous, high resistivity (OFF) state and a crystalline, low resistivity (ON) state. The transformation from the OFF to the ON state can be accomplished thermally after heating to a certain transition temperature or electrically, after passing a transitional current.

The decrease in resistivity and ΔE with time during isothermal annealing is due to the progressive formation of the crystalline conducting phase in the amorphous structure. The isothermal annealing curves (Figs. 5,11) have sigmoidal form and are similar to those of recrystallization of cold worked metals.[12] The crystallization takes place by nucleation and growth. Transmission electron microscopy showed that the growth of the second phase takes place at high speeds once embryos of critical size or nuclei have formed. Thus the effect of temperature and annealing time on electrical properties may be attributed mainly to the nucleation of the crystalline phase.

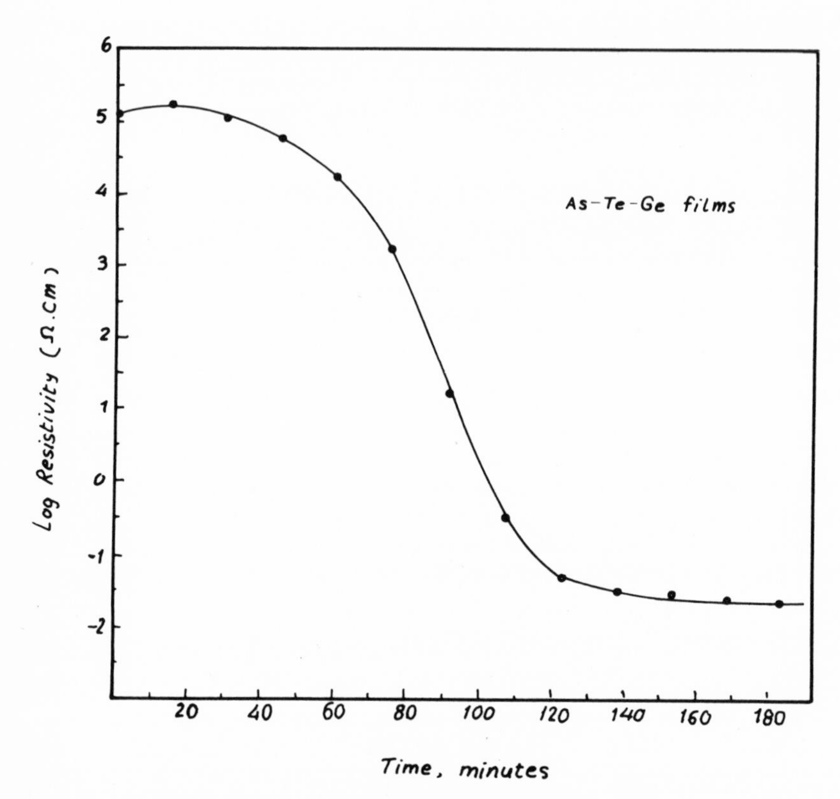

Fig. 11: Effect of isothermal annealing at 451°K on log room temperature resistivity.

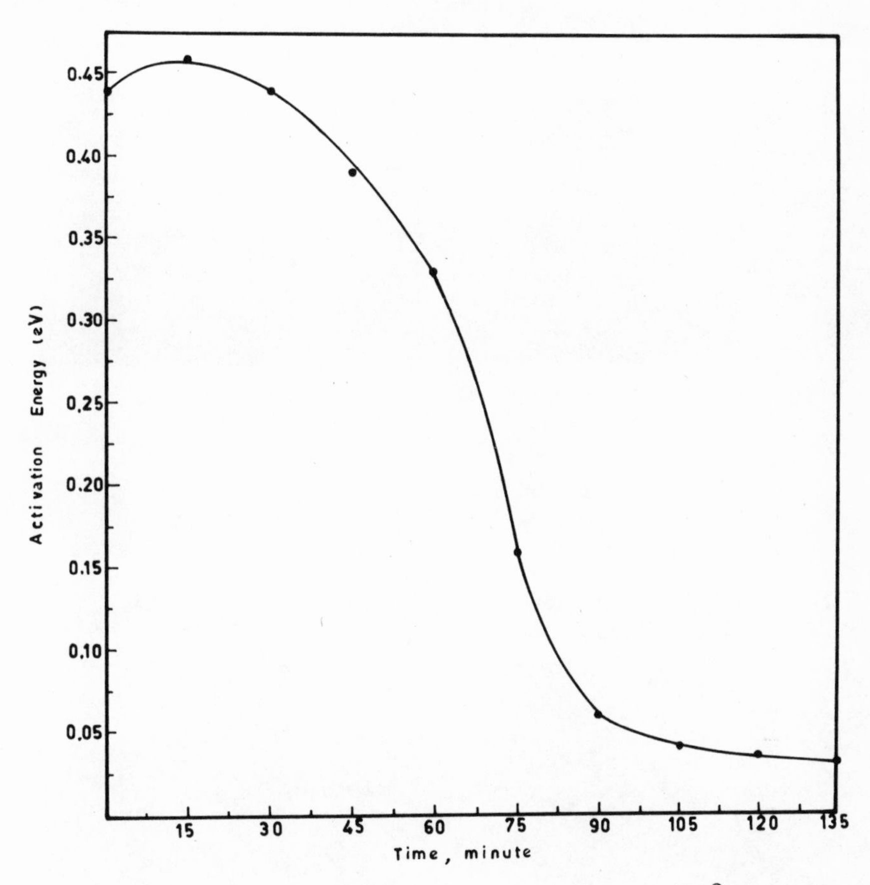

Fig. 12: Effect of isothermal annealing at $451^{o}K$ on the activation energy.

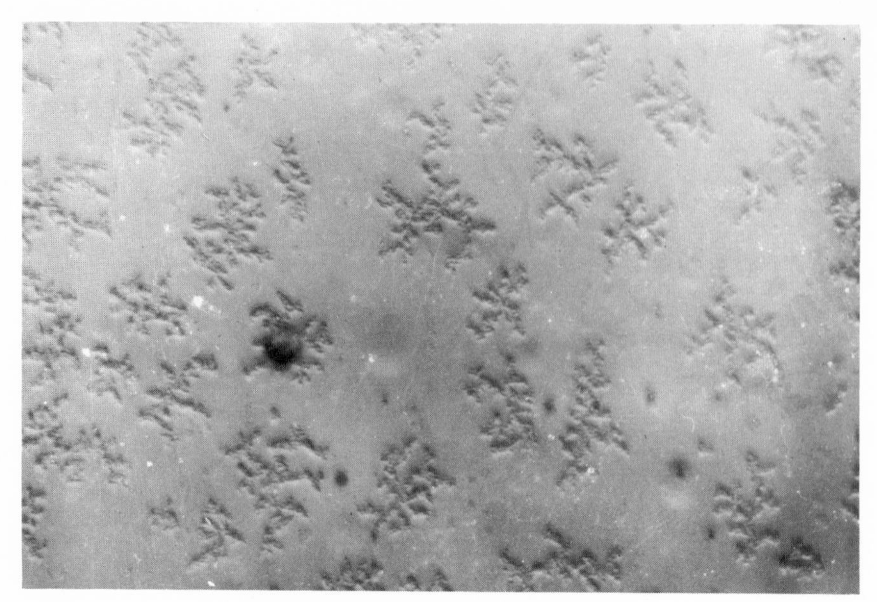

Fig. 13: Photomicrograph of As.Te.Ge film annealed for 60 min. at 451°K. X325.

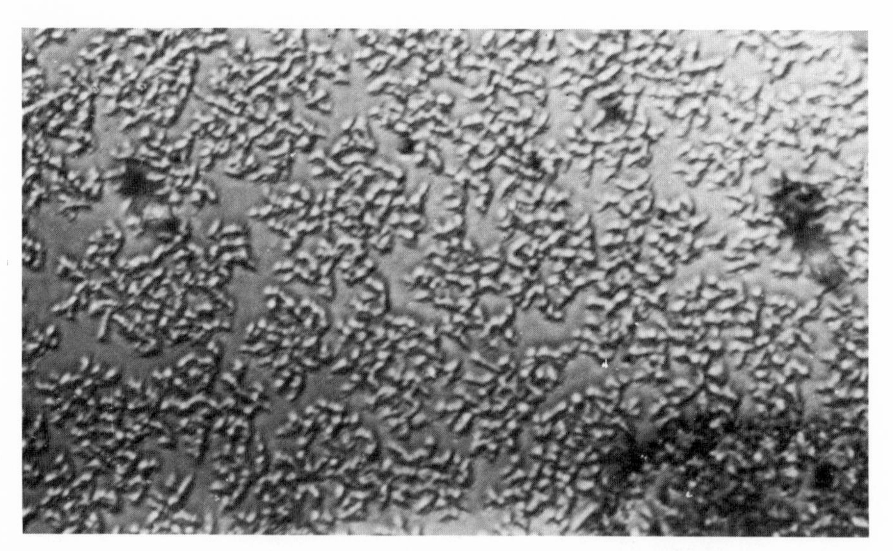

Fig. 14: Photomicrograph of As.Te.Ge film annealed for 90 min. at 451°K. X325.

As the process of nucleation is thermally activated,[12] increasing the annealing time or temperature increases the probability of producing nuclei. The need for thermal activation accounts for the incubation period often observed (Figs. 5,6,11). As the temperature was increased the incubation period decreased, and sometimes was not observed. The sharp drop in resistance at a certain temperature (Fig. 4) or a certain annealing time is due to the rapid formation of the crystalline phase subsequent to the incubation period. Thus, the time required for the conducting or crystalline state to be reached decreases as the temperature is increased.

Similarly, increasing the substrate temperature increases the mobility of deposited atoms and promotes nucleation of the crystalline phase. Thus, crystallization is facilitated with the increase of the substrate temperature. Above a certain temperature the film is deposited in the crystalline or conducting state immediately.

Thus, the crystallization temperature or the transition temperature is not sharply defined and depends on time, and substrate temperature. Sugi et al.[13] also showed that the transition temperature depends on the heating rate.

The establishment of a conducting state electrically can also be attributed to thermally induced crystallization resulting from sufficient Joule heating after passing a current I_t. The effect of temperature on the I-V behaviour can be related to thermally induced crystallization. As the temperature is increased, the resistance decreases due to progressive crystallization. Thus the amount of Joule heating required for switching is decreased. Similarly, the threshold voltage decreases and the threshold current increases.

REFERENCES

1. S.R. Ovshinsky, Phys. Rev. Lett. 21 1450 (1968).

2. H. Fritzsche, S.R. Ovshinsky, J. Non-Cryst. Solids 2 148 (1970).

3. D. Adler, J.M. Franz, C.R. Hewes, B.P. Kraemer, D.J. Sellymer, S.D. Senturia, J. Non-Cryst. Solids 4 330 (1970).

4. A.D. Pearson, J. Non-Cryst. Solids 2 1 (1970).

5. H.J. Stocker, J. Non-Cryst. Solids 2 371 (1970).

6. H. Fritzsche, in Proc. Symp. on Instabilities in Semiconductors, IBM, Watson Res. Lab. (1969).

7. R. Pinto, J. Non-Cryst. Solids 6 187 (1971).

8. E.A. Fagen, H. Fritzsche, J. Non-Cryst. Solids 2 170 (1970).

9. S.V. Phillips, R.E. Booth, P.W. McMillan, J. Non-Cryst. Solids 4 510 (1970).

10. G.V. Bunton, J. Non-Cryst. Solids 6 72 (1971).

11. T. Takamori, R. Roy, G. McCarthy, J. Appl. Phys. 42 2577 (1971).

12. R.W. Cahn, Physical Metallurgy, North Holland (1970) 1129-65.

13. M. Sugi, S. Iizima, M. Kikuchi, K. Tanaka, J. Non-Cryst. Solids 5 358 (1971).

APPLICATION OF AVRAMI'S EQUATION TO THE CONDUCTIVITY CHANGES OF AMORPHOUS SELENIUM

M.K. El-Mously and F.E. Gani

Physics Department, Faculty of Science

Ain Shams University, Cairo, Egypt

The conductivity changes associated with the isothermal annealing of amorphous selenium are due mainly to crystallization. Direct measurements of these changes were conducted, and the experimental data was applied in Avrami's equation in an attempt to evaluate the constant (n), which depends on the crystallization geometry. The crystallization constant (k), as well as the energy of crystallization (E) at different temperatures were calculated on the basis of the constant (n). For an average value of $\bar{n} = 1.38$, the energy of crystallization was found to be 15.1 kcal/mol. The obtained fractional values of the constant (n) suggest that the crystallization of selenium is not a purely one dimensional process.

INTRODUCTION

The crystallization process of Se and Se-S alloys has been studied previously[1,2] by measuring the crystallization parameters, i.e. conductivity (σ), energy of activation (Eg) and density (d) of a sample isothermally annealed for different time intervals. For $S\text{-}Se_{20}$ alloy, the rate of crystallization is relatively small and, hence, the crystallization process can be easily stopped by quenching the annealed sample in air. The experimental data was found to fit Avrami's equation and the calculated value of (k) on the basis of both log σ and (d) lead to the same value for E. In the case of pure Se, this method cannot be successfully applied since the sample needs some interval of time to acquire the temperature of the oven, resulting in some error in measuring the time of annealing. Furthermore, Se has a relatively high crystallization rate which cannot be stopped at a desired time. Additional error may result from considering the constant (n) equal to 1 in Avrami's equation.[3-7]

$$\theta = e^{-kt^n} \quad (1)$$

To minimize these errors as much as possible, another technique has been applied in which the conductivity was measured continuously during the process of annealing; the conductivity was taken as a function of the percentage of crystallization of selenium.[1,2]

The isothermal curves indicated the presence of four different stages, namely:

1. An increase in σ due to normal heating of the sample;

2. Further gradual but less pronounced increase in σ which may be due to nucleation of the crystalline phase;

3. A sharp rise in σ which may be attributed to rapid growth of the crystalline state, a process which is accompanied by the liberation of the heat energy associated with the transition from a nonequilibrium to an equilibrium thermodynamic state;

4. An asymptotic decrease in σ following stage 3, which may indicate that the thermal energy released during this stage is being lost asymptotically from the sample and absorbed by the materials of the oven. At the end of such a process, the conductivity acquires a certain constant value corresponding to the temperature of the oven. This phenomenon was also observed by Chapness.[8]

This report presents a detailed study of the changes in conductivity as a function of time during stage 3 (above) on the basis of Avrami's equation. These studies have been carried out at different temperatures, ranging from the softening temperature T_g to the melting point of selenium. Calculations using an IBM 1130 computer have been carried out in an attempt to show to what extent the experimental data may be analyzed according to Avrami's equation.

EXPERIMENTAL

Samples of 99.99% pure selenium were used. A sample was sealed in an evacuated quartz tube (10^{-3} mm Hg) and heated to 700°C for 6 hours and then quenched in air. The specimen was then placed in a pyrex glass tube with two fixed tungsten electrodes. The pyrex tube was then evacuated to 10^{-3} mm Hg and sealed. In order to obtain good contact between the sample and the lectrodes, the temperature was elevated to 250°C for 30 min. and then quenched rapidly in air, in order to obtain glassy selenium. Using an electrometer of the type VAJ-51 (error less than 2%), the changes of the electrical conductivity were continuously measured during the crystallization

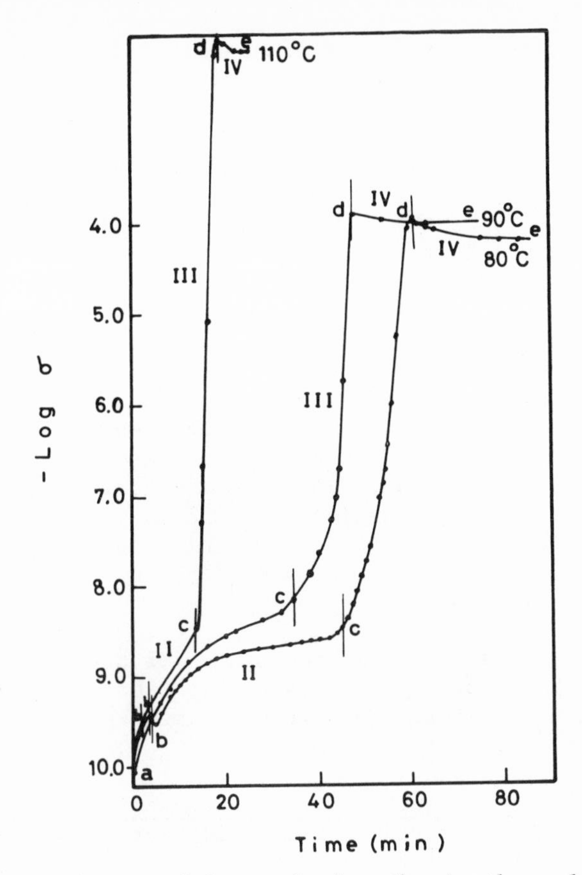

Fig. 1: The change of log σ during the isothermal annealing at 80, 90 and 110°C.

Table 1

Property	Normal Heating		Nucleation		Crystallization		Total change in log σ	Total time min.
	t_h min.	$\Delta \log\sigma_h$	$t_{nucl.}$ min.	$\Delta \log\sigma_n$	t_{crys} min.	$\Delta \log\sigma_{crys.}$		
80	4	0.54	40	0.98	17	4.62	5.86	84
84	3.5	0.58	31.5	0.78	13	4.68	5.90	68
90	3.5	0.68	30	1.23	12.5	4.30	6.06	64
95	3	0.82	17	0.98	10.5	4.50	6.08	40
105	3	0.98	14	1.06	6	4.34	6.26	32
110	2.5	0.89	10.5	1.02	5.5	5.12	6.72	28

process as a function of time. A preheated and automatically controlled oven was used with temperature variation of less than 0.2°C.

RESULTS AND DISCUSSION

The conductivity of the sample thus prepared was first measured at room temperature. The tube was then placed in an oven preheated to the required temperature, and the conductivity of the sample was measured as a function of time. 1-2 volts was applied across the sample and kept constant throughout the exposure time. The sample was examined under identical experimental conditions for temperatures ranging from 80°C to 110°C. Figure 1 shows the experimental results obtained at 80, 90, and 110°C. The results obtained at temperatures 84, 95, and 105°C had the same features as those described above. The total change in the conductivity may be interpreted as being due to the sample's transformation from a low conductivity amorphous selenium to a rather high conductivity trigonal form. The duration and the total change in log σ corresponding to every stage are given in Table 1.

It is obvious that the time necessary for both the process of nucleation t_n and the process of crystallization t_{cryst} decreases with increasing temperature. These two parameters are plotted against temperature in Fig. 2.

Application of Avrami's Equation to the Conductivity Changes. The conductivity changes observed so far, which were considered as corresponding to the crystallization process, can be correlated with Avrami's equation(see eq. 1 above), in an attempt to calculate the crystallization rate k. In equation (1) θ is the amount of material left uncrystallized at a time t, and n accounts for the crystal geometry. According to Fig. 1, θ can be calculated in terms of log σ from the relation:

$$\theta = (\log \sigma_d - \log \sigma_t)/(\log \sigma_d - \log \sigma_c) \qquad (2)$$

where t is the time at any point between the limits c and d. The value of n was computed from the relation

$$\ell n(-\ell n\, \theta) = \ell nk + n\, \ell nt \qquad (3)$$

The calculated values of $\ell n(-\ell n\, \theta)$ were plotted against ℓnt as shown in Fig. 3, which is a series of straight lines, each of which corresponds to a certain temperature. From this graph, the values of n can be deduced directly as the slopes of these lines and are listed in Table 2.

The crystallization rate k can therefore be directly computed, for every time interval, as

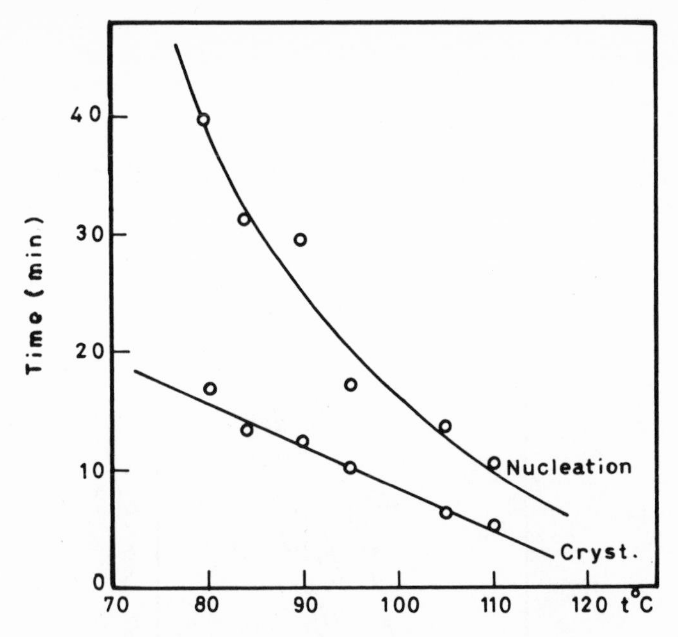

Fig. 2: The dependence of the nucleation time and crystallization time on temperature.

Table 2

Values of n and k for different temperatures

t°C	Range of θ	Values of n	Mean $\bar{n}$ value	$(\bar{k})$, mean k_{-} x 10^4 using n mean
80	90 - 34	1.58		(7.20
84	93 - 32	1.56		(8.15
90	90 - 30	1.36	1.38	(11.73
95	91 - 30	1.28		(15.20
105	90 - 32	1.23		(29.20
110	90 - 35	1.23		(37.40

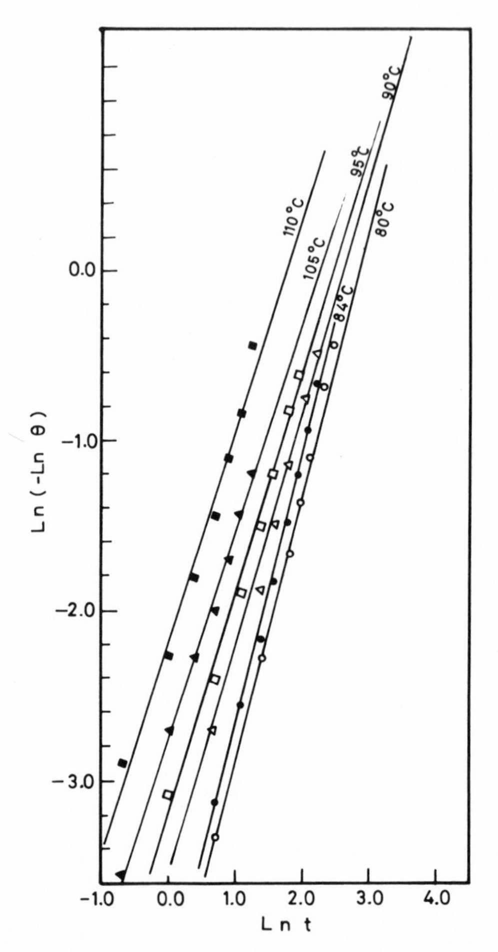

Fig. 3: The plot $\ell n(-\ell n\,\theta)$ against ℓn t(sec) for different crystallization temperatures.

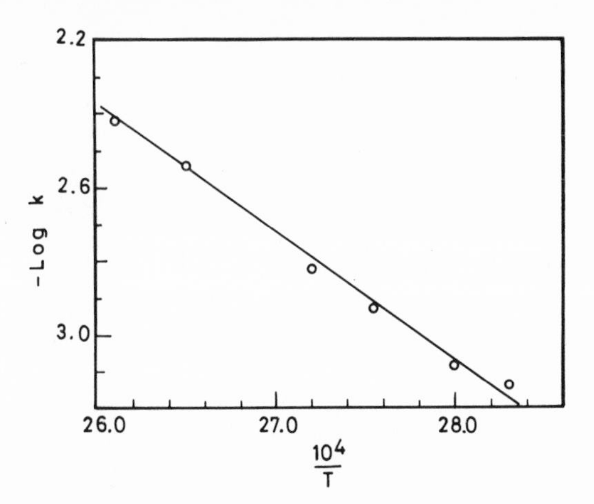

Fig. 4: The dependence of log k (calculated from the conductivity changes) on 1/T for selenium.

$$k = \ell n\ (1/\ \theta)\ /\ t^{n} \qquad (4)$$

For the given range of θ (90% - 34%), k was calcuated for different temperatures using the average value $\bar{n}$ = 1.38. The results as given in Table 2 represent the mean value $\bar{k}$ over the indicated range of θ, for the different temperatures. k is given by the equation

$$k = k_o\ e^{-E/RT} \qquad (5)$$

where R is the universal gas constant and E is the activation energy of the process.

The plot of log k against 1/T would give a straight line whose slope is -E/R (see Fig. 4). The energy of activation, as deduced from the graph, was 15.1 kcal/mol. The activation energy of the same process was calculated and reported by Mamedov and Nurieva[7] using x-ray technique in a temperature range of 70-110°C. Their result was 18.4 kcal/mol, which differs slightly from our obtained value. This small amount of activation energy of crystallization, in comparison with the energy necessary for the destruction of the Se-Se chemical bond (48 kcal/mol)[9], means that the energy of crystallization accounts for the formation of grains from individual nuclei actually present in amorphous selenium and formed during stage 2 (Fig. 1).

ACKNOWLEDGMENT

The authors wish to express their gratitude to Prof. F. El-Bedewi, Head of the Physics Department of Ain Shams University, for his interest and stimulating suggestions.

REFERENCES

1. M.K. El-Mously, Z.U. Borisova, Izv. Akad. Nauk SSSR. Neorg. Mater. 3 92-93 (1967).

2. M.K. El-Mously, M. El-Zaidia, J. Noncryst. Solid (in press 1972).

3. J.N. Hay, Br. Polym. J. 3 74-82 (1971).

4. D. Turnbull, J.C. Fisher, J. Chem. Phys. 17 71 (1949).

5. D. Turnbull, Solid State Physics, Academic Press, New York (1956) 225-306.

6. W.E. Garner, Chemistry of the Solid State, Scientific Publications, London (1955) ch. 7.

7. K.P. Mamedov and Z. Nurieva, Soviet Physics: Crystallography 12-4 605 (1968).

8. C.H. Chapness, R.H. Hoffmann, J. Noncryst. Solids 4 138-148 (1970).

9. R.L. Myuller, Chemistry of Solid State, Izd. Leningrad State University, 18 (1965).

ELECTRICAL PROPERTIES OF QUENCHED SnO_2 FILMS ON GLASS SUBSTRATES

E. W. Wartenberg* and P.W. Ackermann

Institut für Anorganische Chemie

Universitaet Stuttgart

SnO_2 films have been prepared on glass substrates of different linear coefficients of expansion α from different tin-compounds through a hydrolytic reaction by chemical vapor deposition (CVD) under simultaneous quenching. The electroconductivity of the semiconducting films were measured and found to be dependent upon the film forming compound, the film growth rate and temperature, the hydrolytic reaction intrinsic and thermal stress. The thermal stress is responsible for the observed difference in conductivity on glass substrates with different α-values.

INTRODUCTION

Thin films are a matter of great interest in science and technology. In spite of having the same chemical composition as the bulk material they often show different physical behaviour. Optical and electrical properties can differ considerably between bulk and thin film material.

Stannic oxide is an example for such behaviour. The first investigations of the electrical properties of SnO_2-films were made by Bauer.[1] He found an enchanced specific conduction on SnO_2-films in comparison to the bulk material. A study of the electrical properties of stannic oxide single crystals quenched and unquenched was made by Marley and Dockerty.[2] Koch[3] discussed a possible effect of substrate and SnO_2 film on energy gap and electroconductivity. He concluded that the influence of different substrates on the electroconductivity of SnO_2 films is negligible. One objective of our

* Present address: TETRA PAK, Research Dept., Stuttgart.

investigation was to study the electrical properties of stressed SnO_2 films in relation to the substrate. For a systematic investigation of the correlation between stress and electroconductivity we used the CVD monoliquid quenching process, a process of simultaneous coating and quenching.[4,5] By this method fairly high stresses are induced into film and substrate.

EXPERIMENTAL

$SnCl_4$ and H_2O in small concentrations are dissolved in chlorobenzene. When the heated glass is brought into the boiling solution, a gas phase consisting of both solvent molecules of $SnCl_4/H_2O$ is formed around the heated glass. While $SnCl_4/H_2O$ according to the overall hydrolytic reaction

$$SnCl_4 + 2\ H_2O \rightarrow SnO_2 + HCl$$

reacts by a heterogeneous mechanism with the hot glass surface to form the SnO_2 film, the solvent molecules evaporate and cause rapid cooling of the film and substrate, stressing both. Quenching time and stress are dependent upon the heat of vaporization of the solvent and the coefficient of expansion of glass and film. In this process the reaction begins at a substrate temperature of about 400°C. The reaction time in seconds (ΔG) is determined by the duration of the gas phase between the initial substrate temperature (700°) and 400°C.

$$G_{total} - G_{400^oC} = \Delta G$$

G_{total} is the expression for the overall quenching time of the coated substrate. The two terms, G_{total} and ΔG, are inversely proportional to the heat of evaporation of the solvent. The term $d/\Delta G$ corresponds to the average film growth rate, Å: s^{-1}. Four different glasses were investigated: Soda lime glass ($\alpha = 95 \times 10^{-7}$; Schott G20, borosilicate ($\alpha = 49 \times 10^{-7}$); Schott D50, borosilicate ($\alpha = 32 \times 10^{-7}$); Infrasil, silica ($\alpha = 5 \times 10^{-7}$). The film-forming compounds were $SnCl_4$, $SnBr_4$, $Sn(NO_3)_4$, $C_6H_5SnCl_3$.

RESULTS

The electrical conductivity is dependent upon the chemical composition of the tin compound which forms the SnO_2 film. Rising conductivity is observed for SnO_2-films formed from $(SnNO_3)_4 \gg C_6H_5 \cdot \cdot SnCl_3 \approx SnBr_4 < SnCl_4$. No difference could be detected in the resulting SnO_4 structures by x-ray methods. However, FMIR spectra showed a definite decline in Sn-O bond vibration frequency ($\Delta \nu$ = 20 cm^{-1}) with rising conductivity, Figs. 1 and 2. Conductivity rises with rising film growth rate accompanied by rising intrinsic film

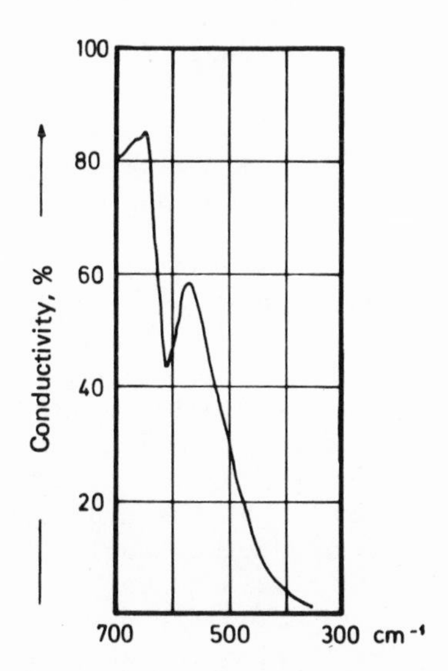

Fig. 1: FMIR spectra of SnO_2 film: Non-conductive.

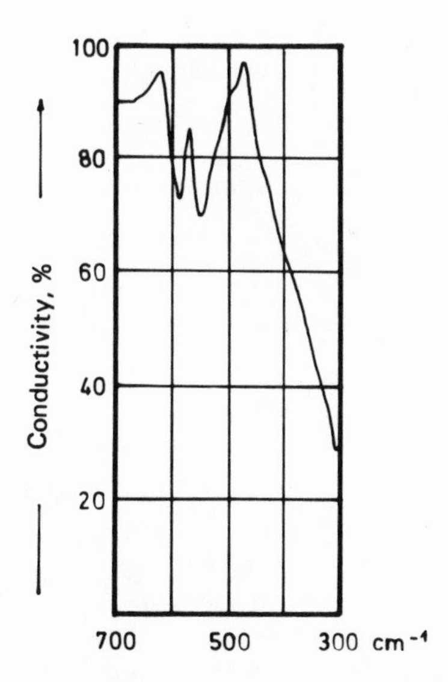

Fig. 2: FMIR spectra of SnO_2 film: Conductive (σ = 350 $\Omega^{-1}cm^{-1}$).

stress, Fig. 3. Differences in the conductivity of the same SnO_2 films were observed when applied to glasses with different chemical composition and different coefficients of expansion (Fig. 4).

The SnO_2 films on glass exhibit two different kinds of stresses, a thermal stress S_{th} and an intrinsic stress S_{in}. The overall intrinsic stress S_{in} has two components, s^{I}_{in} and s^{II}_{in} . s^{I}_{in} is due to

film building temperature and the rapid film growth, which does not allow the formation of a regular crystal lattice. s^{I}_{in} should therefore be directly proportional to growth rate $d/\Delta G$ and reaction temperature. The other component s^{II}_{in} is an additional intrinsic stress induced during quenching by the formation of a thermal stress between film and substrate while the film is formed. This stress can be either compressive or tensile or naught according to $\Delta\alpha$ between substrate and film. It is assumed that the hydrolytic reaction is influenced by stresses which are formed during film growth. Through compression, the complete hydrolysis to SnO_2 is hindered, leading to lattice defects where chlorine takes the position of oxygen. This results in an increase in intrinsic stress as well as in carrier concentration. s^{II}_{in} should therefore be dependent upon the sign and magnitude of $\Delta\alpha$ between film and substrate.

The total intrinsic stress S_{in} and conductivity are related through an exponential function. Increasing stress is followed by increasing conductivity and electron density. From Figs. 5, 6 and 7, the following equation can be deduced:

$$\sigma = A\cdot e^{b\cdot[c\cdot(d-do/\Delta G)^{1/2} + e\cdot\Delta\alpha]} \qquad (1)$$

For $e\cdot\Delta\alpha$, we can substitute the expression

$$K \cdot (\alpha_F - \alpha_S)\ E_F \cdot \Delta T$$

and through further transformations we obtain the equation

$$\sigma = A\cdot e^{b\cdot(s^{I}_{in} + s^{II}_{in})} \qquad (2)$$

which shows the relation between electrical conductivity σ and the total intrinsic stress S_{in}. The same relation applies for the free carrier concentration, Fig. 8. This is reasonable is one assumes that electrical conductivity and intrinsic stress are both caused by lattice defects. From the first equation it can be seen that σ is dependent only upon film temperature and film growth rate, if $\Delta\alpha$ is kept constant. This is shown in Fig. 5, where $\ln\sigma$ is plotted against s^{I}_{in}. Comparison of equations 1 and 2 shows that the first member is equal

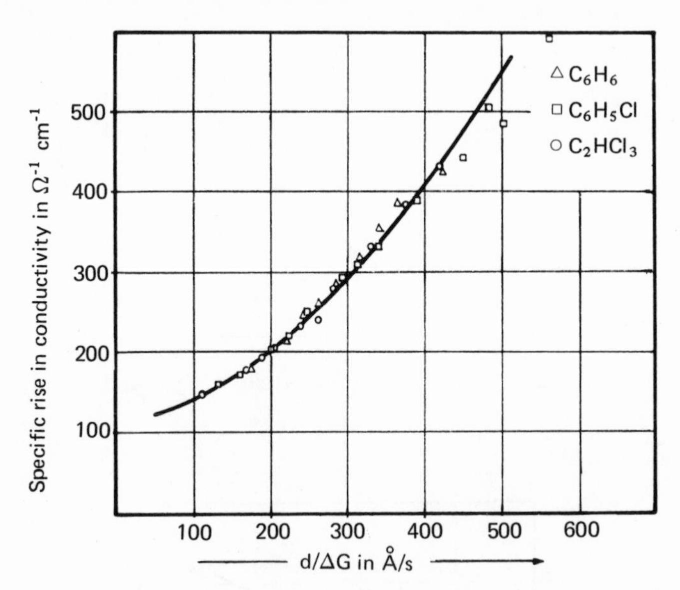

Fig. 3: Conductivity dependence upon film growth rate accompanied by rising intrinsic film stress.

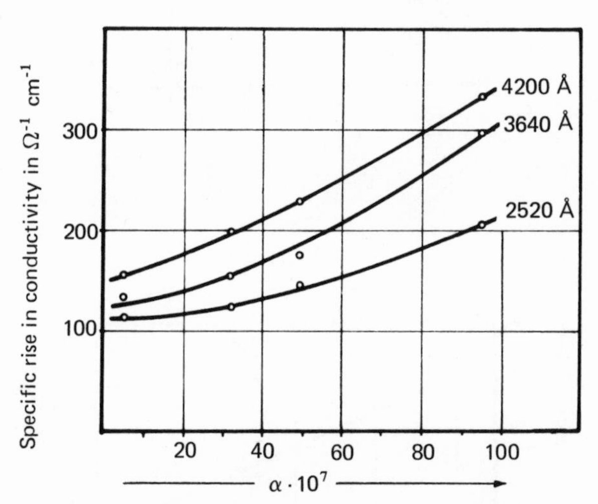

Fig. 4: Conductivity of SnO_2 films on glasses with different chemical compositions and different coefficients of expansion.

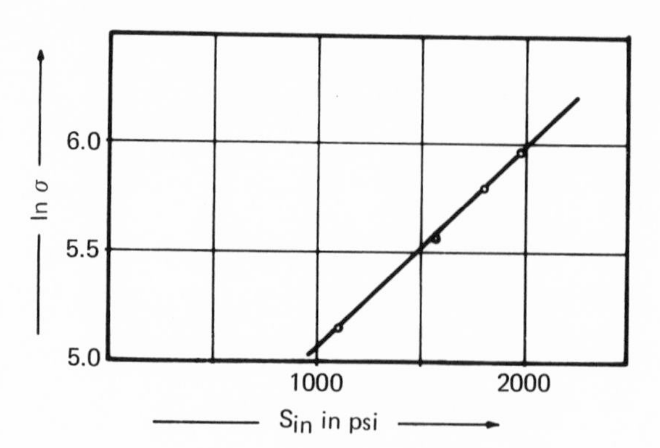

Fig. 5: Relation of total intrinsic stress S_{in} and conductivity.

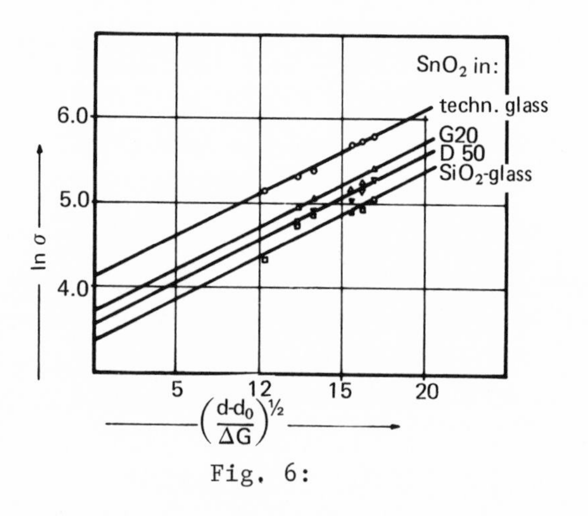

Fig. 6:

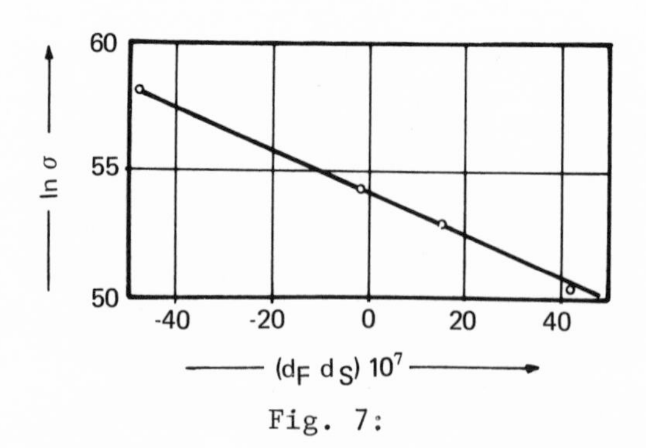

Fig. 7:

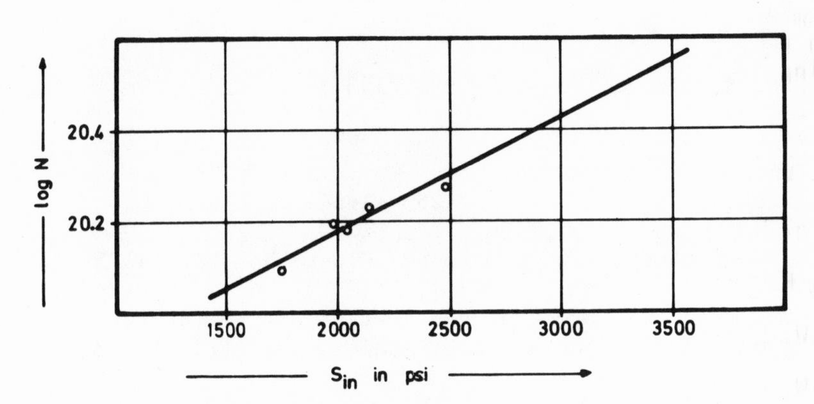

Fig. 8: Relation of free carrier concentration and total intrinsic stress S_{in}.

to s_{in}^{I}. If on the other hand the first expression in the above equation is kept constant a change in σ is dependent only upon the second expression e . $\Delta\alpha$, which is equal to s_{in}^{II}. This relation is seen in Fig. 7 where $\Delta\alpha$ is varied and plotted against ln σ while d / ΔG and T are kept constant.

Film thickness ranged between 1000 Å and 5000 Å. Part of the obtained SnO_2 films have also been doped with antimony and fluorine, resulting in σ - values which were two to three times above the σ - values between 0 and 1500 $\Omega^{-1}cm^{-1}$. Free carrier concentrations were calculated from Hall constants and show values between 1×10^{20} and 3.5×10^{20} cm^{-3}. The electron mobility is in the order of 10 to 20 $cm^2V^{-1}s^{-1}$. The energy gap was calculated from the fundamental edge in the UV-region. The values ranged between 3.97 eV to 4.11 according to the difference in free carrier concentration.

REFERENCES

1. G. Bauer, Ann. d. Phys. 30 433 (1937).

2. J.A. Marley, R.C. Dockerty, Phys. Rev. A140 304 (1965).

3. H. Koch, Phys. Stat. Sol. 7 263 (1964).

4. E.W. Wartenberg, P.W. Ackermann, Glastechn. Ber. 41 55 (1968).

5. E.W. Wartenberg, Proc. Annual Meeting of the Intern. Comm. on Glass, S. Bateson and A.G. Sadler, eds. (1969) 123.

PHYSICAL PROPERTIES OF SPUTTERED AMORPHOUS GERMANIUM THIN FILMS: THE ROLE OF TECHNOLOGICAL PARAMETERS

M. Závětová, S. Koc, and J. Zemek

Institute of Solid State Physics

Czechoslovak Academy of Science, Prague

The role of technological parameters -- decomposition rate, annealing, doping, working gas purity -- in the electrical and optical properties of sputtered a-Ge layers was studied

INTRODUCTION

The influence of technological parameters on the physical properties of a-Ge is stronger than was generally recognized some time ago. Consequently, great care must be taken in the preparation and characterization of samples. a-Ge cannot be prepared in bulk by quenching of the melt, but only in the form of thin layers by various deposition techniques. The most common of these are vacuum evaporation, cathode sputtering, deposition from the electrolyte, and chemical decomposition of the gas.

A great number of papers dealing with aàGe have appeared during the past eight years. Summarizing them, we can conclude that there is a lot of disagreement in the published results. This situation is very well described and discussed in two review papers[1,2] which see the reason for the discrepancies in the different conditions of the preparation of samples.

EXPERIMENTAL

Our samples were prepared by cathode diode sputtering in a stream of especially purified argon. A 50 mm diameter cathode was cast in hydrogen from pure germanium (45 Ω cm). Substrates from optically polished glass, silica, metal or cleared NaCl were kept by cooling to room temperature. Under strictly controlled technological conditions (i.e. deposition rate aprox. 1 μm/hr, constant substrate temperature and deposition geometry, etc.) reproducibility was good from batch to batch. Samples were investigated by electron diffraction and x-ray microprobe to check their structure and chemical composition. We also measured their density by x-ray total reflection,[3] optical transmission in the region of the absorption edge, temperature dependence of the electrical conductivity (usually down to 80 K, in some cases down to 10 K) and the current-voltage characteristics. These physical properties were studied in dependence on the (a) deposition rate (from 0.4 up to 2.0 μm/hr); (b) annealing (up to 500°C after which the samples were completely crystalline); (c) doping (Ag, Au, Ga, Sb, P); (d) oxygen content in argon.

In Fig. 1 the influence of the deposition rate on the density S, electrical conductivity σ and the E_g^{opt}/4/, characterizing the optical gap, is shown. Dependences tend to saturate at the deposition rates around 1 μm/hr.

In Fig. 2 the temperature dependence of the electrical conductivity of the as-deposited and annealed samples is plotted. The curves shift down with increasing temperature of anneal until approximately 400°C where recrystallization results in a sudden increase of conducitivity up to the values characteristic of polycrystalline germanium. The annealing has a similar influence on the spectral dependence of the absorption coefficient. These curves are not only shifted to higher energy with increasing annealing temperature, but the exponential dependence for the as-deposited film also changes to a steeper non-exponential one. The effect is most pronounced in the corresponding low energy parts of the absorption edge. When the annealing temperature exceeds that necessary for recrystallization, the corresponding curve approaches the absorption edge of crystalline germanium. Another effect of the same annealing process can be seen in density measurements. The density of layers increased by 1-2% compared with the density of untreated material.[5]

Figure 3 illustrated the spectral dependence of the absorption coefficient α for pure a-Ge(2) a-Ge (Au)(3) and a-Ge (P) (4). Very similar curves as 3,4 are obtained for a-Ge (Ag) a-Ge(Ga) and a-Ge(sb) In all cases we used heavily doped cathodes (approx. 1% dopant); in the case of a-Ge(O_2) (1) oxygen was present in the argon atmosphere during deposition (approx. 10ppm of O_2). Doping with Ag, Au, Ga,Sb, P shifts the absorption edge almost parallelly to the lower energies, while O_2 present in a-Ge causes not only the shift to the higher energies, but also steepens the edge in its low absorption part.

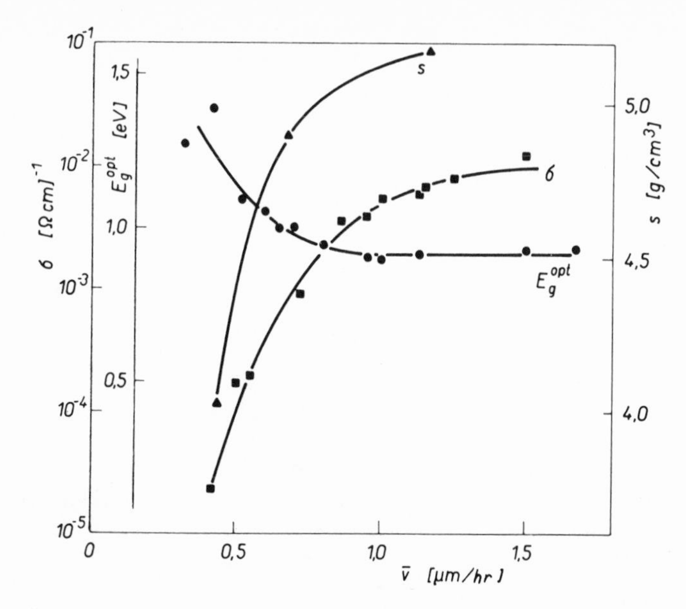

Fig. 1: Dependence of the density S, the electrical conductivity σ and E_g^{opt} of sputtered a-Ge on deposition rate.

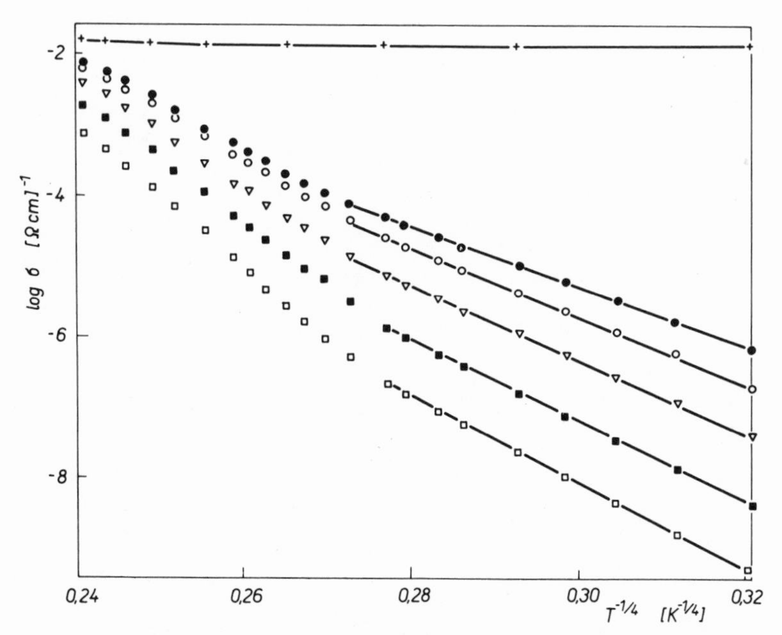

Fig. 2: Temperature dependence of the conductivity for as-deposited and annealed a-Ge.

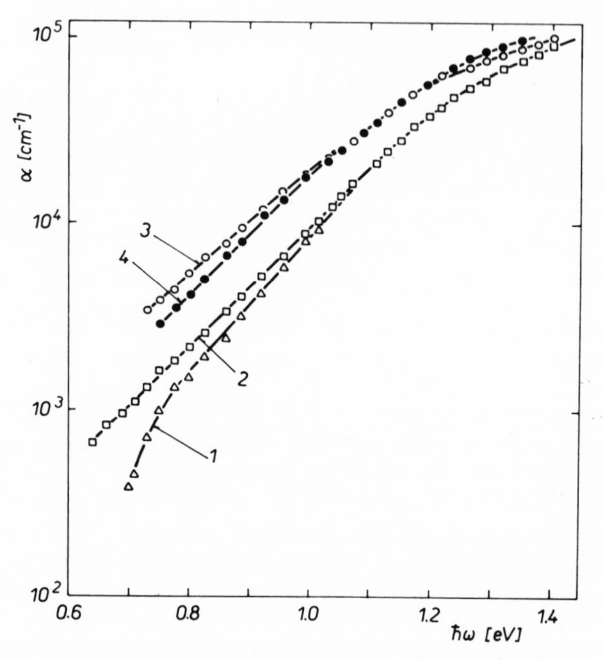

Fig. 3: Spectral dependence of α for pure a-Ge (curve 2) and a-Ge doped with Au[4] and P [4]. Curve 1 corresponds to the sample with deliberate traces of oxygen.

Similar results were obtained from the electrical measurements on doped samples. The curves describing the temperature dependences of the electrical conductivity shift in all cases of doping to higher values of σ. This effect is more pronounced in the low temperature range. In the case of the samples with oxygen content the curve is shifted in the opposite direction.[6]

DISCUSSION

We have tried to explain all our experimental results within one model description. It is possible to explain the changes in the density of the material, due to different deposition rates or annealing, by the assumption that in amorphous Ge, voids are present. From the low density of the layer, prepared with low deposition rate, we estimated the concentration of voids to approx. 10^{20} voids/cm^3, assuming a spherical shape and a diameter of approximately 10 Å. Changes in the electrical conductivity as well as the shift of the absorption edge considerably exceed the effect which voids and their concentration changes could cause by passive presence. To give a true picture of these effects it is necessary to assume an active influence of voids upon their neighborhood, which manifests itself in the electronic structure of amorphous Ge as parallel and antiparallel modulation of the width and position of the gap to the Fermi level.[3] The relative significance of both contributions is not yet clear.

This means that in a-Ge there exist regions with pronounced space charge and other regions with strongly different gap width. Both regions can overlap. Such fluctuations may create localized states. The concept of localized states with tail-like distribution in the gap helps to explain the electrical conducitivity in a-Ge at low temperatures through the hopping mechanism. At high temperatures the carriers thermally excited up to the mobility edge also contribute to the transport of electrical charge. Exponential tails in the optical absorption edge can be explained by the transitions of the electron between the localized states in the gap and the band states. From the increase of density by annealing we conclude that the concentration of voids decreases. The observed shifts of the temperature dependence of electrical conductivity caused by annealing process are connected with the decrease of the density of states around the Fermi level (10^{17} $cm^{-3}eV^{-1}$ for unannealed material; by a half order lower after annealing to 400°C) resulting from the measurements of the current-voltage characteristics[3]. The changes of the absorption edge position show a similar tendency in the density of localized states in the tails. However, the relation between voids and localized states is still an open question.

As to the experiments with doping the low concentration

(< 10^{-2} at. %) of dopants studied has no effect. The influence of impurity levels can manifest itself when their density exceeds the originally high density of localized states in the host material.

During the deposition process the argon and traces of other gases atoms are captured within the growing layer. By comparing the result of optical and electrical observations on sputtered a-Ge layers we can conclude that argon does not significantly influence the studied properties. On the other hand oxygen produces strong effects probably in connection with its ability of chemical sorption. This influence is pronounced at low deposition rates, when probability of the oxygen contamination grows.

CONCLUSIONS

Among the technological parameters studied, the following were found to have profound influence on a-Ge: deposition rate, annealing and traces of active gases. We found the influence of doping below approximately 10^{-2} at.% to be negligible. Argon captured in sputtered a-Ge layers had no observable effect on the properties studied.

REFERENCES

1. S.C. Moss, D. Adler, Comments on Sol. St. Physics 5 47 (1973).

2. D. Adler, S.C. Moss, Comments on Sol. St. Physics 5 63 (1973).

3. S. Koc, M. Závětová, J. Zemek, Thin Solid Films 10 165 (1972).

4. M.H. Brodsky, R.S. Title, K. Weiser, G.D. Pettit, Phys. Rev. B1 2632 (1970).

5. O. Renner, Thin Solid Films 12 S 43 (1972).

6. M. Závětová, S. Koc, J. Zemek, Czech J. Phys. B22 429 (1972).

SOME RECENT RESULTS ON PARAMAGNETIC SPIN-LATTICE RELAXATION IN HYDRATED PARAMAGNETIC SALTS OF THE IRON GROUP

C.J. Gorter and A.J. van Duyneveldt

Kamerlingh Onnes Laboratory

University of Leiden, The Netherlands

The study of relaxation phenomena near the magnetic phase transition has been the aim of many experiments since 1958. In this paper we will first review some early results which have been obtained on manganese chloride and bromide samples. The possibility of using very low frequencies (ν_{min} = 0.2 Hz) at strong magnetic fields (H_{max} = 50 kOe) gave us an opportunity to study the behavior of the relaxation times near the magnetic phase transition. Experiments on manganese chloride and bromide with the external magnetic field parallel to the crystal c axis, showed a maximum in the relaxation time as a function of temperature or external magnetic field. This behavior can be ascribed to the anomalies of the specific heat at the transition temperature. Another interesting sample in this study is $CoCl_2 \cdot 2H_2O$ (external magnetic field parallel to the crystal b axis). This crystal becomes antiferromagnetic at 17.2°K, while below 9.3°K also a ferrimagnetic spin arrangement occurs in magnetic fields between 32.0 and 45.5 kOe. Our susceptibility data did confirm the suggested phase diagram. At the antiferromagnetic-ferrimagnetic transition field of 32.0 kOe a relaxation mechanism can be detected. The relaxation times are remarkably long (1 s at 4.2°K) and can only be measured with a so-called step-field method, in which τ is found from the time dependence of the susceptibility after a sudden change of the external magnetic field. The relaxation times vary strongly with temperature; in the range between 3.2° and 4.2°K an increase of a factor 100 is observed. Though the field dependence of the relaxation time cannot be detected accurately because of the lack of sensitivity, the experiments do show a maximum at the transition field.

HISTORICAL INTRODUCTION

Modern regular investigations on magnetism at low temperature (liquid helium etc.) started essentially in Leiden with the theses of E.C. Wiersma (later professor in Delft) and of one of us (C.J.G.) in early 1932. Since then there have been many investigations in this field.

An interesting step forward, in 1936 by C.J.G., facilitated the discovery and study of the paramagnetic relaxation in rather rapidly oscillating magnetic fields plus a constant field. The start was made at frequencies of the order of 10^6 periods per second. Later the frequencies were broadened gradually from 1 to 10^{10} periods per second. The early experiments were carried out on powders but later, interest concentrated on single crystals in magnetic fields oriented in various directions.

During the first period the investigations were concentrated in Great Britain, the Soviet Union and in the Leiden laboratory, and later also in Groningen and Amsterdam. In Groningen and Amsterdam, Brons, Theunissen and Dijkstra, as well as the semi-theoretician Broer, studied the behaviour of paramagnetic spins in a crystalline lattice under the influence of a constant magnetic field and an oscillating perpendicular high frequency magnetic field.

Before the second world war, C.J.G. suggested nuclear magnetic resonance, and during the war he, in cooperation with his Amsterdam student L.J.F. Broer (now for many years professor of theoretical physics in Eindhoven) tried to discover electronic spin resonance. This latter effect was discovered not much later in Russia by Zavoisky.

Studies of paramagnetic relaxation were carried out in Amsterdam and later mainly in Leiden. The theses of Volger, Bijl, De Vrijer, Van der Marel, Bölger, Van den Broek and Verstelle concentrated on the energy contact from liquid air to liquid helium temperatures at a wide field of frequencies. Around 1962 Locher began studies on spin-spin relaxations in a wide range of external magnetic fields[1] while more recently, De Vries built a very rapid device for the passage and detection of a wide range of frequencies (200 Hz up to 1 million Hz), and modernised the set-up in a relatively short time.[2] During the last few years A.J.v.D. and our students have successfully used such equipment for studying cross relaxations in diluted magnetic salts[3] and relaxation phenomena in antiferromagnetically ordered systems. Drs. Roest and Verbeek's recently published theses record some of this work.

INTRODUCTION TO SOME RECENT RESULTS

Even in very simple regular crystalline structures there are various possibilities. One may have different types of ions and nuclei and those may occupy varying positions and orientations. High temperatures and relatively small energy differences may not give rise to remarkable complications. But radical reduction of the relative temperature may lead to variable entropies and energies and corresponding variations of the situation.

There are two rather large groups of magnetic ions: the iron-group with 19 - 27 electrons per ion and the rare earths with 55 - 67 electrons, having 1 - 9 non-compensated electrons in the 3d shell or 1 - 13 non-compensated electrons in the 4f shell respectively. In the first half of the ions of the iron-group there are up to 5 parallel non-compensated electrons per magnetic ion, and in the first half of the rare earth group, there are up to 7 parallel electrons per ion.

The best known ions in the iron-group are the chromic, manganese, ferric, ferrous, cobaltic, nickel and cupric ions with respectively 3, 4, 5, 6, 7, 8 or 9 electrons in the ionic 3d-shell.

Some of the best known rare earth ions are the tri-valent cerium, bi-valent praseodimium, gadolinium, disprosium and erbium, having respectively 1, 2, 3, 7 and 11 electrons in the 4f-shell.

In Leiden the relaxation phenomena for a few magnetic ions of the iron-group have been studied thoroughly, while so far, not many data are collected on the rare earth group.

We will concentrate on a few rather well studied hydrated cobalt, manganese and chromium salts in the iron-group having 7, 5 and 3 electrons respectively in the 3d-shell. Their energy states have half integral quantum numbers and we must classify them among the so-called Kramers ions. In zero magnetic field the energy levels of these ions are twofold degenerate since the orbit-lattice interaction (an electric field) cannot connect time-reversed states of half integer quantum number.

REVIEW OF SOME RECENT RESULTS

In the first part of this review we will consider spin-lattice relaxation in various cobalt salts.

The ground state of Co^{2+} ions in the samples considered is a so-called Kramers doublet. The next highest doublet is situated at $\Delta \gg kT$, $k\Theta_D$ (Θ_D being the Debye temperature, which is related to

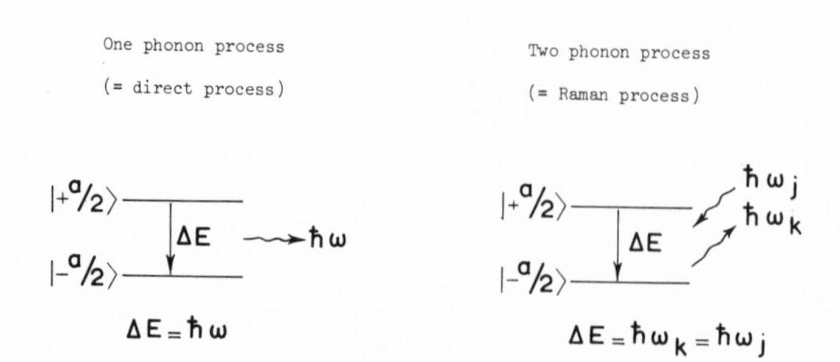

Fig. 1 Schematic diagram, indicating the direct - and Raman relaxation process.

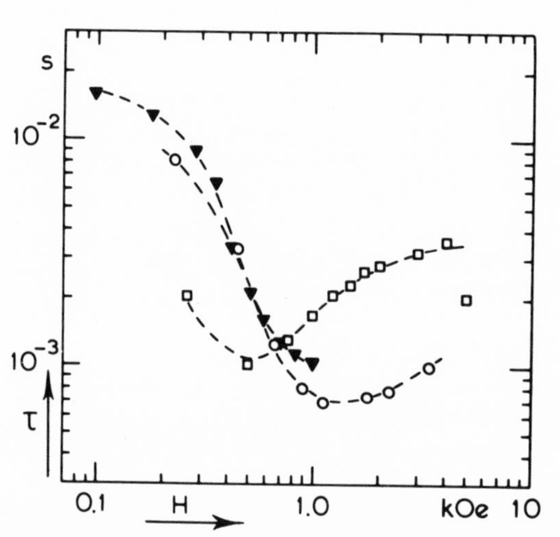

Fig. 2 Relaxation time τ against external magnetic field H at T = 4.2 K; results from early reports.
▼ Haseda,[5] Cobalt ammonium Tutton salt.
o Van den Broek,[6] Cobalt ammonium Tutton salt.
□ Poolman,[7] Cobalt cesium Tutton salt.

the maximum frequency ω_m of the system of lattice oscillations by $k\Theta_D = \hbar\omega_m$) which means that the lattice oscillations cannot excite the cobalt ions into the next highest energy state.

In spin-lattice relaxation processes the magnetic spin-system must exchange energy with the lattice oscillations. This may occur by one or two phonon processes which are schematically indicated in Fig. 1.

To calculate the time constant τ with which the magnetization reaches its equilibrium value after an external disturbance, one must consider the transition probabilities

$$w_{|+a/2\rangle \rightarrow |-a/2\rangle} \quad \text{and} \quad w_{|-a/2\rangle \rightarrow |+a/2\rangle}$$

These quantities equal w if $\Delta E \ll kT$. At the temperature where experimental results are obtained one may state: $\tau = 1/2\,w$. From the general expression for the transition probability one may derive the following (simplified) expression for the spin-lattice relaxation time:[4]

$$\tau^{-1} = \underset{\text{(dir. pr.)}}{ATH^4} + \underset{\text{(Raman prs.)}}{B_1T^9H^o + B_2T^7H^2}$$

This expression is correct if the splitting within the doublet due to the external magnetic field ($g\beta H$) is smaller than kT and if $T \ll \Theta_D$. The relations for the Raman processes have to be considered with more care, as the temperature dependences of these terms vary towards $T^2 \gtrsim \Theta_D$. Figures 3, 4 and 5 demonstrate the direct process term of (1). Figures 6, 7 and 8 illustrate the second term of (1), while Fig. 9 (cobalt salts) shows the influence of the magnitude of the external magnetic field, and thus the third term of (1).

Let us consider the eight diagrams in more detail:

Figure 2 gives three early reports on Cobalt Tutton salts at 4.2 K: Haseda and Kanda's paper[5] in 1956, $Co(NH_4)_2\cdot(SO_4)_2\cdot 6H_2O$; Van den Broek's[6] later similar result, showing an increase of τ above 1 kOe, and the results of Poolman, et al.[7] on $CoCs_2(SO_4)_2\cdot 6H_2O$. The anomalous mounting of the τ vs. H curves above 1 kOe might be due, according to Wharmby and Gill,[8] to a nickel-impurity, which later was confirmed by spectro-chemical analysis to be as much as 3% of the cobalt. The influence of impurities can be overcome by expanding the external magnetic field range towards 60 kOe by making use of superconducting coils. A result of such measurements of the relaxation in samples of

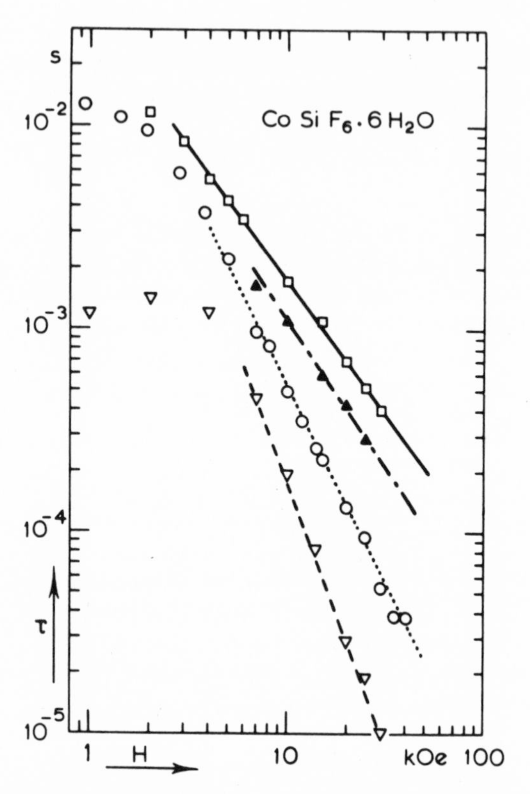

Fig. 3 Relaxation time versus external magnetic field for various powdered samples of $CoSiF_6.6H_2O$ at T = 2.1 K.
□ average crystal diamter 2 mm; ▲ : 0.4 mm; o : 0.1 mm; ∇ : 0.03 mm.

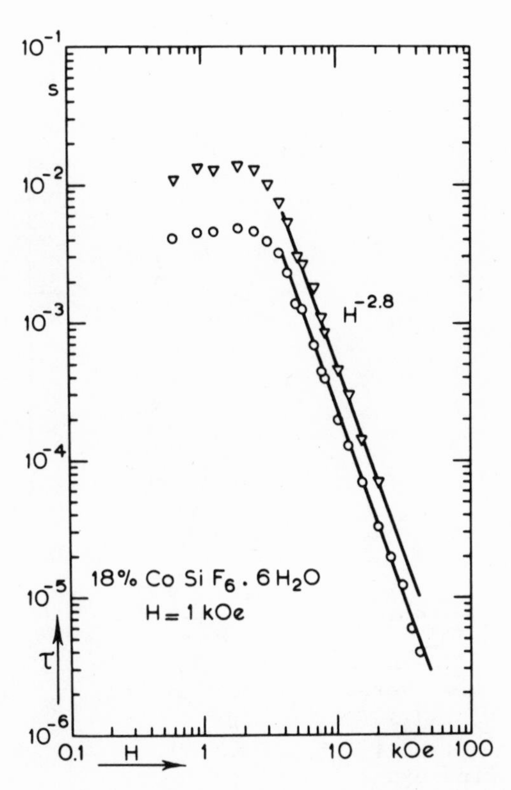

Fig. 4 Relaxation time versus external magnetic field for powdered $Co_{0.18}Zn_{0.82}SiF_6.6H_2O$ (average crystal diameter 0.1 mm). circles: T = 4.2 K; triangles: T = 2.1 K.

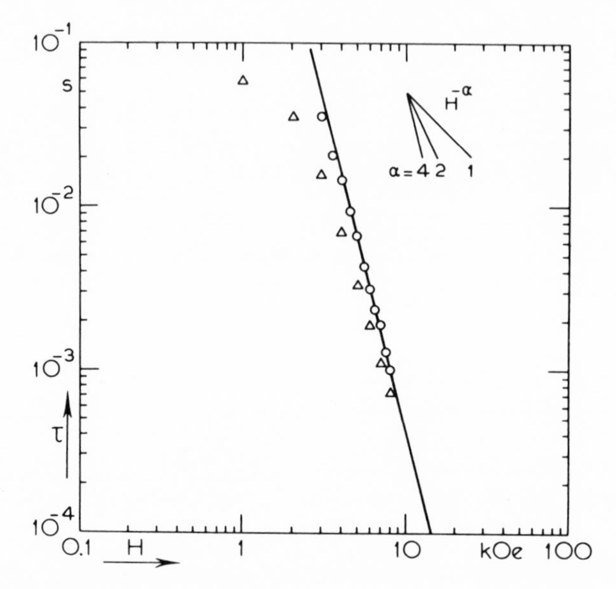

Fig. 5 Relaxation time versus external magnetic field for powdered $Co_{0.02}Zn_{0.98}(NH_4)_2(SO_4)_2.6H_2O$
o : T = 2.1 K; Δ : T = 4.2 K.
The drawn line exposes $\tau \propto H^{-4.0}$.

$CoSiF_6 \cdot 6H_2O$ (phonon bottleneck effect). The effect occurs because the heat contact between the lattice oscillations involved in the direct process and the surrounding helium bath is inadequate to transmit enough energy between the helium bath and the system of cobalt spins. This occurs very strongly in large fields and thus wide splittings. The samples were powders, the average diameter of the crystals in the samples being 2, 0.4, 0.1 and 0.03 mm. Upon decrease of the crystal diameter the slope of the $\ln \tau - \ln H$ curve increases, which indicates that the influence of phonon bottleneck effect diminishes with reduced crystal sizes.

Another means to overcome the phonon bottleneck is shown in Fig. 4, where a powder of diluted cobalt fluosilicate (18% Co in $ZnSiF_6 \cdot 6H_2O$) shows considerably steeper curves: $\tau \propto H^{-2.8}$. In Fig. 5 the relaxation times of an even more diluted sample ($Co_{0.02}Zn_{0.98}(NH_4)_2(SO_4)_2 \cdot 6H_2O$) shows the proper direct process dependence: $\tau \propto H^{-4.0}$ at 2.1 K. At 4.2 K the curve is less steep, in particular at the lower magnetic fields.

Figure 6 gives the relaxation in $CoK_2(SO_4)_2 \cdot 6H_2O$ (left curve and left scale) and in $Co(NH_4)_2(SO_4)_2 \cdot 6H_2O$ (right curve and scale) as a function of temperature at an external magnetic field of 1 kOe. The lines drawn at the lowest temperatures exhibit $\tau \propto T^{-9}$. Near 20 K the relaxation times are proportional to T^{-6} as is to be expected from the exact expression for the Raman process. The calculated lines agree very well with the experimental data. The deviations at the lower temperatures are caused by impurities which are effective in this rather low external field value.

Figure 7 gives the observed relaxation of $Co_3La_2(NO_3)_{12} \cdot 24H_2O$[10], also at one kilo Oersted. The drawn curve exhibits again a nice agreement between the theoretical expression and the experimental data. There seem to be only small irregularities in the liquid hydrogen temperature region. In Fig. 8, the third more or less identical picture, the relaxation times of $CoSiF_6 \cdot 6H_2O$ and $Co_{0.18}Zn_{0.82}SiF_6 \cdot 6H_2O$[9] show the remarkable influence of a slightly changed crystal structure on the observed Raman relaxation times.

Finally, Fig. 9 gives the short relaxation times of samples of $CoK_2(SO_4)_2 \cdot 6H_2O$ and $Co(NH_4)_2(SO_4)_2 \cdot 6H_2O$ at liquid hydrogen temperatures[11] (considering Fig. 6 near the highest temperatures). On the basis of the second and third term of eq.(1) the dependence of τ on external magnetic field can be expressed as $\tau = p + qH^2$. This equation presents good agreement with the data The temperature dependence of q is given by $q \propto T^{-\beta}$ with $2 < \beta < 7$, as is to be expected from the exact expression for the field dependent Raman relaxation process (third term in (1)) at these temperatures. A.J.v.D. pointed out recently that this

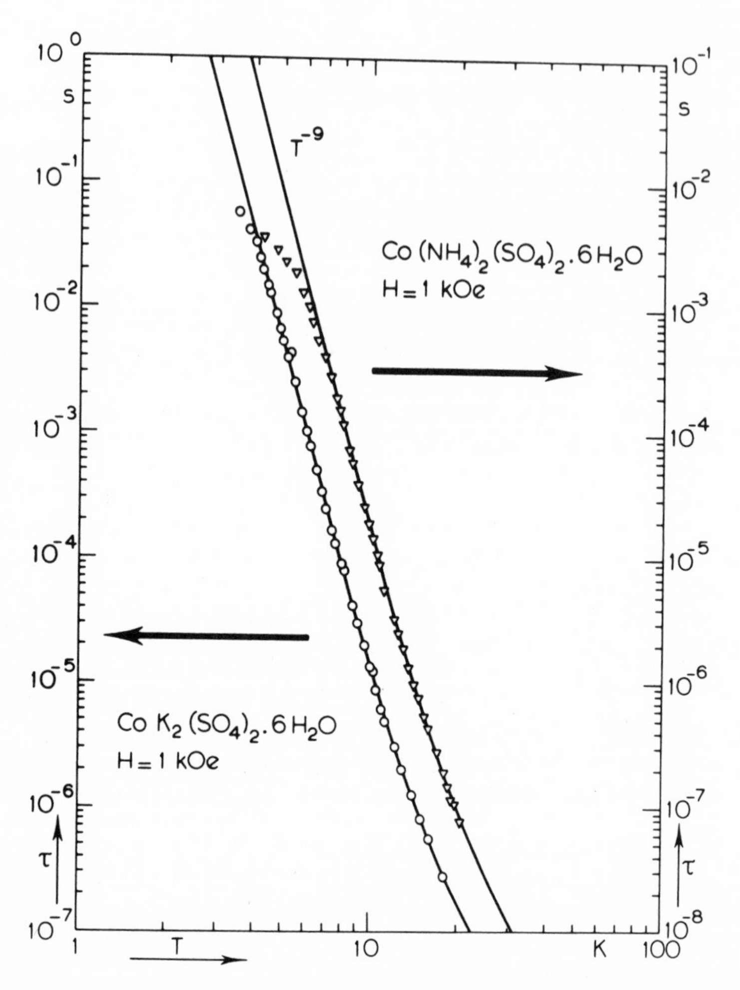

Fig. 6 Relaxation time versus temperature for powdered $CoK_2(SO_4)_2$. $6H_2O$ (circles, left scale) and $Co(NH_4)_2(SO_4).6H_2O$ (triangles, right scale) at an external magnetic field of 1 kOe.

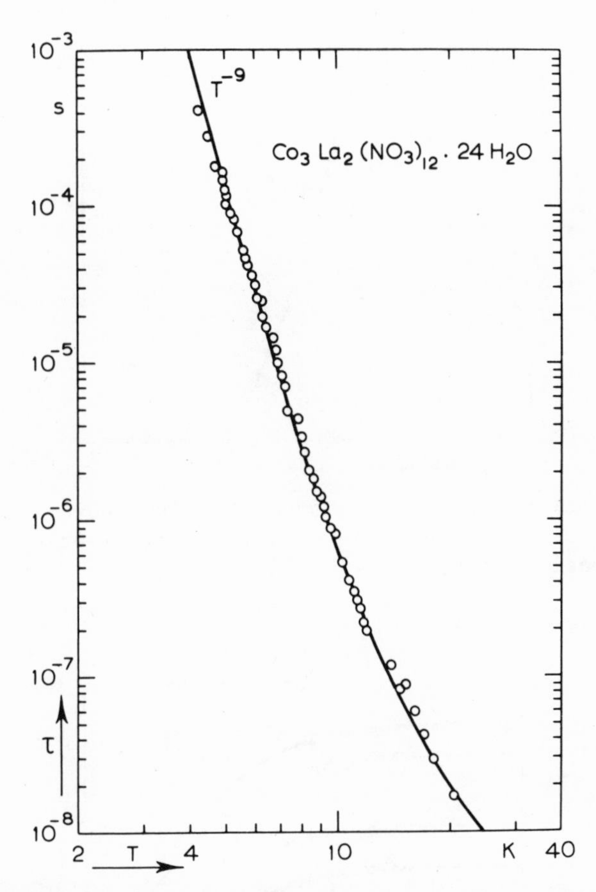

Fig. 7 Relaxation time versus temperature for powdered cobalt lanthanum double nitrate at an external magnetic field of 1 kOe.

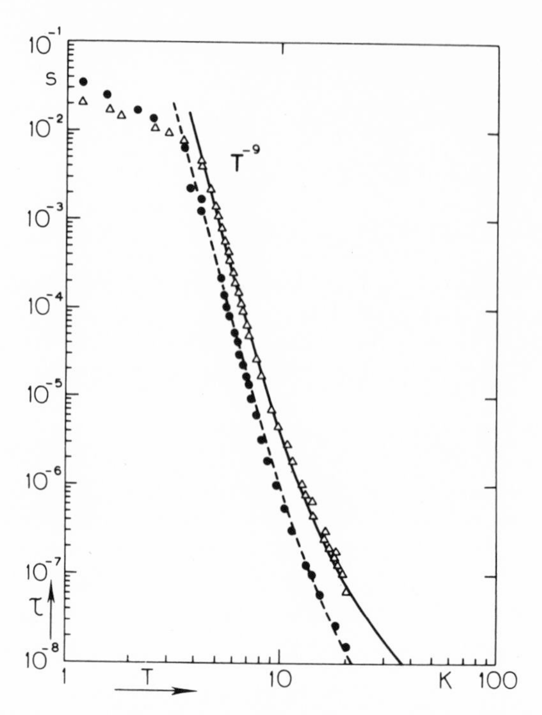

Fig. 8 Relaxation time versus temperature for powdered $CoSiF_6.6H_2O$ (circles) and $Co_{0.18}Zn_{0.82}SiF_6.6H_2O$ (triangles) at an external magnetic field of 1 kOe.

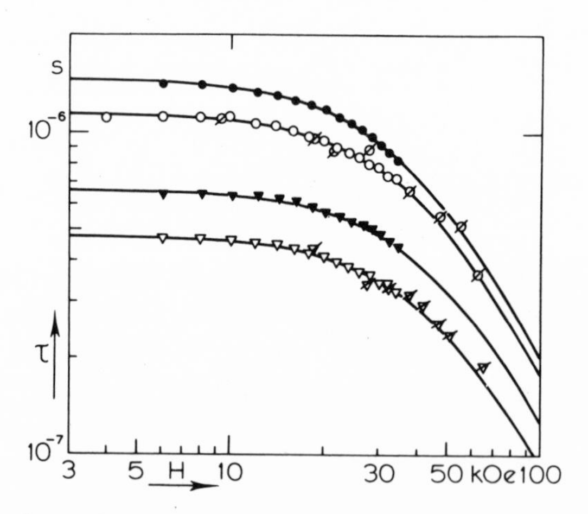

Fig. 9 Relaxation time versus external magnetic field at liquid hydrogen temperatures for $Co(NH_4)_2(SO_4)_2.6H_2O$ (closed symbols) and $CoK_2(SO_4)_2.6H_2O$ (open symbols). circles: T = 14.1 K; triangles: T = 16.0 K.

dependence of the Raman relaxation time on field and temperature was suggested by Kronig in 1939.[12]

We will now discuss some features of the paramagnetic relaxation behaviour in manganese and chromium salts. The magnetic moments for manganese ions are largely due to five parallel electronic spins, the angular moments being zero in the orbital ground state. Consequently, the manganese ions have to be described with a ground state of three closely situated Kramers doublets ($\Delta \sim 0.01$ cm^{-1}). For the chromium ions the situation is slightly more complicated, but the ground state is well described by two Kramers doublets laying approximately 0.1 cm^{-1} apart.

Thus in the temperature range where measurements have been performed, both manganese and chromium salts have to be treated as three or two doublets respectively with an energy separation which is small compared to kT. Therefore the spin-lattice relaxation behaves differently from the case of cobalt.

The simplified expression for τ becomes:[13]

$$\tau^{-1} = \underset{\text{(dir.pr.)}}{ATH^2} + \underset{\text{(Raman pr.)}}{BT^5H^o} \qquad (2)$$

The same restrictions apply for this equation as for eq. (1). Also, the exact expression for the Raman process causes the slope of the log τ - log T graph to vary from -5 at low T towards -2 at high temperatures. Figures 10 and 11 demonstrate the second term of (2) while the direct process becomes clear in Figs. 12, 13, 14, and 15.

Figure 10 shows the Raman relaxation times for $MnSiF_6 \cdot 6H_2O$ (circles, H = 1 kOe) and $Mn(NH_4)_2(SO_4)_2 \cdot 6H_2O$ (triangles, H = 750 Oe). The agreement between the theoretical expressions (drawn lines) and experimental data is good. Above 14 K both curves show the T^{-5} dependence from the second term in (2). The Debye temperature values differ for the salts ($MnSiF_6 \cdot 6H_2O$, Θ_D=140 K and $Mn(NH_4)_2(SO_4)_2 \cdot 6H_2O$, Θ_D=230 K) so the departure towards $\tau \propto T^{-2}$ starts at different temperatures. Relaxation times below 4 K in these small external magnetic fields are influenced mainly by impurities (Figs. 12, 13).

Figure 11 gives the analogous results for chromium cesium alum.[14] Here an external magnetic field of 4 kOe was chosen to avoid complications with the cross-relaxation processes that are operative in fields below 2 kOe in this salt. The T^5-Raman process is clearly observed above 14 K in the concentrated salt (circles) as well as in the diluted (2.6% Cr - 97.4% Al) sample (triangles). The relaxation times below 4 K show a peculiar $T^{-1.5}$ dependence due to the competition between the impurity relaxation

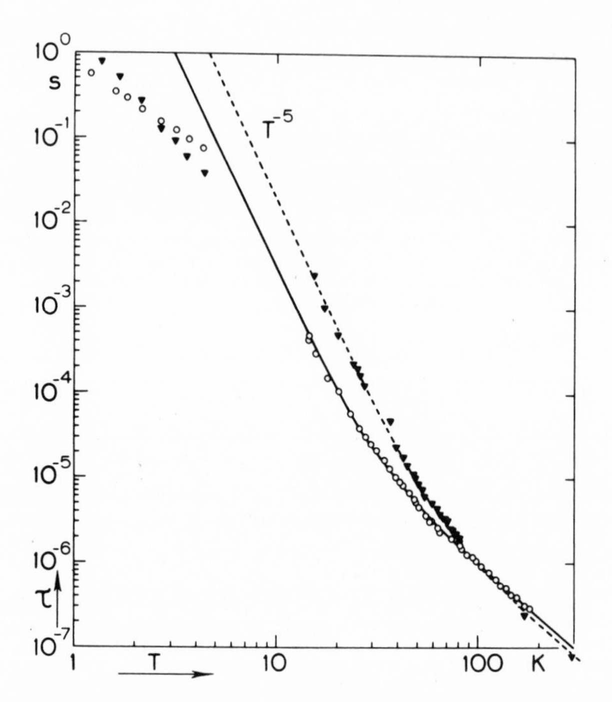

Fig.10 Relaxation time versus temperature for powdered $MnSiF_6.6H_2O$ (circles, H = 1 kOe) and $Mn(NH_4)_2(SO_4)_2.6H_2O$ (triangles, H = 750 Oe).

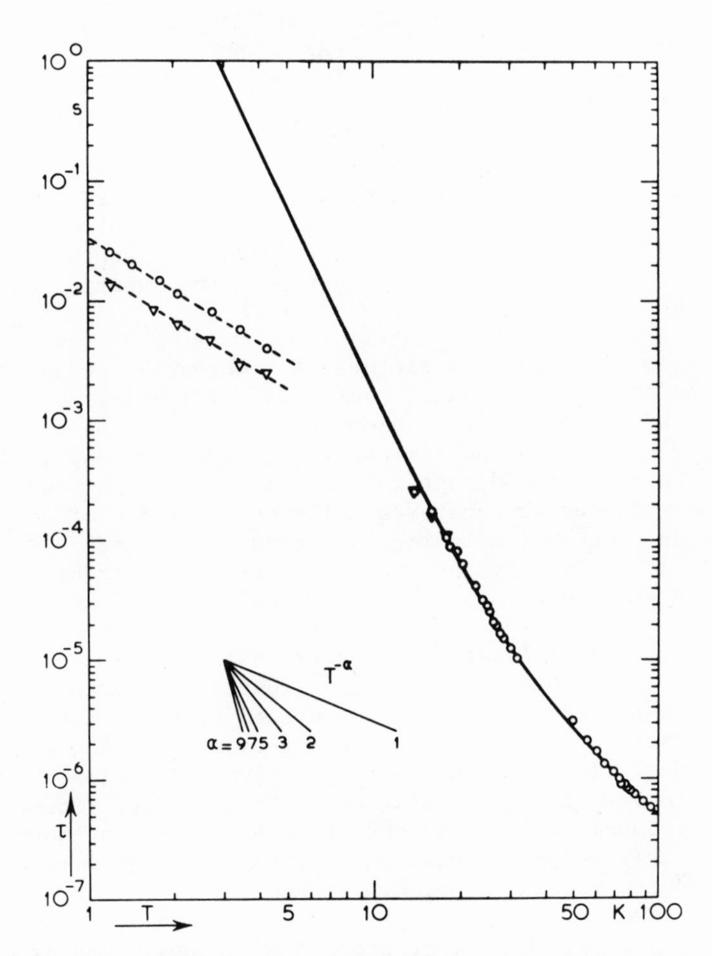

Fig.11 Relaxation time versus temperature for powdered $CrCs(SO_4)_2 \cdot 12H_2O$ (circles) and $Cr_{0.026}Al_{0.974}Cs(SO_4)_2 \cdot 12H_2O$ (triangles) at an external magnetic field of 4 kOe.

and the direct process, as is shown in Fig. 14.

In Fig. 12 the direct relaxation process of powdered $MnSiF_6 \cdot 6H_2O$ is shown.[15] The relaxation time τ is plotted vs. the external magnetic field for T = 2.06; 4.25; 14.1 and 16.0 K. At the two lowest temperatures the H^{-2} dependence of the direct process is reached above approximately 12 kOe, while in low fields an impurity relaxation mechanism is effective. At 14 and 16 K the direct process starts its influence above 10 kOe; the Raman process is observed at the low fields. For this process a slight field dependence according to the Brons-Van Vleck formulae ($\tau = \tau_o \, (b/C + H^2) \, / \, (b/C + pH^2)$ is seen.[19] This field dependence is indeed found for all Raman processes at low fields, but has been omitted in the simple equations (1) and (2). The dotted curves have been obtained from a computer analysis which takes both relaxation processes into account. The agreement with the experimental data is quite good.

Figure 13 gives again a picture of τ versus H for powdered manganese ammonium Tutton-salt. For 2.09 K (Δ) and 4.22 K (O) the H^{-2} dependence is clearly demonstrated. This time at 14.2 K (∇) and 16.0 K (Δ)our highest measuring fields (40 kOe) were sufficient to reach the direct relaxation region with $\tau \propto H^{-2}$ even at these higher temperatures. The computer fit of the direct- and the Raman process (including the Brons - Van Vleck field dependence[19]) is given by the dotted curves and describes the measurements very well.

In Fig. 14 the relaxation times of chromium cesium alum at 4.2 K (O) and 2.1 K (∇) are displayed as a function of external magnetic field. One observes that the fields needed to reach the region of direct relaxation (drawn curves) are higher than in the above cases (about 20 kOe). The influence of impurities is considerably larger. The dotted curve gives a computer analysis where the impurity relaxation mechanism is considered in more detail. A thorough analysis of this mechanism cannot be given here, but will be presented in the thesis of Dr. C.L.M. Pouw.

Also at 14.1 K, the direct process in chromium cesium alum can be observed in the highest measuring fields, as may be seen in Fig. 15. The line represents the computer analysis considering a Raman field dependence (Brons - Van Vleck) in low fields and the direct process in high external magnetic fields.

It is remarkable to compare the numerical values of the direct processes just described. Both manganese samples are described with a constant A (see eq. 2) of approximately 0.2 s^{-1} kOe^{-2} K^{-1}. The chromium cesium alum exhibits A $\sim$ 2 s^{-1} kOe^{-2} K^{-1}, thus showing the influence of the non-zero orbital momentum.

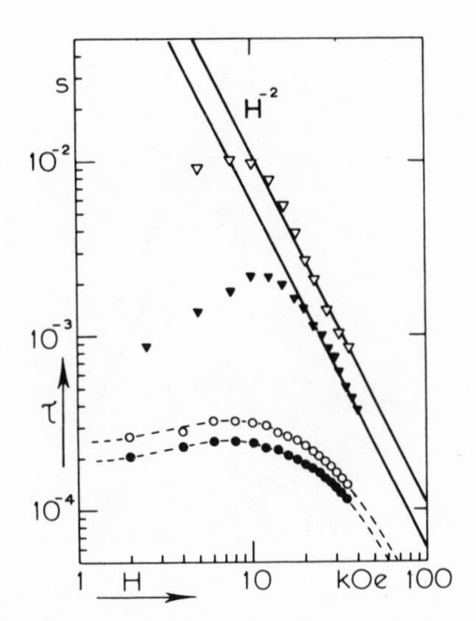

Fig.12 Relaxation time versus external magnetic field for powdered $MnSiF_6.6H_2O$.
∇ : T = 2.06 K; ▼ : T = 4.25 K; o : T = 14.1 K;
● : T = 16.0 K.
Lines are explained in the text.

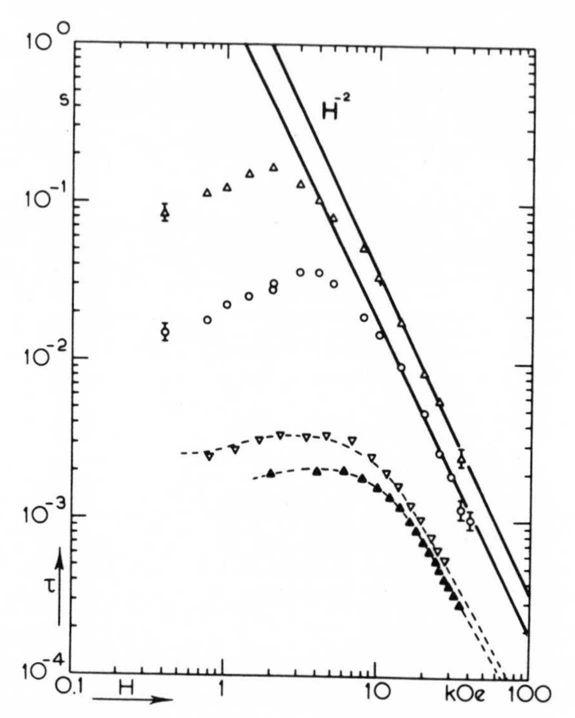

Fig.13 Relaxation time versus external magnetic field for powdered $Mn(NH_4)_2(SO_4)_2 6H_2O$.
Δ : T = 2.09 K; o : T = 4.22 K; ∇ : T = 14.2 K;
▲: T =16.0 K.
Lines are explained in the text.

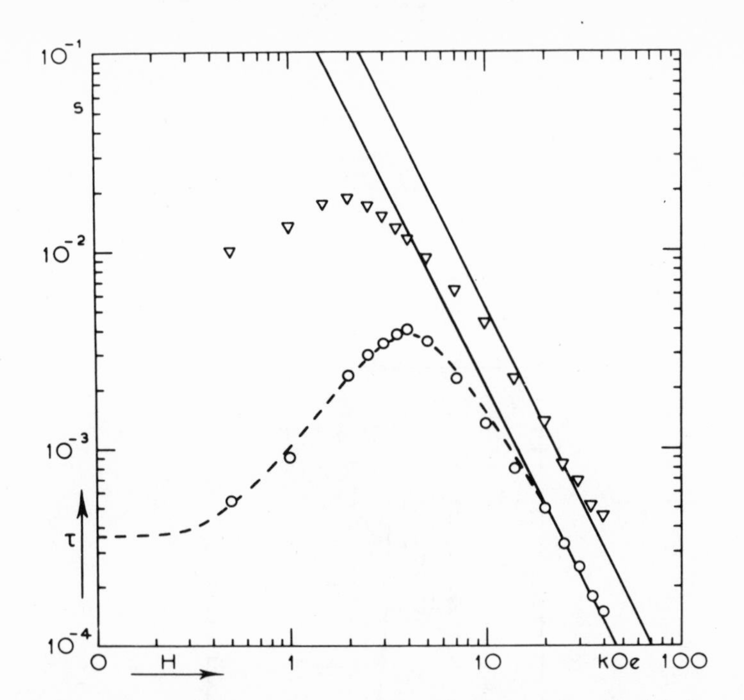

Fig.14 Relaxation time versus external magnetic field for powdered $CrCs(SO_4)_2.12H_2O$..
∇ : T = 2.1 K; o : T = 4.2 K.
Lines are explained in the text.

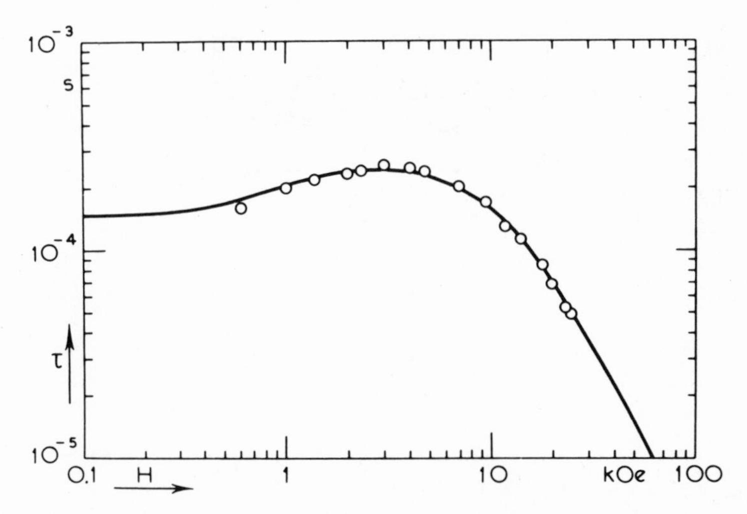

Fig.15 Relaxation time versus external magnetic field for powdered $CrCs(SO_4)_2 12H_2O$ at T = 14.1 K. Line see text.

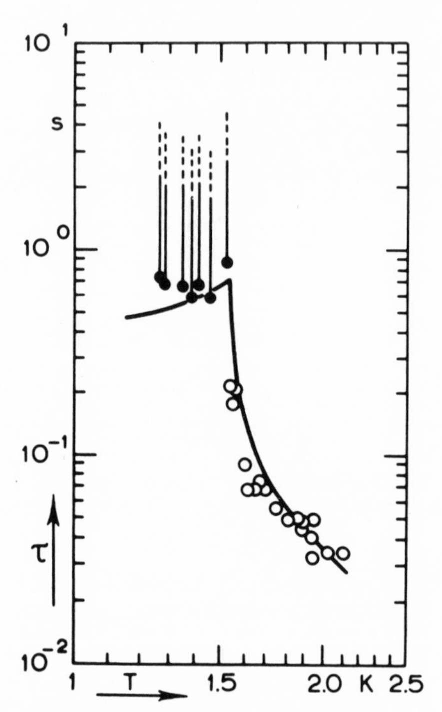

Fig.16 Relaxation time versus temperature for a single crystal (H // c-axis) of $MnCl_2.4H_2O$.

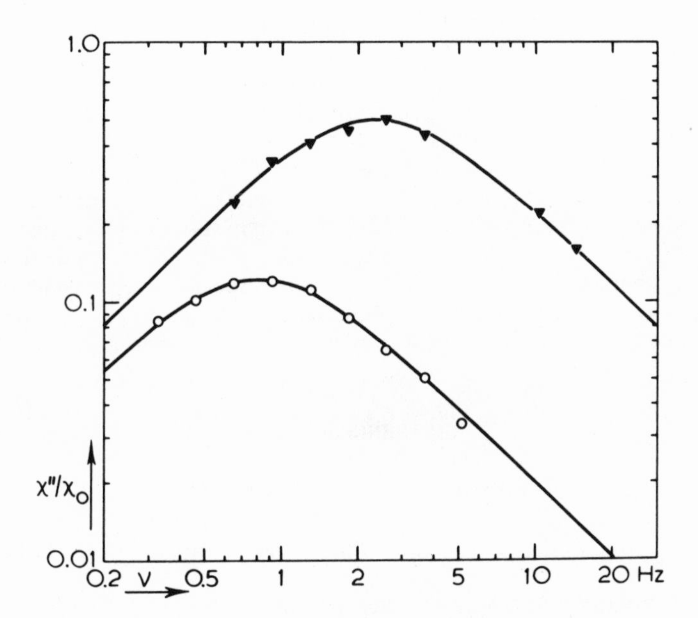

Fig.17 Absorption (χ''/χ_o) versus frequency ν ($=\omega/2\pi$) for single crystals (H // c-axis) of $MnCl_2.4H_2O$ (o : T = 1.3 K; H = 4 kOe) and $MnBr_2.4H_2O$ (▼ : T = 1.8 K; H = 8 kOe). The drawn lines are the Debye-curves.

Finally we will mention some relaxation results obtained on a substance that shows an antiferromagnetic ordering. Figure 16 gives the relaxation times published by A.J.v.D. in his doctoral thesis in 1969. Near 1.5 K (in the external field of 4 kOe) the single crystal of $MnCl_2.4H_2O$ shows antiferromagnetic ordering. The figure shows an increase in the spin-lattice relaxation time at this temperature. Below 1.5 K a slight decrease of τ was observed, but, at that time, the experimental facilities did not allow proper measurements in this region. More recent results indicated that in the antiferromagnetic region, single relaxation times are also observed. This is demonstrated in Fig. 17 where the absorption is plotted versus the frequency for $MnCl_2.4H_2O$ at 1.3 K in an external magnetic field of 4 kOe and for $MnBr_2.4H_2O$ at 1.83 K in a field of 8 kOe. The measurements lay on a Debye curve[16] ($\sim \omega\tau/1+\omega^2\tau^2$) which is the usual case if the susceptibility can be described by one single relaxation time.

Remarkable and unexpected antiferromagnetic relaxation behaviour has been observed in various manganese salts[17] and in $CoCl_2.2H_2O$,[18] but it is far beyond the scope of this paper to discuss these results in more detail. We hope that the survey given above of the latest experimental phenomena in the field of magnetic spin-lattice relaxation did show you some of the features that inspired so many Dutch physicists.

REFERENCES

1. P.R. Locher, C.J. Gorter, Physica 27 997 (1961).
2. A.J. De Vries, J.W.M. Livius, Appl. Sci. Res. 17 31 (1967).
3. A.J. Van Duyneveldt, H.R.C. Tromp, C.J. Gorter, Physica 38 205 (1968).
4. R. Orbach, Proc. Roy. Soc. A264 458 (1961).
5. T. Haseda, E. Kanda, Physica 22 647 (1956).
6. J. Van den Broek, L.C. Van der Marel, C.J. Gorter, Physica 25 371 (1959).
7. P.J. Poolman, J.J. Wever, G.J.C. Bots, L.C. Van der Marel, B.S. Blaisse, Proc. XIVth Colloque AMPERE (1966) 225.
8. D.O. Wharmby, J.C. Gill, Physica 46 614 (1970).
9. J.A. Roest, A.J. Van Duyneveldt, A. Van der Bilt, C.J. Gorter, Physica 64 306 (1973).

10. J.A. Roest, A.J. Van Duyneveldt, H.M.C. Eijkelhof, C.J. Gorter, Physica 64 335 (1973).

11. A.J. Van Duyneveldt, C.L.M. Pouw, W. Breur, Phys. stat. sol. (b)55 K63 (1973).

12. R. de L. Kronig, Physica 6 33 (1939).

13. M. Blume, R. Orbach, Phys. Rev. 127 1587 (1962); R. Orbach, M. Blume, Phys. Rev. Lett. 8 478 (1962).

14. H.M.C. Eijkelhof, C.L.M. Pouw, A.J. Van Duyneveldt, Physica 62 257 (1972).

15. A.J. Van Duyneveldt, C.L.M. Pouw, W. Breur, Physica 27 205 (1972).

16. J.H. Barry, D.A. Harrington, Phys. Rev. B4 3068 (1971).

17. A.J. Van Duyneveldt, J. Soeteman, C.J. Gorter, Commun. Kamerlingh Onnes Lab., Leiden, No. 397b (to be pub. 1973).

18. A.J. Van Duyneveldt, J. Soeteman, Phys. stat. sol. a16 K17 (1973).

19. F. Brons, thesis Groningen (1938). J.H. Van Vleck, Phys. Rev. 57 426 (1940).

ESR AND OPTICAL STUDIES OF INDUCED CENTRES IN BINARY SILVER AND ALKALI BORATE GLASSES*

F. Assabghy, E. Boulos, S. Calamawy, A. Bishay, and
N. Kreidl

American University in Cairo and University of Missouri
Rolla, U.S.A.

A series of binary borate glasses containing up to 35 mol% Ag_2O was X-irradiated and the induced centres were studied by means of ESR and optical absorption spectroscopy. The results showed that the Ag^+ traps an electron or a hole to form Ag^o or Ag^{++} respectively. The induced ESR spectra of these centres consisted of two hyperfine Ag^o lines (g = 2.20 and g = 1.83) and an axially symmetric line for Ag^{++} ($g_\perp$ = 2.04 and $g_\parallel$ = 2.31). This is in line with what has previously been reported for low silver content glasses.

It was observed that increasing the Ag_2O content was accompanied by a gradual decrease and final disappearance of the complex ESR spectra induced in borate glasses. This is due to the ability of Ag^+ to trap holes which are mainly responsible for the complex borate spectra.

The ESR and optical data correlate to confirm that the following two mechanisms occur during irradiation:

$$Ag^+ \xrightarrow{e-} Ag^o$$

and

$$Ag^+ \xrightarrow{e+} Ag^{++}$$

The two centres absorbing at 6 eV and 4 eV, characteristic of the base borate glass, compete with the Ag^o and Ag^{++} for the

* Research sponsored by U.S. NSF Granted G5 36216.

electrons and the holes respectively. This is believed to give rise to the observed maximum in the ESR intensity and minimum in the UV absorption at a composition of about 5 mol% Ag_2O.

INTRODUCTION

The study of silver centres in glass has been to a large extent restricted to glasses containing small amounts of silver. This is because the various applications of silver such as in photochromic, photosensitive, dosimeter, and semiconductor glasses, require but traces of the ion. The basis for all these applications relates to the electronic states taken up by the silver ion and the radiation induced changes in these states. When somewhat larger amounts of silver are introduced in the glass it is then likely to assume a determining structural role. Indeed there is a large body of work indicating that silver behaves in much the same way, structurally, as alkalis.[1,2] The ESR and optical spectra associated with Ag^{o} and Ag^{++} induced by radiation is then likely to reflect the structural changes caused by the introduction of silver in the glass. The system Ag_2O - B_2O_3 has been chosen because of the ease of glass formation in this system up to about 35% mol Ag_3O content.

Extensive ESR and optical studies have been made on various glass systems activated by small amounts of silver. Yokota and Imagawa,[3] Kreidl and Lell[4] have investigated the ESR and optical spectra, respectively, of irradiated phosphate glasses. They have shown that the silver ion Ag^+ acts as a trap for both electrons and positive holes leading to the formation of Ag^o ($4d^{10}$ $5S^1$) and Ag^{++} ($4d^9$), respectively, a situation perhaps analogous to the processes taking place in halide crystals.[5,6]

Silver has two naturally occurring isotopes, Ag^{109} (48.1%) and Ag^{107} (51.9%), both of which have a nuclear spin of I = 1/2. The resulting ESR spectra of Ag^o is then likely to exhibit a hyperfine structure. The ESR spectra of Ag^o with the resolved hyperfine structure has been observed in phosphate glasses, [3,7] in pyrex and sodium tetraborate glasses,[8] in borate, phosphate and silicate glasses[9] and in alkali silicate glasses.[10]

EXPERIMENTAL

Sample Preparation. The borate glasses were prepared from reagent grade boric acid, silver nitrate and sodium carbonate. Melting was conducted in an electric muffle furnace in alumina crucibles at about 1000°C. The glasses were annealed at about 350 to 450°C and then stored in black walled dessicators.

The same samples used in ESR studies were polished carefully and used for optical measurements. Glass films were also prepared from the same melts and used for measurements in the U.V. region.

ESR and Optical Studies. ESR measurements were made at X-band frequencies using a Varian Model V-4502 Spectrometer.

Most measurements were carried out at room temperature, but a number of liquid nitrogen runs were also performed.

All samples were X-irradiated (a tungsten target was used) at room temperature for one hour using a G.E. XRD-6 machine at a dose rate of about 3×10^5 rad/hour.

A Beckman DK-2A spectrophotometer was used for optical absorption measurements for the visible and UV range of the spectrum of the glass samples before and after irradiation.

RESULTS AND DISCUSSION

Figure 1 shows a typical induced ESR spectrum for a binary silver borate glass. The glass contains 30 mol% Ag_2O. The two outer lines are the characteristic Ag^o lines. The lack of resolved isotopic structure is not unexpected in view of the relatively large concentration of silver in the glass.[7,10] However, on going to liquid nitrogen temperatures a marginal increase in the isotopic resolution is observed in some cases.

The central part of the spectrum is the complex borate resonance. This part of the spectrum has been investigated by a number of workers.[11,12,13,14] It appears that five different centres are responsible for this complex spectrum. However, the greatest part of the intensity is attributed to two types of hole centres. The first of these is anisotropic and interacts with the B^{11} ($I = 3/2$) nucleus to give 5 lines plus a shoulder and is predominant in low alkali glasses. The second centre is axially symmetric and interacts with the B^{11} nucleus to give four lines and is predominant at high alkali contents (above 20 mol%).

The assignment of the Ag^{++} resonance has been made by Yokota and Imagawa[3,7] and will be discussed later.

Figure 2 shows that the complex borate spectrum decreases appreciably on increasing Ag_2O content. The spectrum for a sodium borate glass is included in the figure for comparison. The five lines plus shoulder spectrum is still identifiable in the spectrum for the 2 mol% Ag_2O although some of the resolution is lost. With 5 mol% Ag_2O, the complex borate induced spectrum is substantially

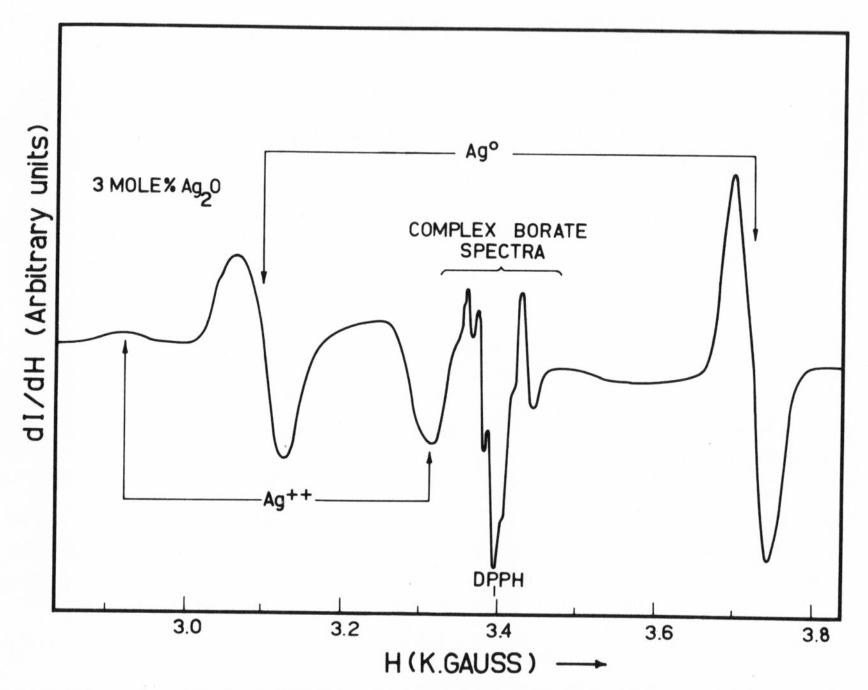

Fig. 1: Induced ESR spectra of silver borate glass.

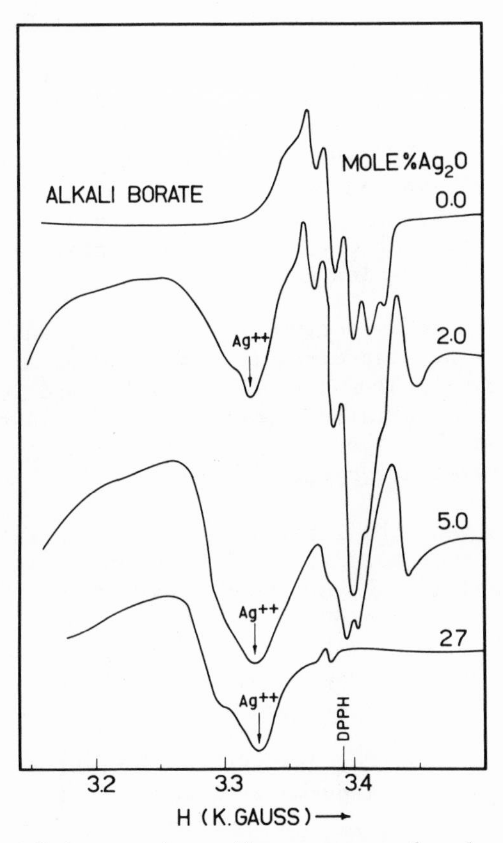

Fig. 2: Effect of increasing silver on complex borate spectra.

decreased and at 27 mol% Ag_2O no trace of the five or four line spectra remains.

This state of affairs is clearly different from what occurs in the case of alkali borate glasses in which, with increasing alkali content, the resonance changes from a five line spectrum to a four line spectrum with some gain in resolution and intensity.

The above observation must be regarded as being associated with the capability of silver to trap holes, which is in line with the observed decrease in the complex borate spectrum and the increase in the resonance attributed to the Ag^{++} species up to about 17 mol% Ag_2O (Fig. 3). At concentrations between 15 and 20 mol% Ag_2O, the five or four line spectra can be safely considered to be absent and the remaining observed spectrum consists mainly of the resonances for the two species Ag^o and Ag^{++}.

In Figure 4, the Ag^o spectrum is constructed and when subtracted from the observed spectra, the spectrum of Ag^{++} is obtained. The $g_{\perp} = 2.04$ and $g_{\parallel} = 2.31$ resonance is considered to be that of a powder pattern for an axially symmetric spectrum associated with the d^9 system of the Ag^{++} species. However, further work is needed to explain the residual absorption at $g \sim 1.88$.

Figure 5 shows that the intensity of the resonance absorption for the Ag^o species increases at first with increasing mol% Ag_2O, reaching a pronounced maximum at about 4 mol% Ag_2O. This is followed by a general decrease in the intensity of Ag^o lines. Intensity data above 15 mol% Ag_2O are not shown in view of indications of line shape variations in this range. Measurements with somewhat greater refinements are presently being made.

An attempt at correlating the ESR and optical data was undertaken. Figure 6 shows the induced UV absorption for 3 samples of different silver contents. Two induced optical bands at about 5.5 - 6.00 eV and 4.0 - 4.5 eV are observed in the sample containing 2.0 mol% Ag_2O. The 6 eV band can be attributed to an electron trap whereas the 4 eV band can be attributed to a hole trap.[13] Both bands are drastically reduced by increasing the silver content. In fact, on going from 2.0 to 5.0 mol% Ag_2O a gain in transmission is observed (i.e. negative induced absorption).

Figure 7 shows the change in induced absorption at about 6.0 eV with increasing Ag_2O. If the assumption is made that the competing mechanism for the centre absorbing at 6 eV is Ag $\overset{e^-}{\rightarrow}$ then it can be seen that there is complete agreement with the ESR intensity changes observed for the Ag^o species.

Figure 8 shows that if the competing mechanism for the centre

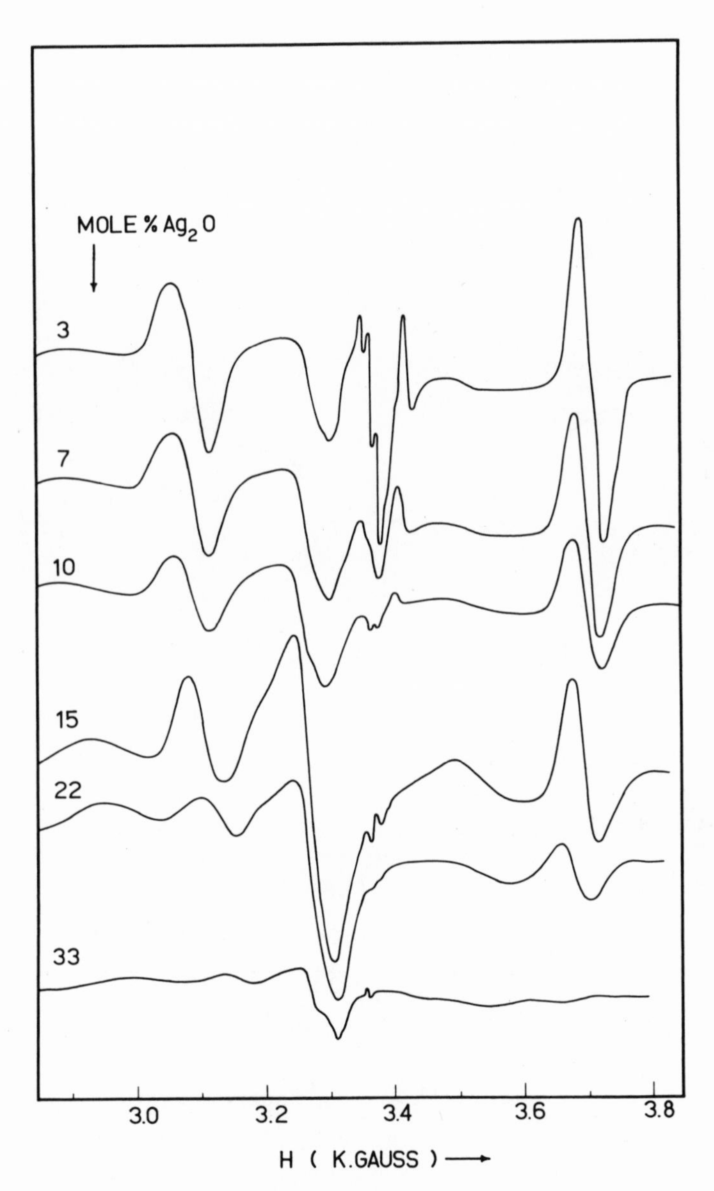

Fig. 3: Effect of increasing silver.

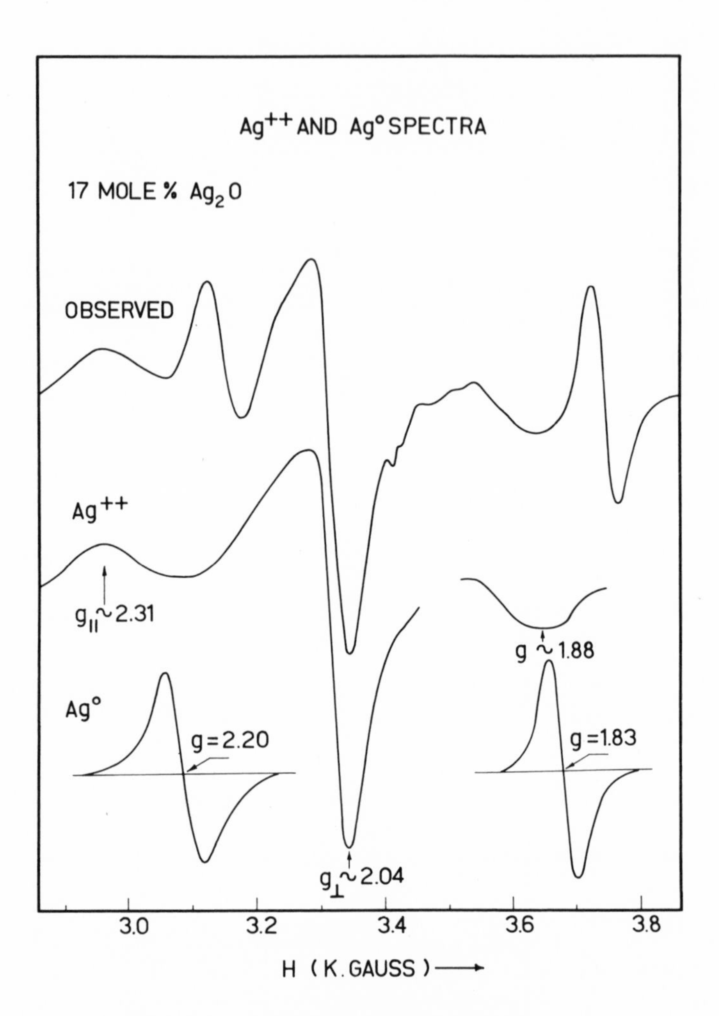

Fig. 4: Ag^{++} and Ag^{O} spectra.

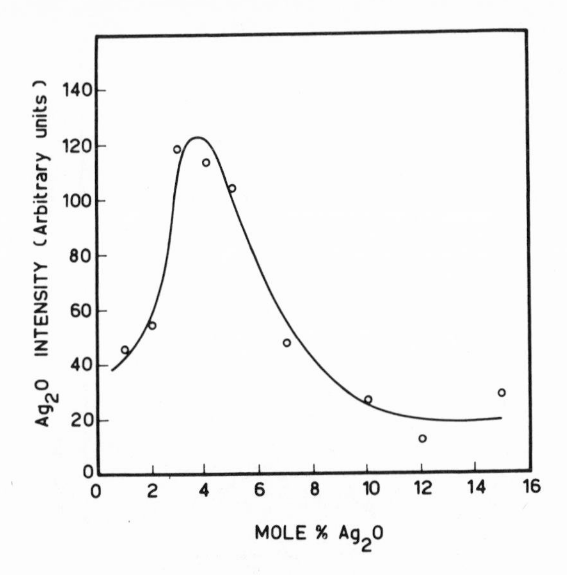

Fig. 5: Effect of increasing silver content on Ag^O signal.

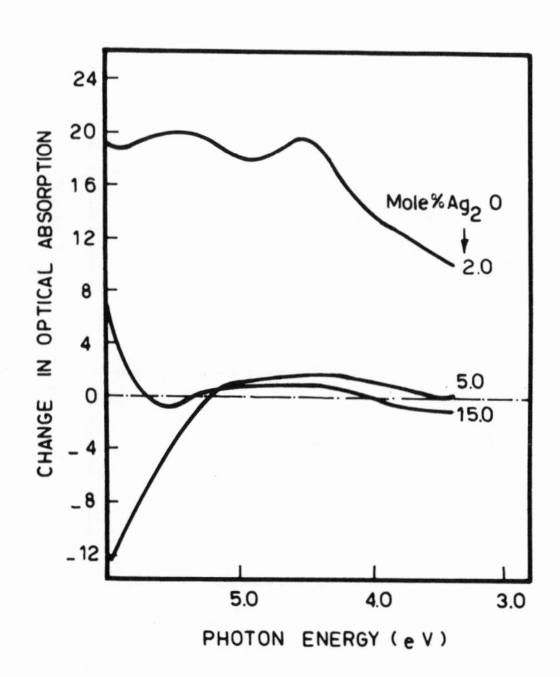

Fig. 6: Induced UV absorption.

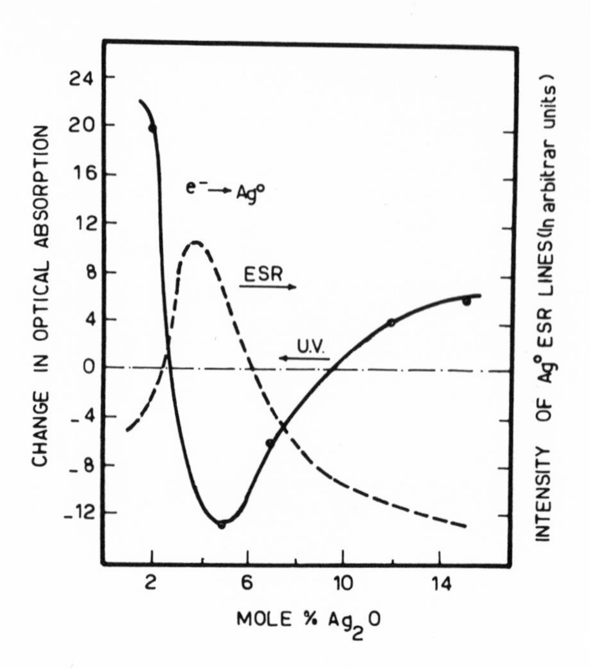

Fig. 7: Induced absorption at 6 eV.

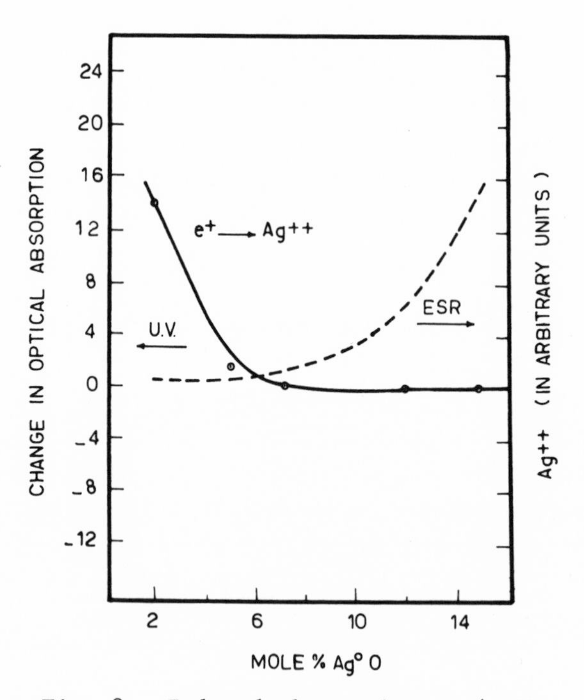

Fig. 8: Induced absorption at 4 eV.

absorbing at 4.0 eV is assumed to be $Ag^+ \overset{e+}{\rightarrow} Ag^{++}$ then again, agreement with the ESR intensity changes for the Ag^{++} species is obtained.

As the silver concentration is increased above 19 mol% Ag_2O, coloration is observed prior to irradiation. Subjecting the glass to irradiation or increasing the silver content above 19 mol% causes the coloration to deepen. It should be noted, however, that no ESR signals are observed above about 22 mol% Ag_2O where in fact the coloration is strongest. Clearly an explanation is needed, and it is being sought.

CONCLUSION

In conclusion, irradiating binary silver borate glasses containing up to 35 mol% Ag_2O by means of X-rays results in the Ag^+ trapping an electron or a hole, thus forming Ag^o or Ag^{++} respectively. The induced ESR spectra of these centres consisted of two hyperfine Ag^o lines ($g = 2.20$ and $g = 1.83$) and an axially symmetric line for Ag^{++} ($g_\perp = 2.04$ and $g_\parallel = 2.31$). This is in line with what has previously been reported for low silver content glasses.

It was observed that increasing the Ag_2O content was accompanied by a gradual decrease and final disappearance of the complex ESR spectra induced in borate glasses. This is due to the ability of Ag^{++} to trap holes which are mainly responsible for the complex borate spectra.

The ESR and optical data correlate to confirm that the following two mechanisms occur during irradiation.

$$Ag^+ \overset{e-}{\rightarrow} Ag^o$$

and

$$Ag^+ \overset{e+}{\rightarrow} Ag^{++}$$

The two centres absorbing at 6 eV and 4 eV, characteristic of the base borate glass, compete with the Ag^o and Ag^{++} for the electrons and the holes respectively. This is believed to give rise to the observed maximum in the ESR intensity and minimum in the UV absorption at a composition of about 5 mol% Ag_2O.

REFERENCES

1. E.N. Boulos, N.J. Kreidl, J. Am. Ceram. Soc. 54 368 (1971).
2. E.N. Boulos, N.J. Kreidl, J. Am. Ceram. Soc. 54 318 (1971).
3. R. Yokota, H. Imagawa, J. Phys. Soc. of Japan 23 1038-1048 (1966).
4. E. Lell, N.J. Kreidl, Proc. Cairo Solid State Conf. 1966, Interaction of Radiation with Solids, A. Bishay, ed., Plenum, New York (1967), 199.
5. C.J. Delbecq, W. Hayes, M.C.M. O'Brien, P.H. Yuster, Proc. R. Soc. London A271 243 (1963).
6. C.J. Delbecq, A.K. Ghosh, P.H. Yuster, Proc. Cairo Solid State Conf. 1966, A. Bishay, ed., Plenum, New York (1967) 387.
7. R. Yokota, H. Imagawa, J. Phys. Soc. Japan 20 1537-38 (1965).
8. L. Shields, J. Chem. Phys. 45 2332 (1966).
9. R.A. Zhitnikov, N.I. Mel'nikov, J. Sov. Phys.-Solid State 10 80 (1968).
10. H. Imagawa, J. Non-Cryst. Solids 1 335-38 (1969).
11. S. Arafa, A. Bishay, J. Amer. Ceram. Soc. 53 390-96 (1969).
12. Sook Lee, P.J. Bray, J. Chem. Phys. 39 2863 (1963).
13. A. Bishay, J. Non-Cryst. Solids 3 54-114 (1970).
14. P.C. Taylor, P.J. Bray, Bul. Amer. Cer. Soc. 51 234-39 (1972).

ESR AND STRUCTURAL STUDIES ON PSEUDO-ALEXANDRITE

S. Arafa*
The American University in Cairo, Egypt

S. Haraldson and A. Hassib**
University of Uppsala, Sweden

Structural analysis using x-ray, mass spectroscopy and ESR measurements were made on "Alexandrite" stones obtained from markets in Egypt, Germany and Sweden. The results show that all the samples studied are not synthetic Alexandrite but are synthetic crystals of α - Al_2O_3 doped with varying concentrations of vanadium, chromium and iron. The crystals show similar optical properties to Alexandrite crystals (the chromium bearing chrysoberyl, $BeAl_2O_4$:Cr) and are therefore termed "pseudo-Alexandrite."

ESR measurements were made on powdered samples and single crystals at room temperature and -160°C. The spectra indicate the presence of Cr^{3+} and Fe^{3+} ions. The resonance due to vanadium can be observed only at very low temperatures. The symmetry of the dopants sites is discussed. Heating at temperatures higher than 100°C causes a color change from red to green in the samples which disappears in a few seconds after cooling to room temperature.

The effect of ionizing radiations as well as the temperature dependence of the observed spectra are reported, and some comments made concerning the potentialities of ESR technique as a non-destructive test for investigating archaeological materials.

---

*Work conducted at the Solid State Unit, Institute of Physics, Uppsala University, Sweden, during a leave of absence.

**Present address, Physics Dept., Khartoum University, Sudan

INTRODUCTION

The samples used in this investigation were all obtained from markets in Egypt, Germany and Sweden with the understanding that they were Alexandrite gem stones. Preliminary tests by lane camera and ESR indicate clearly that the samples are single crystals, but not of Alexandrite, the chromium-bearing variety of chrysoberyl structure, $BeAl_2O_4$: Cr. They are synthetic crystals of α- Al_2O_3, corundum, doped with a variety of dopants. The impurities detected in the samples studied were vanadium, chromium and iron.

The samples show optical properties similar to those of Alexandrite (e.g. pleochroism) and so were named pseudo-Alexandrite. The structure of both crystals (Alexandrite $BeAl_2O_4$:Cr and pseudo-Alexandrite α-Al_2O_3: V_1Cr,Fe) may be viewed as a hexagonal close packing of oxygen ions with aluminium ions at interstitial sites along the crystalline c-axis. It is interesting to note that the average Al-O bond length of 1.914 Å for Alexandrite[1] is almost identical with the value of 1.915 for corundum.[2] In both types of crystals the Al^{3+} ions are surrounded by a distorted octahedral distribution of O^{2-} ions. Also, there are non-identical sites occupied by the Al^{3+} in both structures. Corundum crystals doped with chromium are called ruby crystals while those doped with iron are called sapphire crystals. The ESR spectra of the pseudo-Alexandrite crystals are presented. Temperature variation and effect of irradiation on those spectra are also presented and discussed in relation to its structure.

Crystal Structure. Corundum crystallizes in the rhombohedral (trigonal) system, with the structure shown in Fig. 1. The space group assigned is $D_{3d}6$ (R $\bar{3}$ c).[3] The rhombohedral angle is α = 55°17' and the length of the unit cell is a_o = 5.42 Å. The unit cell contains two Al_2O_3 (Z = 2). The exact coordination of an Al^{3+} is shown in Fig. 2. Electrostatic forces between the Al^{3+} and O^{2-} ions reduce the size of the shared triangles and produce a slight distortion in the hexagonal close packing. This contraction results in a slight enlargement and rotation about the c-axis of the adjoining triangles. Newnham and De Haan[2] gave a value of 3.9° for this angle of rotation. Three O^{2-} ions of the octahedron lie in a plane 1.37 Å from the Al^{3+} sites, but the other three are in a plane only 0.80 Å from the Al^{3+}. The aluminium ions are all octahedrally coordinated but occupy sites of two different symmetries (Al_I and Al_{II}).

Figure 3, from Geschwind and Remeika,[4] shows a portion of the structure of corundum. The Al^{3+} sites are all physically equivalent; however, there are two types which are magnetically inequivalent. All Al^{3+} sites between adjacent planes of oxygen are magnetically equivalent to these in the next set of planes, i.e. (b) and (c),

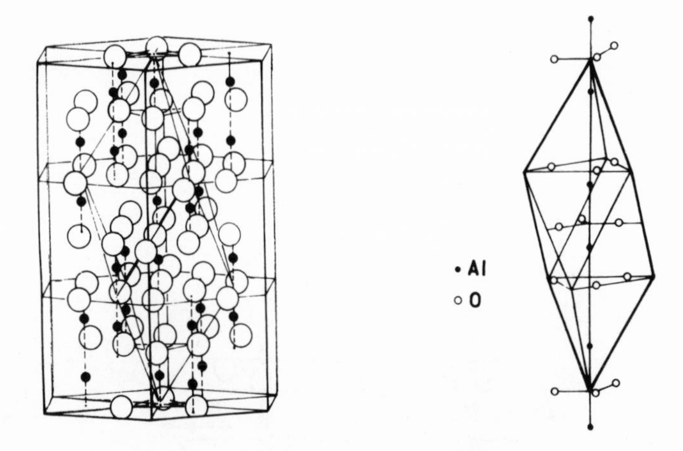

Fig. 1: A unit cell of α -Al_2O_3.

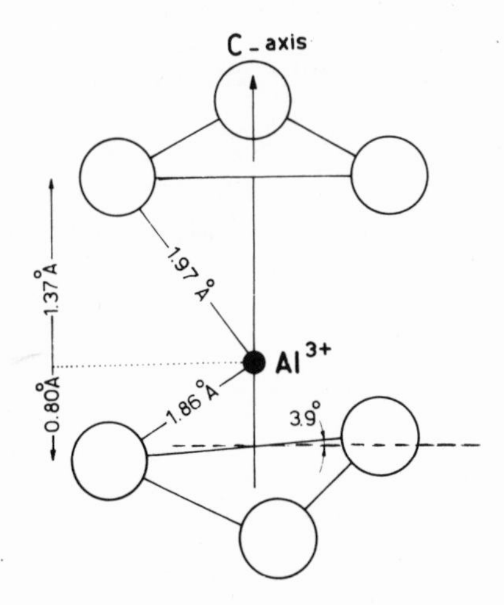

Fig. 2: Coordination of Al atom in α -Al_2O_3.

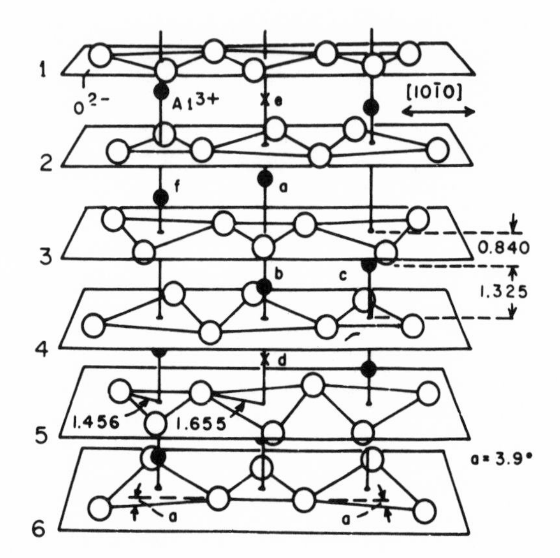

Fig. 3: The different sites occupied by Al in corundum.

while they are magnetically inequivalent to those in the next set of planes, i.e. (a) and (f), in that their ϕ axes are not rotated from each other by nearly 60°.

One feature of interest is the fact that the Al_{11} octahedron is appreciably larger than that of Al_{1}. The point symmetry is C_3 at the Aluminium site, C_{3i} at the void, D_3 at the centre of an Al_2O_3 molecule and C_2 at an oxygen site. Different dopants such as V, Fe or Cr substituted for Al might therefore prefer one site to the other, where differences in crystal fields are also expected, thus influencing the optical obsorption and ESR spectra of the crystal.

EXPERIMENTAL

The samples designated E_1, E_2, E_3, G and S in this study were obtained from Egypt, Germany and Sweden respectively.

The samples were analyzed by mass spectroscopy. Sample E_3 showed concentrations of impurities as follows: vanadium 0.7%, chromium 0.2% and iron 0.09% by weight.

ESR experiments were carried out at about 10 GHZ, using a V-4502 X-band spectrometer with 100 KHZ modulation. The spectra were recorded in the temperature range 77 K^{o} to 550 K^{o} temperature accessory. The magnet could be rotated through 50° about a vertical axis and the crystal could be rotated through 360° about the same axis, thus enabling an accurate orientation of the specimen with respect to the magnetic field.

RESULTS AND DISCUSSION

ESR Measurements. Figure 4 shows the ESR spectra obtained for pseudo-Alexandrite specimen E_3, at room temperature and at X-band frequency, with the magnetic field at various angles from the b-axis of the crystal. The circles indicate the three allowed transitions for Cr^{3} ions. The values of the magnetic field for the various transitions can be computed from the following Hamiltonian spin:

$$\mathcal{H} = g_{11}H_ZS_Z + g\ (H_xS_x + H_yS_y) + D\ [S_Z^2 - 1/3(S+1)]$$

The parameters of the spin Hamiltonian are the same as those measured before for ruby:[5] S $^{3}/2$, $g_{\parallel}$ = 1.9840, $g_{\perp}$ = 1.9867 and D = 5.747 GHZ.

Figure 5 shows the ESR spectra of the same specimen at other

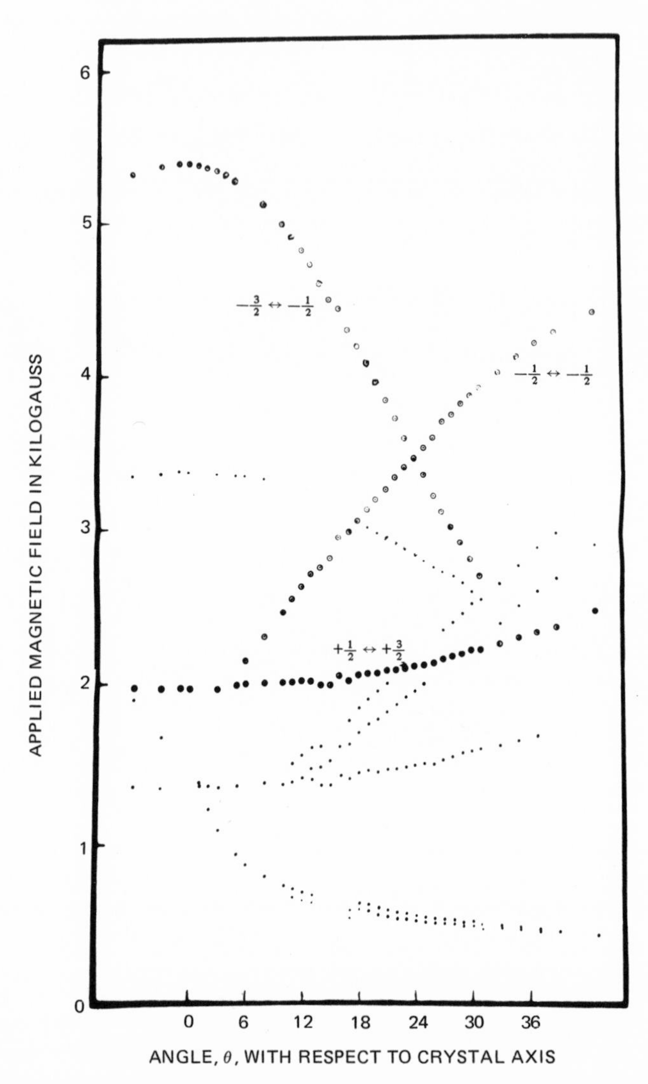

Fig. 4: The position of the observed ESR lines at various angles from the b-axis of the pseudo-Alexandrite crystal.

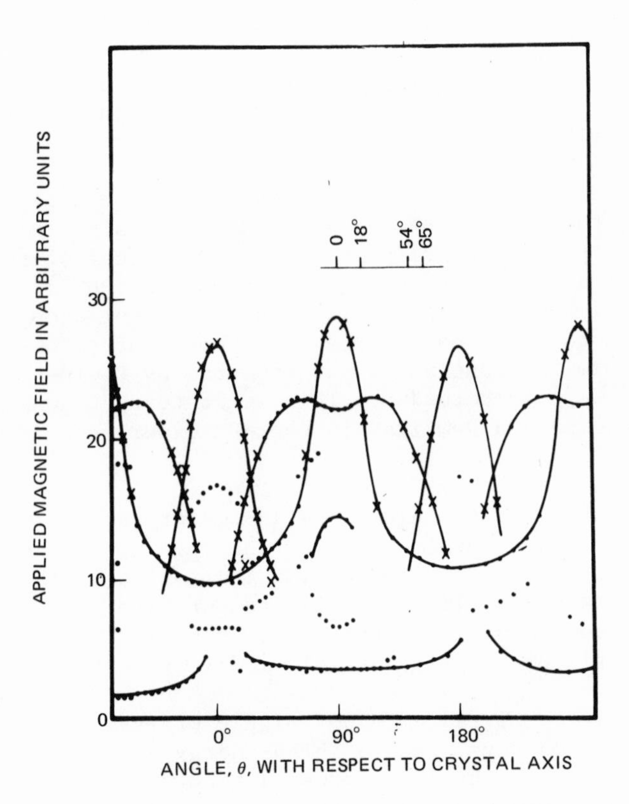

Fig. 5: The angular variation of the observed ESR spectrum for the pseudo-Alexandrite crystal.

orientations. Several other weak lines are observed, which may be due to forbidden transitions. In a strong crystal field, the selection rule $\Delta M_s = \pm 1$ is not obeyed and thus "forbidden" transitions are usually observed. The weak lines may also be the allowed and the forbidden transitions for Fe^{3+} ions present in the structure. These lines, however, are very weak, which makes it difficult to assign them to the different transitions. Furthermore, the concentration of Fe^{3+} is very small compared to the Cr^{3+} concentration. Closer examination of the angular variation of the lines showed that each line splits into two. The cross points in Fig. 5 are in fact the centre of gravity of two lines. The slow change of field position with angle near extrema, combined with finite line width, prevented the two components of each line from being observed with the magnetic field oriented near the magnetic axes.

This indicates that the magnetic axes are, for a first order approximation, the same as the crystal axes. It also indicates that the two components correspond to the resonances of the Cr^{3+} ion in two non-equivalent sites, as described in the literature.[4] The intensity of the split lines is about equal for each of them. This means that there is no preferential choice of sites for the Cr^{3+} ions. A similar conclusion was reported by Lewiner et al.[6] in the case of Fe^{3+} in corundum. This is contrary to what was observed by Geschwind and Remeika[4] in the EPR experiment in Gd^{3+} doped corundum.

No ESR spectra for vanadium ions in the crystals studied were observed. Lambe and Kikuchi[7] reported the spin resonance properties of various oxidation states of vanadium in α -Al_2O_3. The spectra for V^{3+} were obtained at liquid helium temperature. An analysis of the optical and magnetic properties of vanadium corundum was given on the basis of the ligand-field model of Macfarlane.[6]

Irradiation Studies. Irradiation of single crystals of α-aluminium oxide by U.V., gamma-rays, reactor high-energy electron radiation or by a combination of these produces lattice defects. These defects can be manifested in a variety of ways, including electron spin resonance.

The modified properties of irradiated Al_2O_3 have been studied extensively by many techniques. However, none of them have provided conclusive identification of the underlying centres. Gamble et al.[9] studied the ESR of gamma-ray irradiation at 77 K. As a result of the irradiation a single asymmetric anisotropic resonance line about 50 G wide with $g_{\parallel} = 2.012 \pm 0.002$ and $g_{\perp} = 2.008 \pm 0.002$ (where $\parallel$ and $\perp$ refer to the orientation of the magnetic field with respect to the crystal c-axis) were observed. All of the unirradiated cyrstals showed prominent background spectra of Fe^{3+} resonances. The induced ESR line has been analysed as a superposition of three Gaussian lines with the isotropic g-values: $g_1 = 2.020 \pm 0.003$, $g_2 = 2.006 \pm 0.003$ and $g_3 = 2.006 \pm 0.003$. The component lines were tentatively attributed to two types of centres.

The structure of trapped hole defects in Al_2O_3, has been discussed in a theoretical paper by Bartram et al.[10]

Figure 6 shows the ESR spectrum of a pseudo-Alexandrite crystal, after room temperature x-ray irradiation for 80 mins. Three groups of 8 lines each were observed in the induced ESR spectrum. This indicates the formation of some V^{2+} ions in the crystal by irradiation according to the following reactions:

$$V^{3+} + e \rightleftarrows V^{2+} \quad (1)$$

$$Fe^{3+} + e \rightleftarrows Fe^{2+} \quad (2)$$

$$Cr^{3+} \rightleftarrows Cr^{4+} + e \quad (3)$$

The ESR spectra of Cr^{4+} and Fe^{2+} cannot be observed at room temperature. The hyperfine lines are of about width 15 Gs and separation 120 Gs. There was no indication of the formation of induced color centers. The presence of Cr seems to suppress the formation of hole centers. For V^{3+} no signal is observed at room temperature or liquid nitrogen temperature. The hyperfine structure 8 lines, with a width of about 20 G and 110 G separation were observed for V^{3+} at liquid helium temperature.

The spin resonance of the irradiated crystals at room temperature gave a group of lines (8 lines) in the region of $g \simeq 2$. The center of the group remained in the same position as the angle between the c-axis and the static magnetic field varied.

Temperature Variation. Heat treatment produced changes from pale violet to green at temperatures around 550 K°. On cooling, the sample regains its original colour in a few minutes.

ESR spectra were measured at different temperatures in the range 77 K° to 550 K°. No detectable changes were observed in the ESR spectrum except that the intensity of the lines decreased as the temperature increased.

Since there was no detectable change in the ESR as a result of heating, we believe that the change in the color is due to lattice distortion resulting from the expansion of the lattice, specially along the c-axis, by heating. This results in a shift in the absorption peak. As the temperature decreases, the lattice relaxes and the crystal regains its original color.

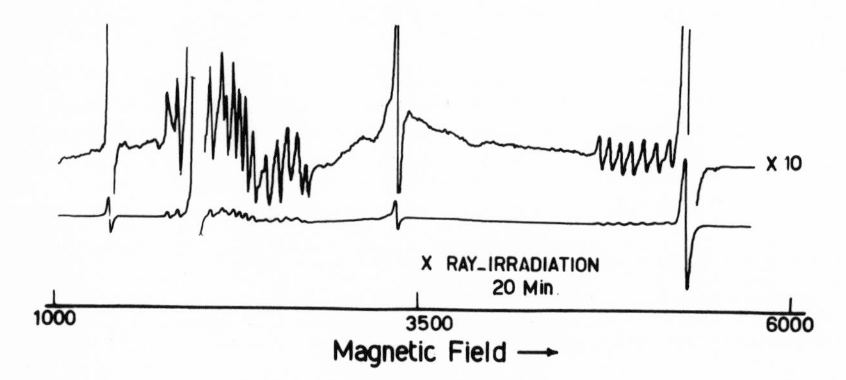

Fig. 6: The ESR spectrum of the pseudo-Alexandrite crystal after x-ray irradiation.

APPLICATION TO ARCHAEOLOGY

What then are the differences between synthetic and naturally grown single crystals? Apart from the difference in shape and size one can generally say that the naturally grown crystals have lesser impurities and lesser defects. Both types can be studied successfully by ESR technique, as may some archaeological materials, especially gem stones. Precise data from ESR can tell something about the method of preparation of such materials, the quantity of different impurities and also the different treatments these substances have been subjected to. It may help in dating these materials' preparation and use. Another suitable field and application of ESR is to pottery and glazing materials. The impurities present and the induced centers can be studied by ESR and other complementary techniques. Some problems on glazing, pottery and gem stones are currently under investigation at the Solid State and Materials Research Center, The American University in Cairo.

ACKNOWLEDGMENTS

The authors wish to acknowledge the interest and encouragement of Professor O. Beckman, Head of the Solid State Group, the Institute of Physics, Uppsala University. Thanks are also due to M. Richardson for his assistance in X-ray mass spectroscopic analysis. S. Arafa and A. Hassib are indebted to the Swedish International Development Authority (SIDA) for fellowships awarded.

REFERENCES

1. E.F. Farrell, J.H. Fang, R.E. Newnham, Amer. Mineralogist 48 804 (1963).

2. R.E. Newnham, Y.M. de Haan, Z. Krist. 117 235 (1962).

3. W.J. Moore, Seven Solid States, W.A. Benjamin Inc., New York (1967).

4. S. Geschwind, J.P. Remeika, Phys. Rev. 122 757 (1961).

5. E.O. Schulz-Du Bois, J. Bell Syst. Tech. 38 271 (1959).

6. J. Lewiner, P.H.E. Meijer, J. Res. Nat. Bur. Stand. 73A 241 (1969).

7. J. Lambe, C. Kikuchi, Phys. Rev. 118:1 (1960).

8. R.M. Macfarlane, J. Chem. Phys. 40 273 (1964).

9. F.T. Gamble, R.H. Bartram, C.G. Young, O.R. Gilliam, P.W. Levy, Phys. Rev. 134(3A) 1964.

10. R.H. Bartram, C.E. Swenberg, J.T. Fournier, Phys. Rev. 139 (3A) 1965.

MAGNETIC AND X-RAY CRYSTALLOGRAPHIC STUDIES OF A SERIES OF NITROSODISULPHONATES AND HYDROXYLAMINE-N,N-DISULPHONATES

B. D. Perlson and D. B. Russell
University of Saskatchewan, Saskatoon Campus

R.J. Guttormsom and B. E. Robertson
University of Saskatchewan, Regina Campus, Canada

A series of nitrosodisulphonate and two series of hydroxylamine-N,N-disulphonate salts with monovalent cations have been prepared and characterised. The crystal structure of one of the rubidium hydroxylamine-N,N-disulphonates has been determined to be triclinic with space group p1 and shown to contain the anion $\{[ON(SO_3)_2]H\}^{5-}$. *This salt, on irradiation with* Co^{60}_{γ} *was found to contain an S=1 species with* $|D|/hc = 0.048\ cm^{-1}$ *and* $|E|/hc = 0.002\ cm^{-1}$. *Two crystalline modifications of potassium nitrosodisulphonate and two crystalline modifications of rubidium nitrosodisulphonate were found to have thermally accessible triplet states.*

INTRODUCTION

It has long been thought that the monoclinic modification of potassium nitrosodisulphonate (Fremy's salt) is diamagnetic in the solid state[1] presumably due to dimerisation of nitrosodisulphonate ions, $ON(SO_3)_2^{=}$, on crystallisation.[2] On the other hand solid tetraphenylstibonium nitrosodisulphonate has been prepared by Chu et al[3] and shown to be a typical free radical salt.[4] This difference in the magnetic properties of these two salts was felt to be due primarily to cation size, and this study was undertaken to see how cation size affects the degree of interaction and thus the magnetic properties of these radical ions in the solid state. In addition it was decided that it would be of interest to compare the magnetic properties of these salts with those of nitrosodisulphonate ions known to be pro duced in the corresponding Co^{60}_{γ} irradiated hydroxylamine-N,N-disulphonates[5,6] where the radical ions produced would, on average, not

Table I Compounds Prepared and Characterized

a) Hydroxylamine-N,N-disulphonates

1. $Na_2(SO_3)_2NOH$
2. $Na_2(SO_3)_2NOH$. 2/3NaOH . $1/3H_2O$
3. *$K_2(SO_3)_2NOH$. 1/2KOH . $1/6H_2O$ $[K_5\{[(SO_3)_2NO]_2\}.3/4H_2O]$†
4. *$K_2(SO_3)_2NOH.2H_2O$
5. $Rb_2(SO_3)_2NOH$. 1/2RbOH. H_2O (I) $[Rb_5\{[(SO_3)_2NO]_2H\} . 3H_2O]$†
6. $Rb_2(SO_3)_2NOH$. 1/2RbOH . H_2O (II)
7. $Rb_2(SO_3)_2NOH$. 1/2RbOH
8. $Rb_2(SO_3)_2NOH$. $RbNO_2$
9. $Cs_2(SO_3)_2NOH$. 1/3CsOH . $1/4CsNO_2$

b) Nitrosodisulphonates

1. *$\{K_2(SO_3)_2NO\}_2$ (I) (monoclinic form)
2. *$\{K_2(SO_3)_2NO\}_2$ (II) (triclinic form)
3. $\{Rb_2(SO_3)_2NO\}_2$ (I)
4. $\{Rb_2(SO_3)_2NO\}_2$ (II)
5. $Cs_2(SO_3)_2NO$. 1/3CsOH . xH_2O where x ~ 1/2
6. Rb_8Na_6 . $\{(SO_3)_2NO\}_3$. $\{(SO_3)_2NOH\}_3$. $(OH)_2.xH_2O$ where x=1 or 2
7. Cs_8Na_6 . $\{(SO_3)_2NO\}_3$. $\{(SO_3)_2NOH\}_3$. $(OH)_2.xH_2O$ where x=1 or 2
8. *$\{(C_6H_5)_4Sb\}_2(SO_3)_2NO$
9. $\{(C_6H_5)_4As\}_2(SO_3)_2NO$ (I)
10. $\{(C_6H_5)_4As\}_2(SO_3)_2NO$ (II)
11. $\{(C_6H_5)_4As\}_2(SO_3)_2NO$. H_2O

* These compounds have been described previously.

† From crystallographic data.

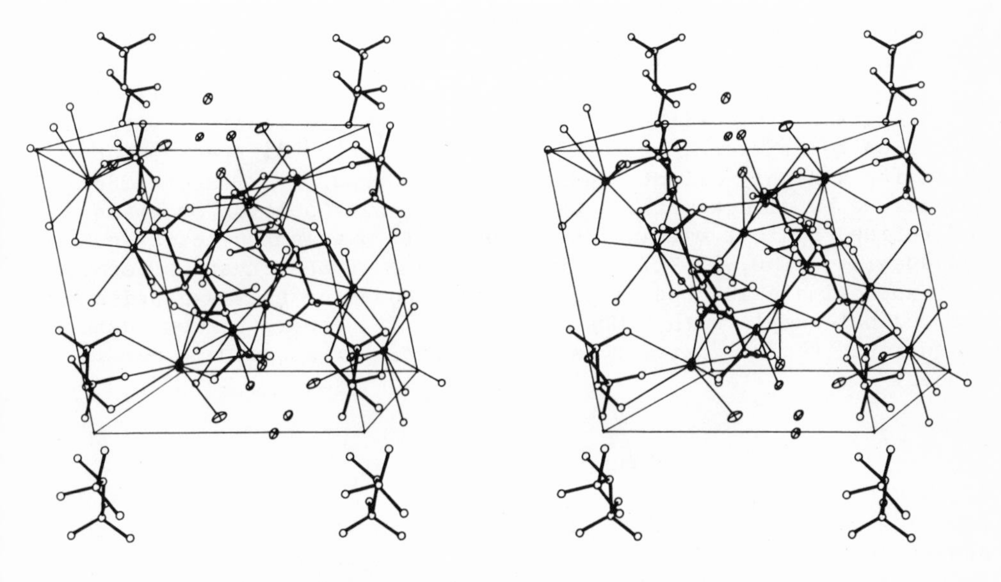

Fig. 1: Stereoscopic view of $Rb_5\{[ON(SO_3)_2]_2H\}.3H_2O$. Stereograms may be viewed with the naked eye or with the aid of a simple lens system available from W.A. Benjamin Inc., 2465, Broadway, New York 25, N.Y. The atoms making up the hydroxylamine-N,N-disulphonate groups are shown as spheres and the bonds in those groups are shown as heavier lines. The water oxygen atoms are shown as closed ellipsoids and the Rb^+ ions are shown as ellipsoids with one octant removed. Rb-O distances of less than 3.59Å are shown as light lines. The outline of the unit cell is also shown.

be expected to interact to any great extent. Accordingly a series of nitrosodisulphonate salts and two series of hydroxylamine-N,N-disulphonate salts have been prepared and characterised (Table 1).

EXPERIMENTAL

The crystal structure of one of the rubidium hydroxylamine-N,N-disulphonates has been determined by x-ray diffraction. The compound, formulated as $[HON(SO_3)_2]^{2-}.2Rb^+.1/2RbOH.H_2O$, is triclinic with space group $P\bar{1}$ and lattice constants a = 9.357(6)Å, b = 11.109 (7)Å, c = 11.206(5)Å, α = 102.03(3)°, β = 99.04(4)°, γ = 115.53(4)°. The integrated intensities of 4858 independent reflections with 2θ values up to 55° were collected with an automated diffractometer, utilizing a highly oriented graphite monochromator. The structure was solved by direct methods using the Germain Woolfson programs,[7] and refined by full matrix least squares to a weighted least squares residue, R_2, of 0.071. The irregular shape of the crystals permitted only approximate absorption corrections, based on the assumption of a spherical sample with $\mu(Mok\alpha)$ = 50.1 cm^{-1}. The refinement assumed anisotropic thermal parameters but the hydrogen atoms were not included.

RESULTS AND DISCUSSION

The structure is shown in Fig. 1, which was prepared with the aid of the Oak Ridge program ORTEP.[8] The water molecules are grouped around the center of the a,b face and form rods of water molecules running the a direction. The distances between the water oxygen atoms and the distances to other oxygen and nitrogen atoms indicate that the water molecules may enter into some weak hydrogen bonds but the existence of hydrogen bonds cannot be confirmed in the absence of a knowledge of the position of the hydrogen atoms. The cations do not have well defined coordination spheres, and even though distances greater than the sum of the ionic radius of Rb^+ and oxygen are included, the structure shows large voids with no neighbours to the cations.

The anions appear as two crystallographically independent $\{[ON(SO_3)_2]H\}^{5-}$ groups, one associated with the center of cymmetry at [000] and the other at [01/2 1/2]. These groups are separated by the Rb^+ ions and water molecules. The distance between the nitroso oxygen atoms strongly suggests the existence of symmetric hydrogen bonds. It would appear that one of the hydroxylamine-N,N-disulphonate hydrogen atoms has neutralized the hydroxyl ion in the formation of the dimer, leaving one hydrogen atom bonded symmetrically to each half of the dimer. The presence of a hydrogen bond is also inferred from the geometrical parameters of the hydroxylamine-N,N-disulphonate groups as shown in Fig. 2 and Fig. 3. The N-S bonds are 0.06Å longer than the values found by Howie, Glasser and Moser in the triclinic modification of potassium nitrosodisulphonate.[9] The N atoms in the present study are removed from the plane of the oxygen and sulphur atoms by 0.51A and 0.48A for the case of the dimers at [000] and [0 1/2 1/2] respectively. Also the N-O distance is increased from 1.28A in the nitrosodisulphonate to 1.43A here. This suggests that the presence of the hydrogen bond to the oxygen atom promotes the formation of a lone pair of electrons on the nitrogen atom and leads to approximate sp^3 hybridization of this atom. A slight tendency towards sp^3 hybridization was observed for the nitrosodisulphonate with the nitrogen atom roughly 0.10A out of the plane of the sulphur and oxygen atoms. The N-O bond lengths are similar to that in the hemihydrochloride of coccinellin[10] which also shows a symmetric hydrogen bond between N-O groups. The S-O distances are similar to those in the nitrosodisulphonate and are consistent with the analysis of $R(SO_3)_2$ groups given by Cruickshank.[11]

The errors given in Fig. 2 and Fig. 3 are based on a statistical analysis of the observed data are are probably low by a factor of 2. The differences in chemically equivalent bond lengths are therefore probably not significant.

Preliminary crystallographic data on the triclinic form of potassium-N,N-hydroxylamine disulphonate has shown that the unit cell contains six independent $\{[ON(SO_3)_2]_2H\}^{5-}$ groups, all in a similar orientation with respect to the unit cell edges and separated by about 6.3Å.

Triclinic rubidium hydroxylamine-N,N-disulphonate, irradiated with Co^{60}_{γ} at 77°K, exhibited room temperature spectra characteristic of a free radical species and a S=1 species, Fig. 4. The free radical spectrum consisted of three equally spaced absorption lines of equal intensity, and was shown to be due to the nitrosodisulphonate radical anion.[6,12] The S=1 species has zero field splitting parameters[13] of $|D|/hc = 0.048\ cm^{-1}$ and $|E|/hc = 0.002\ cm^{-1}$. Each line of the S=1 species was split into five hyperfine lines of relative intensity 1:2:3:2:1 due to interaction with two equivalent N^{14} nuclei. The magnitude of the hyperfine splitting was approximately half that observed in the free radical at all orientations, suggesting that the S=1 species results from the interaction of two nitrosodisulphonate radical anions. The EPR spectra given by a Co^{60}_{γ} irradiated single crystal of the triclinic form of potassium hydroxylamine-N,N-disulphonate were extremely complex. At least three distinct kinds of S=1 species and two free radical species were shown to be present.

The monoclinic and triclinic modifications of potassium nitrosodisulphonate were found to exhibit EPR spectra characteristic of S=1

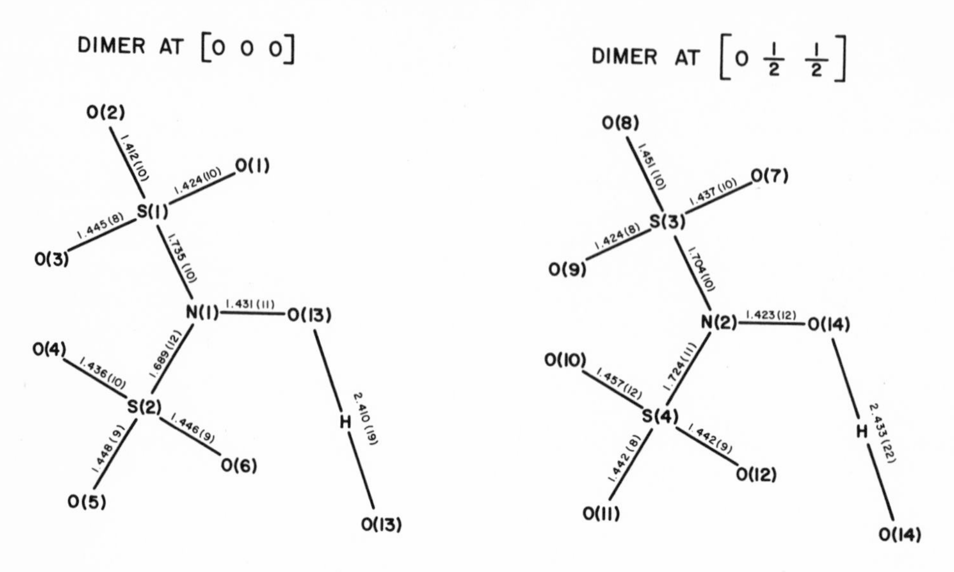

Fig. 2: Bond lengths (Å) in hydroxylamine-N,N-disulphonate groups. Numbers in parentheses are estimated standard deviations.

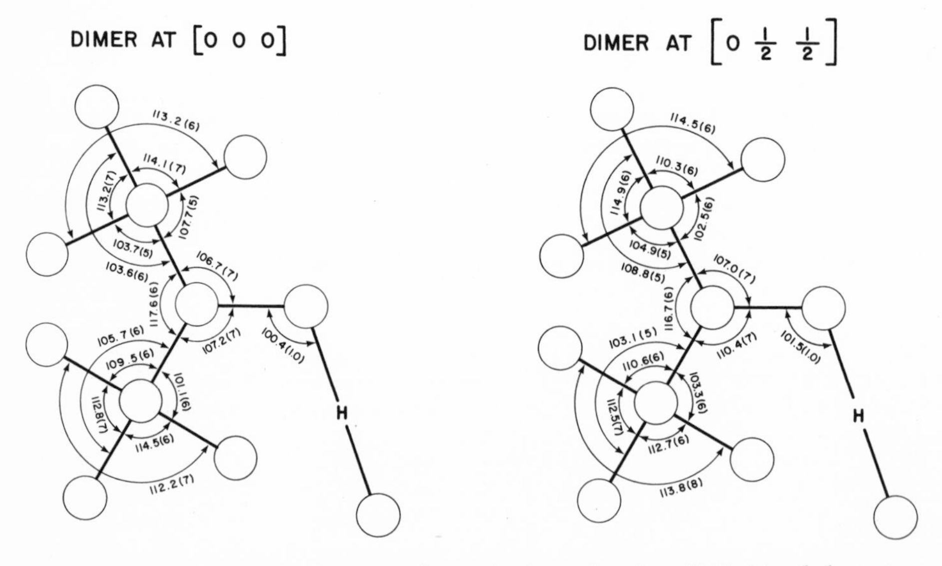

Fig. 3: Bond angles (degrees) in hydroxylamine-N,N-disulphonate groups. Numbers in parentheses are estimated standard deviations.

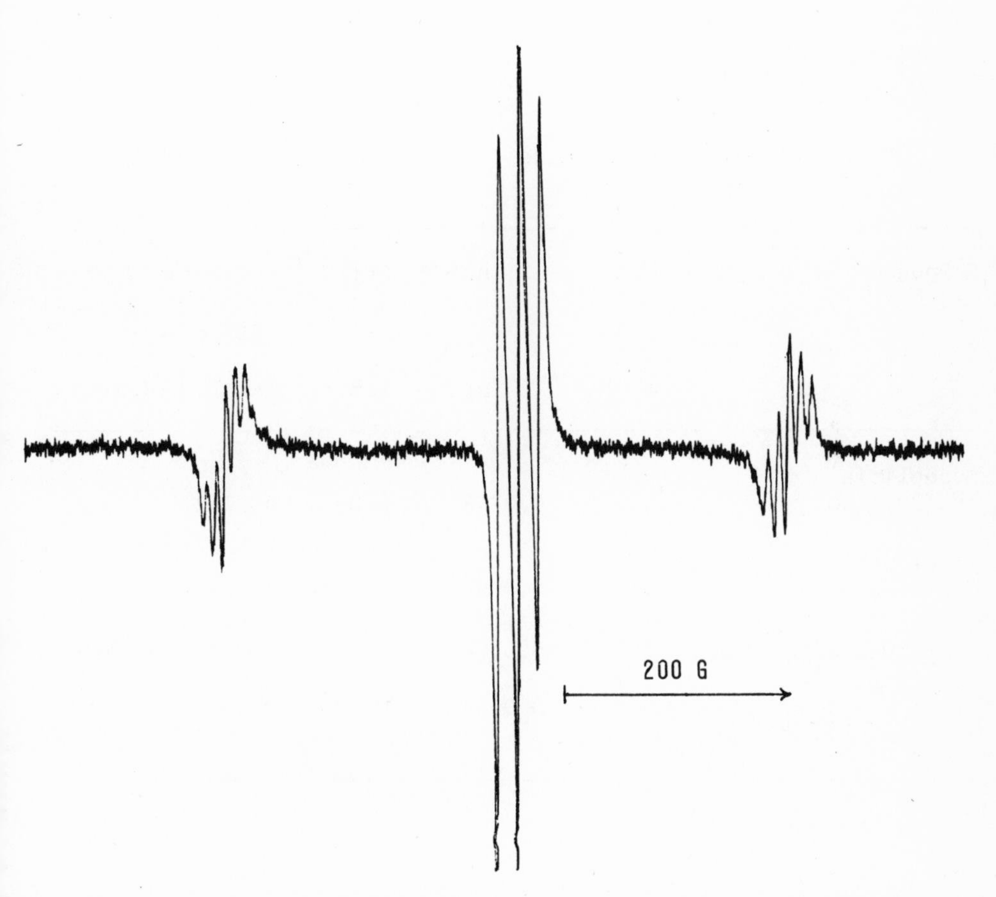

Fig. 4: Representative EPR spectrum given by a single crystal of Co^{60}_{γ} irradiated rubidium hydroxylamine-N,N-disulphonate.

Table II Zero Field Splitting Parameters for the S=1 Dimeric Nitrosodisulphonate Ions in $\{M_2(SO_3)_2NO\}_2$ (M=K,Rb)

Compound		Temp (^oK)	$\|D\|/hc$*	$\|E\|/hc$**	$\{(\|D\|/hc)^2 + 3(\|E\|/hc)^2\}^{1/2}$	
				(cm^{-1})	($\Delta m=1$)	($\Delta m=2$)
Monoclinic $\{K_2(SO_3)_2NO\}_2$		363	0.0748	0.0044	0.0752	0.0776
Triclinic $\{K_2(SO_3)_2NO\}_2$		234	0.0686	0.0037	0.0689	0.0700
$\{Rb_2(SO_3)_2NO\}_2$	(I)	275	0.0666	0.0037	0.0669	0.0688
$\{Rb_2(SO_3)_2NO\}_2$	(II)	216	0.0654	0.0035	0.0657	0.0683

* All values ± 0.0005

**All values ± 0.0002

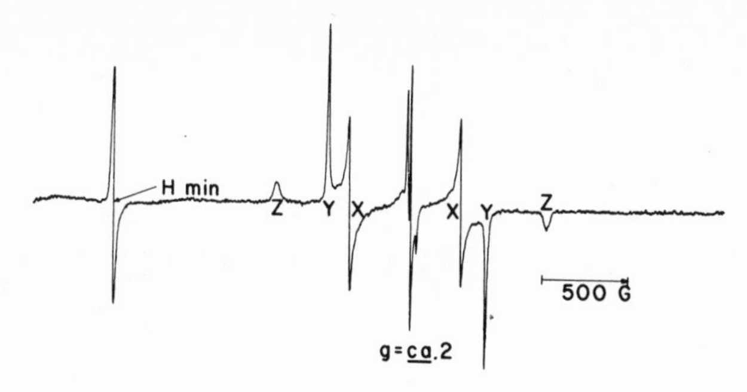

Fig. 5: EPR spectrum of a powdered sample of monoclinic potassium nitroso-disulphonate at 363 K.

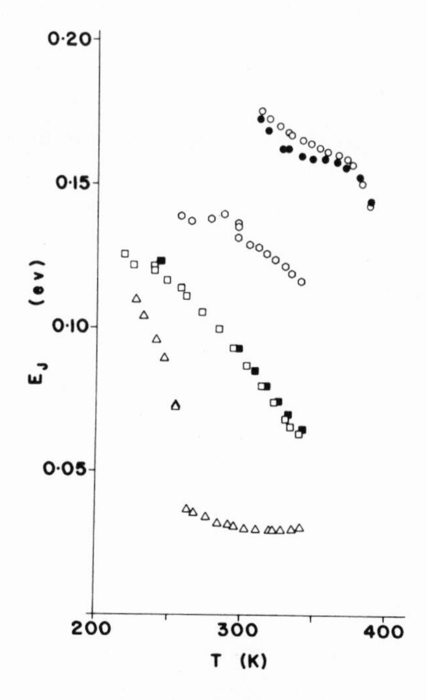

Fig. 6: Variation of measured singlet-triplet energy gaps, E_J, with temperature where ● and O represent two samples of monoclinic potassium nitrosodisulphonate, ⬡ represents $[Rb_2(SO_3)_2NO]_2$,I ■ and □ represent two samples of triclinic potassium nitrosodisulphonate and Δ represents $[Rb_2(SO_3)_2NO]_2$,II.

species.[14,15] The crystal structure of the triclinic potassium nitrosodisulphonate[9] strongly suggests that the S=1 species is formed by the interaction of the $^-$N-O groups of two nitrosodisulphonate ions. In addition, two distinct crystalline modifications of rubidium nitrosodisulphonate give similar EPR spectra.[16] A typical powder spectrum is shown in Fig. 5. The zero field splitting parameters of the S=1 dimeric ions are shown in Table II. Changes in intensities of these spectra with temperature showed that the S=1 state is thermally accessible from a ground singlet state. This was confirmed by Faraday static magnetic susceptibility measurements. These measurements also suggest that the apparent singlet-triplet energy gap,[17] E_J, decreases with temperature (Fig. 6), possibly due, in part, to increasing separation of nearest neighbour nitrosodisulphonate ions. These compounds are, to our knowledge, the first examples of inorganic salts not containing a transition metal atom with thermally populated triplet states.

Measurements of the temperature dependence of the static magnetic susceptibility in the range 77-300K and EPR studies have shown that the Cs^+, $(C_6H_5)_4Sb^+$ and $(C_6H_5)_4As^+$ nitrosodisulphonates are typical free radical salts.

ACKNOWLEDGMENT

The authors thank the National Research Council of Canada for financial support, Mr. S. Van der Heijden for assistance with the drawings, and Dr. J. Rutherford for helpful discussions.

REFERENCES

1. (a) R.W. Asmussen, Z. Anorg. Chem. 212 317 (1973).

   (b) W.A. Moser, R.A. Howie, J. Chem. Soc. (A) 3039 (1968).

2. A. Hantzsch, W. Semple, Ber. 28 2744 (1895).

3. T. Chu, G. Pake, D. Paul, I. Townsend, S. Weissman, J. Phys. Chem. 57 504 (1953).

4. D.L. Fillmore, B.L. Wilson, Inorg. Chem. 7 152 (1968).

5. J.R. Morton, personal communication.

6. P.T. Hamrick, H. Shields, T. Gangwer, J. Chem. Phys. 57 5029 (1972).

7. G. Germain, M.M. Woolfson, Acta Cryst. B24 31 (1968).

8. C.K. Johnson, ORTEP, Report ORNL-3794, Oak Ridge National Laboratory, Oak Ridge, Tennessee (1965).

9. R.A. Howie, L.S.D. Glaser, W. Moser, J. Chem. Soc. (A) 3043 (1968).

10. R. Karlsson, D. Losman, Chem. Comm. 1972, p. 629.

11. D.W.J. Cruickshank, J. Chem. Soc. 1961, p. 5484.

12 S.J. Weissman, D. Banfill, J.A.C.S. 75 2534 (1953).

13. M.S. de Groot, J.H. van der Waals, Mol. Phys. 35 1002 (1963).

14. B.D. Perlson, D.B. Russell, Chem. Comm. 1972, p. 69.

15. Details to be published elsewhere.

16. Details to be published elsewhere.

17. Calculated using the Van Vleck formulation. (Van Vleck, The Theory of Electric and Magnetic Susceptibilities, 1932, p. 235)

DEVELOPMENT OF MATERIALS FOR ENERGY-RELATED APPLICATIONS*

J. S. Kane

Lawrence Livermore Laboratory, University of

California, Livermore, California, USA

In this paper I will discuss several areas in energy systems where materials research and development will lead to improvements in the efficiency of energy utilization. I will consider improvements in current technology as well as new technologies.

The theme of this paper will be the application of materials science and technology to develop new energy sources, and to make current energy systems more efficient. Although the per capita energy requirements and details of energy use may vary in different nations, the need for new or at least improved technology is universal.

It is probably true that there is no energy "shortage" in the world, nor need there be in the future. The basic question is: Will energy be available at a price society can afford, and can it be generated, transported and consumed with minimal impact on our common environment?

The world today depends largely on conventional sources of energy. By this, I mean almost entirely on fossil fuels, to a smaller extent on hydro-electricity, and on a now small, but growing, nuclear capacity. This pattern will change as the following, inevitable events take place:

---

*Work performed under the auspices of the United States Atomic Energy Commission

1. Available hydroelectric sites will be used up.

2. Costs of fossil fuels will rise as readily available resources are consumed, and as expectations of citizens rise.

3. The capital costs of "new" energy sources will cause energy costs to rise. Environmental restrictions will raise costs from some sources more than from others.

It is obvious that we, the technologically advanced nations, must develop a strategy to develop efficient production and use of energy, and that this plan must address both short and long range questions. By short range I mean approximately the next 25 years, which is probably much sooner than any major, new technology can be introduced. The long range question implies using one of the "inexhaustible" sources of energy: breeding nuclear fission, thermonuclear fusion, solar energy, or deep-lying geothermal energy. This is not to imply that other sources will not be important, for example wind, tidal, and thermal and chemical gradients in oceans and rivers; but the latter are localized, and probably not available in sufficient quantity to make a major impact on a global scale.

I will give you my opinions upon where we can most fruitfully apply our materials efforts to yield the greatest short and long term benefits.

The first point I will urge you to remember is that in dealing with the production, conversion and transportation of energy, we are involved with an immensely large and costly system. It is the overall system that should be optimized, and in choosing an individual component for work, we should choose only those components where improvement will yield the largest benefit for the overall system.

I will divide my subject matter into three categories, which I have defined as follows:

SOURCES: The physical or chemical process from which the energy is obtained.

CONVERSION: The process by which the original form of the energy is changed to a more useful, convenient, or transportable form. In this paper I have assumed that this change is to electricity or hydrogen gas. There are other possibilities, but these two are the most probable forms that require the least modifications in our existing distribution systems.

TRANSPORTATION: The shipment of energy from one point (usually the source) to its point of consumption, or point of further conversion.

The second classification of topics I will make is even more arbitrary. I have chosen to mention further only a limited number of areas, ignoring other equally interesting and important ones. My choices are:

	SELECTED FOR DISCUSSION	*NOT TO BE DISCUSSED*
SOURCES	Fossil fuels Solar energy Thermonuclear energy	Fission nuclear energy Geothermal energy Tidal, wind, thermal gradients in oceans
CONVERSION	Thermal cycles Solar photovoltaic Thermal decomposition of water Hydrogen-air fuel cells	Electrical generation Electrolysis
TRANSMISSION	Hydrogen pipelines Superconducting or cryogenic electrical transportation	High voltage, AC or DC Cryogenic liquid transportation

MATERIALS PROBLEMS

SOURCES

Fossil Fuels. In fossil fuel utilization, improvements are needed for increased efficiency and in reducing pollutants, especially sulfur. Any long range approach to fossil fuels must focus on coal, since this is by far the most abundant resource. It is remarkable, however, that the way we burn coal has changed little during the last few hundred years, other than to replace the man with a shovel with a power driven machine. Novel ways are needed to convert coal to an easily transportable form (e.g. CH_4, H_2, liquid hydrocarbons or alcohols) and simultaneously to remove the sulfur, the most objectionable pollutant. One technique is to convert the mined coal to "power gas," a low energy content mixture of hydrogen, hydrocarbons, carbon dioxide and nitrogen. This gas, while not suitable for some uses, is completely satisfactory for efficient, advanced thermal cycles for generating electricity.

Materials work is crucial in each step of this process. For instance, the conversion could utilize MHD, or more likely high temperature gas turbines as "topping" cycles whose exhaust then powers a conventional steam cycle.[1] Another approach is to somehow gasify the coal underground, without mining.[2] The resulting CH_4 or hydrogen could be separated from CO_2 and distributed by pipeline, perhaps in existing systems.

Solar Energy. I will not consider uses such as reflectors for cooking, home heating, etc., although these may become quite important.

The chief problem with solar energy is the low areal energy density. The challenge is to collect this energy and convert it to a transportable form at sufficiently low cost. No one has yet conceived a method of doing this at a capital cost anywhere near that of competitive sources, such as fission nuclear, or conventional fossil fuel. It takes only a cursory consideration to become convinced that schemes requiring arrays that track the sun, or require optical quality components are not apt to become economic in the foreseeable future.

A much better approach is to use planar absorbers that can achieve reasonable efficiency without focusing optics. Two techniques that meet this requirement are direct conversion to electricity by means of photovoltaic cell arrays, and collection of thermal energy with subsequent conversion to electricity via thermal cycles and rotating electrical generation. Both ideas are inefficient: the photovoltaic cells because low energy photons do not contribute, and the thermal collectors because their low temperature implies a Carnot cycle of low theoretical efficiency. For both, then, it is essential that the cost per unit area of collector be made as low as possible. The thermal collection approach depends mainly on developing very inexpensive materials that will attain as high a temperature as possible. Also, if a thermal conversion cycle (probably Rankine) is used, it will doubtless be necessary to use a low-boiling fluid, such as isobutane, rather than water.

The photovoltaic arrays must also maximize efficiency against cost per unit area. Single crystal silicon cells are currently the most efficient means of direct conversion, but to interconnect cells made from individual crystals would be unfeasible for the large areas needed. What will be necessary are techniques for manufacturing cells of very large area so the number of parallel connections can be minimized. It might be possible to use vapor deposition to obtain large area, thin-film cells such as CdS. The major task is to increase the efficiency and lower the cost. It is also worth mentioning that the basic processes occurring in such cells are very poorly understood.

A problem that inevitably accompanies solar energy utilization is the need to store energy in some form during dark periods. Suggestions have been made to use thermal storage or pumped hydrostorage. Both ideas have great practical difficulties.

If hydrogen gas were to be the product of solar energy collectors, either from electrolysis of water or from thermal decomposition of water, the storage would be somewhat higher, using the same methods currently used for storing methane. I will return to thermal decomposition of water later, since I believe this is an extremely important topic.

To couple electrolysis cells with photovoltaic arrays seems an ideal combination, since electrolysis is a low voltage, DC process. Perhaps a large, modular panel could be vapor-deposited, and connected to its individual electrolysis cell. Modular arrays have great economies in manufacture, maintenance, repair, etc.

Thermonuclear Energy. To date, the feasibility of obtaining energy from thermonuclear reactions has not been demonstrated, although efforts are being conducted in many nations in both magnetic and inertial (laser heated) confinement.[3] Even if the scientific feasibility of either approach is demonstrated, there will be a large effort required to bring the concept to the point of economic, predictable performance. If our experience with fission reactors is any indication (and I believe fusion reactors will be far more challenging technologically), most of the difficulties will lie in the area of materials. For illustration let me list five of the problems that must be solved before thermonuclear energy will become economic. I have assumed that the DT reaction will be used; therefore, T must be "bred" by neutron capture in ^6Li.

1. Corrosion problems in Li-containing "blanket."
2. Diffusion, migration and permeation of T in structure.
3. Radiation damage problems in structure.
4. Development of economic superconductors for magnetic confinement.
5. High-temperature materials for efficient thermal conversion.

As more detailed studies are made, new problems will emerge. My colleague, Professor Richard Borg, will describe one such problem that he is studying and one that I believe to be very fundamental. It illustrates the fact that much basic solid-state science as well as technology will be required for the development of practical, economic thermonuclear energy.

CONVERSION

Thermal Cycles. I have mentioned the need for more efficient thermal conversion (liquid metal Rankine, high temperature Brayton, MHD). Without exception, current efficiencies are limited by the high temperature capabilities of materials. Work is being done, but much more is required. It is worth pointing out that rejecting waste heat is expensive, especially where environmental limitations are imposed. This is an additional incentive for increased efficiency.

Current fossil fuel conversion technology is chiefly limited to steam cycle efficiencies, about 40%. In the past, fossil fuels have been so cheap that development of more efficient cycles, which are inherently more complex and expensive, was not warranted. Also, the relatively low-temperature water cooled nuclear reactor has stifled incentive to increase conversion temperatures. This situation will change, however, as the liquid metal cooled breeder reactor is developed, since its design will push the steam cycle to its supercritical limit. An even greater impetus toward higher temperature conversion will be the high temperature, helium cooled reactor. It would be intolerable if conversion cycles were to place a limitation on the efficiency of nuclear electrical generation. Much work on improved materials for combined cycles, novel cycles, and above all, high temperature cycles will be required.

Thermal Decomposition of Water. It is well known that new sources of electricity do not solve all our energy requirements. For example, most vehicles need a fuel with greater capability than batteries can provide.

As long as fossil fuels are available, hydrocarbons will be our best portable fuel, but in the long run we will need another. Hydrogen has been proposed by several authors[4] as the most suitable. A problem of high priority, therefore, is the interconversion between heat and hydrogen. This can be achieved with present technology by first converting the heat to electricity, with a loss of perhaps 60%, then electrolyzing water, at an efficiency of perhaps 85%. This is extremely wasteful of energy.

It should be possible to decompose water into its constituents directly by heating. For a single step process, the temperature required is unacceptably high, again because no materials exist that are suitable. By using a multi-step process, in which no materials are consumed except water, it should be possible to decompose water at temperatures well within the capabilities of high temperature nuclear reactors, either fission or fusion. In fact, even lower temperatures, such as could be obtained from solar thermal panels or geothermal sources, may eventually be sufficient. Although at first the problems encountered in obtaining hydrogen from water appear to be predominantly chemical, I have no doubt materials will play an essential role in their ultimate solution.

Electricity from Hydrogen. One can always obtain electricity from hydrogen by combustion followed by conversion via a thermal cycle. A far more attractive alternative is the hydrogen-air fuel cell, which is not limited to Carnot cycle efficiencies. Efficient fuel cells are therefore an essential component of a hydrogen-based energy economy.

Fuel cells are already in use, and hydrogen is in many ways an ideal fuel. The theoretical efficiency is very high (83%), they are quiet, generate few pollutants, and are not hazardous.

One of the chief advantages of fuel cells is that they are modular; the cost per kilowatt capacity does not vary strongly with station size. There are advantages to dispersing the conversion in smaller units close by the consumer. Electrical line losses are reduced, and use of waste heat for space heating or cooling is possible.[5]

The materials problems in fuel cells are chiefly those of corrosion, and of devising economic but highly efficient electrode surfaces.

TRANSPORTATION OF ENERGY

In the 21st century fuel economy I foresee, energy will be transported predominantly in two ways -- via electrical transmission lines, and via pipelines containing hydrogen and perhaps hydrocarbons or alcohol. Again, both electrical and gas transportation will involve materials development.

The most dramatic improvement in electrical transmission would be the large scale development of cryogenic or superconducting lines. If the transition temperature for superconductors could be raised into the liquid hydrogen range, a far more practical system is possible than that based on scarce helium. There is a further, equally great need for technology for fabricating reliable and economical superconductors. Current methods are prohibitively expensive.

If high pressure hydrogen pipelines become necessary, it will be essential that steels behaving satisfactorily under these conditions be available. In my opinion, no field-welded steel in current pipeline use can be guaranteed fracture-safe when filled with high pressure hydrogen gas.

SYSTEM CONSIDERATIONS

I have touched on only a few of the many energy-related materials areas that I believe warrant your attention. I would like to discuss briefly new system approaches that should be studied. Consider a future time production-conversion-transportation-consumption system, where energy is produced remotely from solar, nuclear or thermonuclear heat sources, transported, and finally consumed for the purpose of home heating and cooling, electrical uses in homes, commerce and industry, and for public and private transportation. What would be the optimum overall system for this purpose? I must admit I can't really answer this question, and instead I will show how this would be done with today's technology, and contrast this with what I believe to be very attractive options that should be developed, or at least investiged.

CURRENT TECHNOLOGY

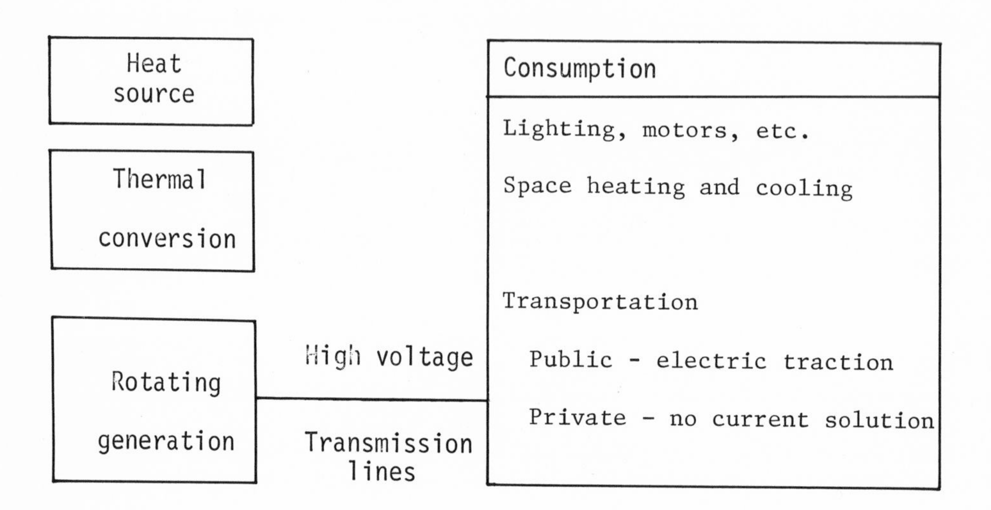

The following comments apply to this system:

1. 60% of thermal energy is wasted in conversion, at efficiencies operative today.

2. Line losses are high over long distances.

3. Private transportation is not patterned around electricity; batteries have low energy/unit mass.

Contrast this with an alternate system, which I believe could be developed over the next decade or so:

POSSIBLE FUTURE TECHNOLOGY

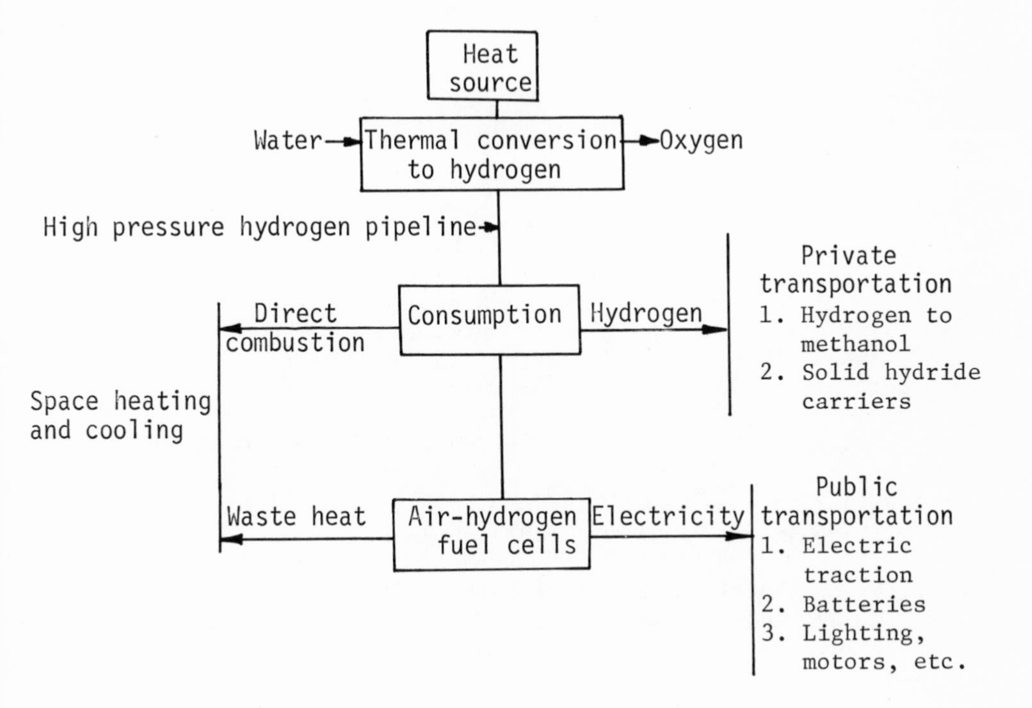

The above system appears attractive to me. The two areas that will require the greatest attention are the thermal dissociation of water and the hydrogen-to electricity fuel cells. Neither of these is Carnot-limited. I hope some of you will be stimulated to conceive other, better systems.

The water-to-hydrogen conversion must avoid the waste of large amounts of thermal energy at the source, where it is usually impossible to utilize. Transportation by high pressure pipeline is fully competitive with electrical transmission at distances greater than a few hundred miles. And if efficient fuel cells can be developed in sizes economical for relatively small areas (such as a few urban blocks) the waste heat can be utilized for space heating and cooling. One advantage of conversion at the point of consumption is that hydrogen/electricity can be consumed in whatever 'mix" is most desirable at any time.

I have tried to point out in very general terms a few of the areas where I believe materials science and technology can be best applied to energy systems. As scientists, technologists and engineers we all appreciate the importance of a clean, economical energy in affording a decent life to the people of the world. What greater contribution could we make than to help insure that a continued supply will be available, both in our lifetime and in that of our children?

REFERENCES

1. F.L. Robson, A.S. Giramonti, G.P. Lewis and G. Gruber, "Technological and Economic Feasibility of Advanced Power Cycles and Methods of Producing Nonpolluting Fuels for Utility Power Stations," report prepared for the National Air Pollution Control Administration, U.S. Dept. of Health, Education and Welfare, December, 1970.

2. G.H. Higgins, U.S.A.E.C. Report UCRL-51217, Rev. 1 (1972).

3. "Fusion Power, an Assessment of Ultimate Potential," U.S.A.E.C. Report WASH-1239, February, 1973.

4. Cf. C. Marchetti, eurospectra 10:4 (1971) 117-129; C.P. Gregory, D.Y.C. Ng, and G.M. Long, Electrochemistry of Cleaner Environments, J. O'M. Bockris, ed., Plenum Press, New York (1972) 226-279.

5. W.J. Luckel, L.G. Eklund, S.H. Law, "Fuel Cells for Dispersed Power Generation," I.E.E.E. Transactions of Power Apparatus and Systems, Vol. PAS 92, No. 1, Jan-Feb. (1973) 230-235.

PROBLEM AREAS IN THE FIRST-WALL MATERIALS OF A CONTROLLED THERMONUCLEAR REACTOR

A.C. Damask*

Queens College, City University of New York

Flushing, New York

There are at present several design schemes for controlled thermonuclear reactors (CTR) which involve magnetic containment of a D-T reaction. In these designs thermal energy will be extracted through a heat-transfer wall to a coolant which is coupled to a steam power conversion system. The coolant will be liquid lithium or lithium salts to permit the continuous generation of tritium.

The heat-transfer wall, called the first wall, must have certain thermal design characteristics. It must be thin enough to permit rapid heat transfer, it must be strong to serve as a vacuum chamber, it must be resistant to corrosion by liquid lithium and it must possess these characteristics at an operating temperature of about 800°C, the temperature required for proper thermodynamic efficiency of the heat generating system. The first wall will also be subjected to a rather severe radiation environment which will severely limit the choice of materials. The D-T reaction will produce large fluxes of 14.1 MeV neutrons, 3.5 MeV α-particles, and bremsstrahlung peaking at about 0.5Å. The anticipated effects of this environment with respect to such phenomena as corrosion, transmutation, hydrogen and helium embrittlement, swelling, sputtering, defect accumulation and ductility changes are reviewed. There are as yet no data on the effects of radiation of these particular energies and fluxes. Existing radiation damage data is used to anticipate problem areas and to call attention to experiments which must be done before reasonable performance of a material may be predicted.

---

*Consultant at Brookhaven National Laboratory, Upton, New York.

INTRODUCTION

The use of controlled thermonuclear fusion as a source of power is extremely attractive from both the economic standpoint and the environment standpoint. Economically, the fuel cost approaches zero and, environmentally, there is no radioactive ash. These two advantages, as is often the case, are theoretical while the practical achievement is frought with difficulties. The economics require a competitive plant cost while environmental considerations require a well-controlled radiation level in the container. Both of these conditions are dependent on the materials which are used. At this state of knowledge we are forced to assume that controlled fusion will be possible and that materials exist which can contain it. It is the purpose of this paper to describe the anticipated environment for the materials and on the basis of our present knowledge try to predict which materials would be the best initial choice and the types of experiments which are required to test them.

Reactions and Reactors

Of the many known nuclear fusion reactions only a few seem to have any potential use for power reactors. Since the positive charges of nuclei mutually repel each other only those with the largest fusion cross sections for a given energy are currently considered and of these the deuterium-tritium interaction $D + T \rightarrow He^4$ (3.52 MeV) + n(14.06 MeV) is the most favorable. Since there are an equal number of electrons in a hot D-T gas or plasma the collisions of these electrons with the ions will produce bremsstrahlung or X-rays. In the reactors under consideration this will peak in the range of 0.5-1Å.

There are minimum necessary conditions for a fusion reactor which yields a net power production. These, developed by J.D. Lawson,[1] show that there exists a minimum confinement time τ for a reaction hydrogenic gas of ion density n. The product $(n\tau)$ is a significant parameter and is a function of the efficiency of recovery of the energy release and the temperature. The various designs of plasma confinement are the results of attempts to achieve the Lawson criterion.

From fusion power considerations an ideal mean-particle energy is about 10^4 eV, which is the order of 10^8 °K. Clearly containment must be by magnetic means. However, at some point beyond the magnetic container, there must be a conventional type of material wall which will serve as both a vacuum jacket and a heat-transfer wall.

The present concept of this wall, called the "first-wall" is that it be made of metal of about 0.5 cm thickness, for proper thermal conductivity.[2] Since the production of tritium at some other location for use in the reactor is extremely expensive, a desirable design feature is the incorporation of a continuous source of tritium. This is planned through the use of a blanket of lithium or lithium salts in contact with the outside of the first wall. Tritium will then be produced by the neutrons from the D-T reaction via the reactions:

$$n + Li^6 \rightarrow T + He^4$$

and

$$n + Li^7 \rightarrow T + He^4 + n$$

The liquid lithium blanket could then be circulated through radiation shields into coils in a conventional steam power generator system while at the same time the tritium gas is being extracted to be fed into the reactor.

From the standpoint of economically competitive construction and operating costs it is desirable to have the inlet temperatures, i.e. the liquid lithium, at as high a temperature as possible. There will also have to be insulators for electrical leads into the reactor and one of the reactor designs, a pulsed type, requires that the reactor side of the first wall be insulated.

The First Wall Environment. We can now summarize the basic requirements of the first wall:

1. It must be strong enough at high temperature, about 800°C, for a 0.5 cm wall to resist buckling under stress of high vacuum.

2. It must not have much residual radioactivity after shutdown.

3. There must not be sufficient transmutation to change its characteristics appreciably.

4. It must resist corrosion by molten lithium.

5. It must not undergo hydrogen embrittlement (caused by direct injection from the plasma or by diffusion of tritium from the lithium).

6. It must not undergo helium embrittlement.

7. It must withstand surface corrosion effects of the α-particle and the bremsstrahlung.

8. It must withstand an accumulated neutron irradiation flux of about 10^{23} nvt without serious dimensional change or loss of mechanical strength or ductility. This latter requirement means

that each atom is expected to undergo 10-100 displacements during the operating life of the wall.

Preliminary Choice of Materials. The requirement of high thermal conductivity for good transfer restricts the material of the first wall to a metal or an alloy. The further requirement, that it be strong at 800°C, restricts the choice further to refractory metals such as vanadium, zirconium, niobium, molybdenium or their alloys.[3] It is conceivable that if none of these can be made to work satisfactorily stainless steel could be used but with an operating temperature of only 600°C. This would greatly increase the cost per kilowatt but at least in some future time when fossil fuels become prohibitively expensive it is a possibility.

An initial evaluation of the choices indicates that Mo, V and Nb or their alloys such as TZM (titanium, zirconium, molybdenum), V-Ti, V-Ti-Cr or Nb-Zr have the best chance of meeting the general requirements.[3] Of these Mo and Nb retain good strength up to 1000°C while V loses strength above 800°C. Molybdenum has a disadvantage in that nonbrittle welds cannot be made at the present state of our technology. This embrittlement apparently occurs because of the introduction of impurities during welding. This difficulty is, in principle, not insoluble.

Materials Problems. Transmutation by neutron capture will occur in these metals. For example, there will be a rate of transmutation of niobium to zirconium of about 1% per year. Therefore, even initial selections of candidate metals or alloys will have to have their properties studied with a full range of expected transmutation impurities.

The resistance to corrosion by liquid lithium or lithium salts must be dealt with either by further limiting the choice of material or by appropriate doping. For example, the oxygen solubility of the refractory metal plays a key role. Experiments have shown that niobium, tantalum and vanadium all have high oxygen solubility and that they are deoxidized by exposure to lithium at 600°C or above. In the case of niobium and tantalum this oxygen loss may be accompanied by lithium penetration to form a complex oxide phase along the grain boundaries which can create sites for fracture nucleation.[4,5]

Preliminary studies have shown that the oxygen thresholds for lithium penetration may in some cases be varied by appropriate alloying. If this problem is not solvable it has been proposed that helium gas surround the first wall to prevent contact with the lithium. Such a scheme would also simplify the magnetohydrodynamic design complications which will be encountered in trying to pump a liquid metal coolant across a magnetic field. However, helium gas next to the first wall will introduce metal embrittlement problems discussed below.

Much data has been taken on "swelling" or dimensional changes of irradiated metals and some principles are emerging. Irradiation by heavy particles displaces atoms from their sites. If the temperature is sufficiently high to permit migration of these displaced atoms, called interstitials, and the vacant lattice sites they left behind, called vacancies, then a variety of fates await them, depending on their relative concentration and energy. A vacancy and an interstitial can encounter one another and thereby restore an atom to a lattice site, vacancies and interstitials can encounter species of their own type and form clusters, or some other types of lattice defect can preferentially stabilize some clusters and others will dissolve in their favor to create rather large vacancy clusters called voids, Fig. 1.[6] Interstitials form small clusters or platelets and escape to the surface. At even higher temperatures some voids appear to be better stabilized than others, in some cases even ordered, Fig. 2,[6] and serve as nuclei for even larger voids called bubbles. This results in swelling proportional to dose up to about 100 dpa.[7]

The sequence of these phenomena has been observed in most metals and alloys and the temperature of their occurrences have been related to T_m, the melting point of the metal as follows:

Clusters	$<$	$0.2\ T_m$
Voids	$\sim$	$0.3 - 0.5\ T_m$
Bubbles	$\gtrsim$	$0.5\ T_m$

The planned operation temperature of the CTR is in the range of $0.3 - 0.5\ T_m$ for the refractory metals under consideration. As seen in Fig. 1, for the total flux anticipated, which can cause about 10 - 100 displacements of each atom per year, volume changes of the order of a few percent can be caused by voids in the metals under consideration. There has been some cause for optimism in this area because, first, there seems to be a saturation to the swelling at about 5 - 10% of volume and, second, a few alloys of stainless steel exhibit no swelling whatsoever.[8] Although these are not suitable for a first wall because of a lower operating temperature it is hoped that an understanding of the mechanism will lead to techniques of doping or fabrication to inhibit swelling in other metals.

One of the mechanisms of stabilization of voids is the precipitation of helium. Helium is formed in the metal as a transmutation by-product; it is injected by the α-particle escaping

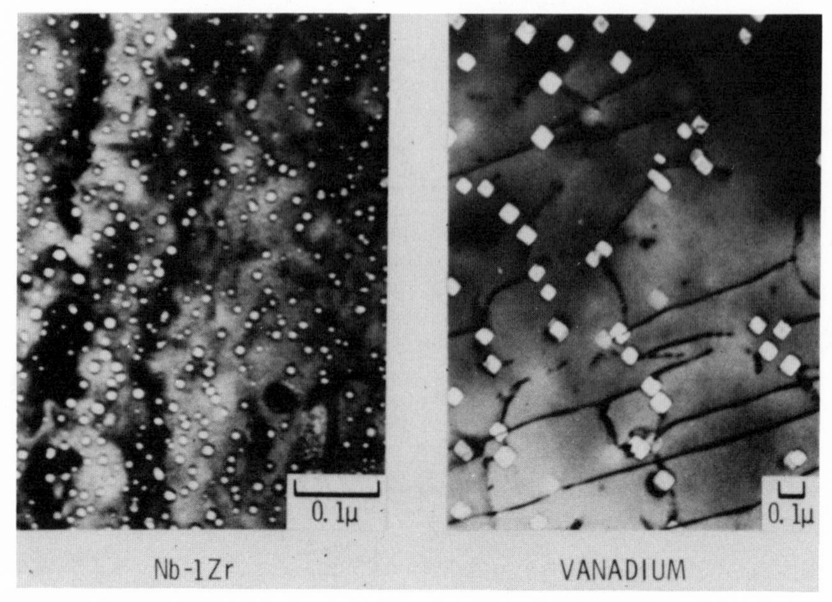

Fig. 1: Voids induced by 7.5 MeV tantalum bombardment to ∿30 dpa at 800°C.

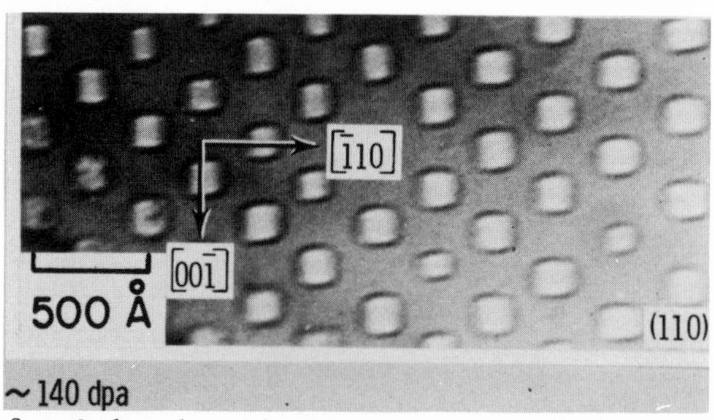

Fig. 2: Ordered voids in niobium bombarded at 800°C with 7.5 MeV Ta^{++} ions.

from the plasma and diffuses in from the lithium where it is a transmutation by-product in the tritium production. Since helium is insoluble in metals it diffuses rapidly until it reaches an internal or external surface. The internal surfaces are vacancy clusters or grain boundaries. It stabilizes a cluster because of the large energy required to go back into the lattice. Thus, helium is a significant mechanism for void development. If, however, the helium precipitates in the grain boundaries it prevents metallic bonding between the atoms which it separates. With further accumulation helium bubbles serve as crack nucleation sites during grain boundary sliding, which operates on a deformation mechanism at high temperature. The result is premature intergranular cracking and failure.[9,10]

The phenomenon of hydrogen embrittlement of metals is well-known although the mechanism is not yet agreed upon. In fact, there may be more than one mechanism for this phenomenon. Hydrogen diffuses rapidly through most pure, fault-free metals. Its solubility varies in different metals and with temperature. The solubility does not follow the usual vant Hoff relation but depends upon the external partial pressure of hydrogen[11] as well as the ability of the metal to form hydrides. Since hydride stability decreases with increasing temperature, it is often found that hydrogen solubility also decreases with increasing temperature. Without bothering to argue the mechanism it can be said that Nb and V are embrittled by hydrogen[12,13] at room temperature,[14] e.g. 400°C. Since hydrogen will be present at all times in and about the first wall, in the form of deuterium and tritium, this is an important phenomenon for further study with respect to the anticipated operating temperature and hydrogen partial pressure.

In some pulsed operation designs it is necessary to have an insulating layer on the inside of the first wall. This requirement presents a new set of problems. An insulating material must be found which remains an insulator at high temperature and high radiation fluxes while not flaking away from the wall during operation and subsequent accumulated damage. Increased conductivity occurs at high temperature from a current of thermally produced ions. In addition, irradiation will knock electrons up to the conduction band and also displace additional ions. Both sintered polycrystalline materials and glasses must be considered. The former have advantages in that most of solid state knowledge is about crystals, while it is known from glass technology how to produce a glass that will wet and therefore stick to a metallic surface. Additional studies must be made on the effects of accumulated radiation damage on both the glass and the glass-metal interface. For example, sputtering experiments have shown that in

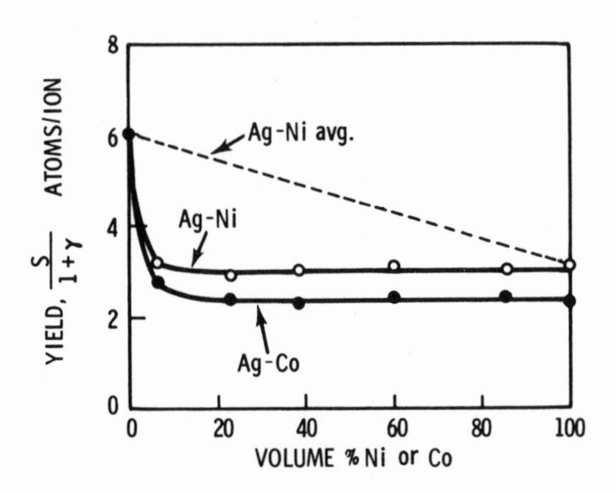

Fig. 3: Sputtering yields for two-phase Ag-Ni and Ag-Co targets sputtering with 1500 eV krypton ions. Computed average yields for two-phase Ag-Ni targets based on sputtering of Ni and Ag is shown by dashed line. Ref. 16.

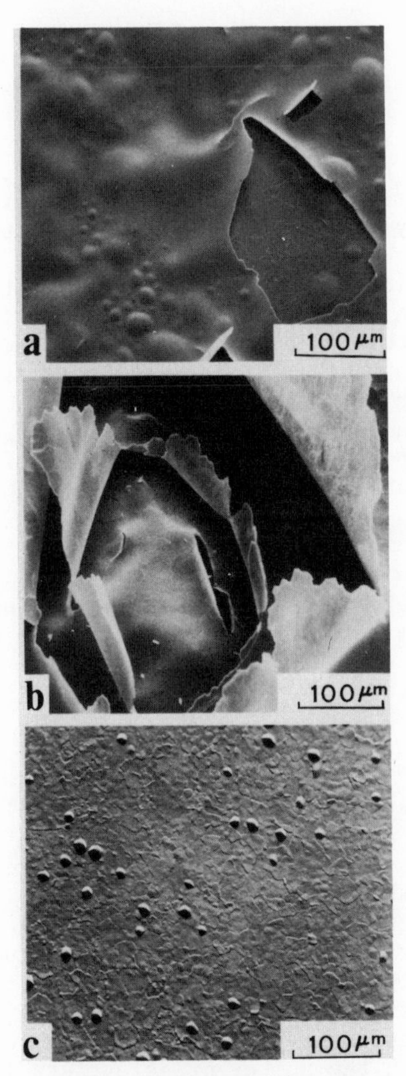

Fig. 4: Annealed polycrystalline vanadium irradiated with 0.5 MeV $^{4}He^{+}$ to a dose of 1.0 C/cm^2. (a) At room temperature; (b) at 600°C; (c) at 900°C.

a component subjected to sputtering, radiation degradation occurs by the loss of lighter atoms, e.g. $Fe_2O_3 \rightarrow FeO \rightarrow Fe$ under proton bombardment.[15]

The sputtering phenomenon just mentioned is an example of the class of surface erosion effects which will be experienced by the first wall with or without an insulating layer. Surface effects may be expected from bombardment by α - particles, neutral atoms formed by charge exchange at the plasma boundary, ions leaking out of confining fields, protons, deuterons as well as energetic photons, e.g. bremsstrahlung, synchrotron radiation and soft X-rays. The erosion caused by such an environment takes place in a variety of ways. For instance, sputtering usually takes place by momentum transfer in a collision between the incident particle and the lattice atoms in such a way that a lattice atom is ejected from the surface. Detailed studies of sputtering from a variety of metals and alloys have shown that it is possible to reduce the amount of sputtering by appropriate alloying[16] (Fig. 3). The surface of the wall can be pitted in the following way. It has been observed that H^+, D^+, He^+, etc., upon penetrating solids, can form gas bubbles in the solid which subsequently migrate to the surface as blisters which then burst and pit the surface, with a corresponding loss of material. Examples of such blisters[17] are seen in Fig. 4. The energetic photons can erode the surface either directly by conversion to high energy electrons, photo-decomposition of surface compounds, photo catalysis of leakage ions with the surface, etc. All of these studies are still in their infancy.

CONCLUSION

The brief summary given above represents only the anticipated problems which must be solved or circumvented before fusion reaction by plasma is possible as an energy source. The list already seems formidable and, when one considers the extremely limited choice of materials which are conceivable candidates the problems may be insurmountable. However, the study programs must proceed with a positive attitude and the recognition that the researchers in plasma physics feel that they won't be ready for a first wall for at least twenty years. In 1953 both our knowledge of the solid state and the number of scientists in the field was small and yet tremendous technological advances were made in the succeeding twenty years. Now, both our knowledge and number of skilled scientists is much greater, so our view of possible technological advances in the next twenty years should be anything but defeatist.

REFERENCES

1. J.D. Lawson, Proc. Phys. Soc. London B70 6 (1957).

2. Fusion Reactor First Wall Materials, U.S. Atomic Energy Commission Wash.-1206 (1972).

3. T.E. Tietz, J.W. Wilson, Behavior and Properties of Refractory Metals, Stanford Univ. Press, Stanford (1965).

4. J.R. Di Stefano, Penetration of Refractory Metals by Alkali Metals. ORNL-TM-2836 (1970).

5. J.R. Di Stefano, A.P. Litman, Corrosion 20 392 (1964).

6. G.L. Kulcinski, in Ref. 2.

7. G.L. Kulcinski, J.L. Brimhall, H.E. Kissinger, in Radiation Induced Voids in Metals, J.W. Corbett and L.C. Ianniello, eds., U.S. Atomic Energy Commission (1972) 449.

8. J.J. Laidler, ibid. 174.

9. D. Kramer, J. Nucl. Met. 25 121 (1968).

10. D. Kramer, Iron Steel Inst. 207 1141 (1969).

11. E. Veleckis, R.K. Edwards, J. Phys. Chem. 73 683 (1969).

12. R.H. Van Fussen, T.E. Scott, O.N. Carlson, J. Less-Common Metals 9 437 (1965).

13. D.H. Sherman, C.V. Owen, T.E. Scott, Trans. TMS-AIME 242 1775 (1968).

14. W.T. Chandler, R.J. Walter, in Refractory Metal Alloys, I. Machlin, R.T. Begley, E.D. Weisert, eds., Plenum, New York (1968).

15. G.K. Wehner, in The Lunar Surface, J.W. Salisbury and P.E. Glaser, eds., Academic Press (1964).

16. S.D. Dahlgren, E.D. McClanahan, J. Appl. Phys. 43 1514 (1972).

17. M. Kaminski, S.K. Das, Appl. Phys. Letters 21 433 (1972); S.K. Das, M. Kaminsky, J. Appl. Phys. 44 2520 (1973); S.K. Das, M. Kaminsky, J. Nucl. Mat. Aug. (1974); M. Kaminsky, S.K. Das, Appl. Phys. Lett. 23 293 (1973).

SIMULATION OF 14-MeV NEUTRON DAMAGE TO POTENTIAL CTR MATERIALS*

R. J. Borg

University of California, Lawrence Livermore Laboratory, Livermore, California, U.S.A.

The reaction $D(T,n)^4He$ occurs either as the primary plasma reaction or as a side reaction in nearly all the controlled thermonuclear reactor (CTR) design schemes. Neutrons produced by this reaction possess an energy of 14.1 MeV and are capable of causing an enormous amount of damage, the extent and character of which is still mostly unknown.

Current CTR design concepts anticipate fluxes of 10^{14} to 10^{15} n/cm^2/sec which, except for EBR-II, are quite outside our present technological experience. Hence, from the standpoint of both energy and intensity CTR neutron sources are expected to create materials problems in an area that is still virtually unexplored. The difficulties in researching these problems are compounded by the obvious lack of an experimental neutron source of comparable strength. This lack forces the materials scientist to employ simulation rather than direct measurement as a means of obtaining the desired information.

In this paper we briefly review the major CTR designs being considered with special reference to the problem of neutron damage. Subsequently, we present a review of the aims and methods of the simulation program in progress at the Lawrence Livermore Laboratory. This program centers about the Rotating Target Apparatus, which is currently the world's most powerful source of 14-MeV neutrons capable of continuous production of approximately 10^{12}n/cm^2/sec. This machine is described in detail as are the preliminary results of our attempt

*Work performed under the auspices of the United States Atomic Energy Commission.

to correlate and compare 14-MeV damage to V, Nb, Mo, and Cu with proton and reactor neutron damage.

Most of the world's conventional sources of energy are either fully developed, approaching a foreseeable state of exhaustion, or else utilization imposes an undesirable environmental hazard or economic handicap. Nevertheless, the world demand for energy is expected to increase everywhere as a result of the increasing population of our planet and as a consequence of the universal aspiration for an ever higher material standard of living. The ubiquitous need for more energy has currently stimulated the proposal of several unconventional technologies as possible remedies for this demand. Some of these proposals are merely further extensions of currently existent systems, e.g., fast breeder reactors, geothermal heat, or in situ coal gasification. Such schemes certainly contain their share of challenges and uncertainty, although much of the initial research and development is already completed. There are, however, other technological approaches which are even newer but which hold equal, if not superior, promise for mitigating the so-called energy crisis. In the latter category, controlled thermonuclear reactor (CTR) are high on the list.

As a source of energy, controlled thermonuclear fusion has many attractive features. It constitutes a substantially reduced radioactive hazard compared to conventional or fast breeder reactors and is generally more acceptable from an ecological viewpoint. The economics for fuel production are extremely favorable, and the potentially usuable fuels are themselves of almost unlimited abundance. However, we have scarcely begun to solve the problems attendant upon a CTR; in fact, it is almost certain that many serious problems remain unidentified. Nevertheless, in recent years we have made considerable progress toward solving the problem of plasma confinement, and the next major research phase in the development of this complex machine is beginning to emerge. This phase, or area of activity, bears the general title of "CTR Technology" as distinct from CTR Plasma Physics. Of the many problems and areas for research which have been specified thus far, we will deal here solely with one resulting from radiation damage incurred by the inner wall of a CTR as a result of bombardment by high-energy neutrons.

It would not be appropriate to review here all the various confinement geometries which are now being investigated at various laboratories around the world, since this has been done elsewhere.[1] The essential features with regard to radiation damage which appear almost regardless of design are shown schematically in Fig. 1. A very hot plasma is confined by means of magnetic fields which are produced by superconducting coils. An exception to this general

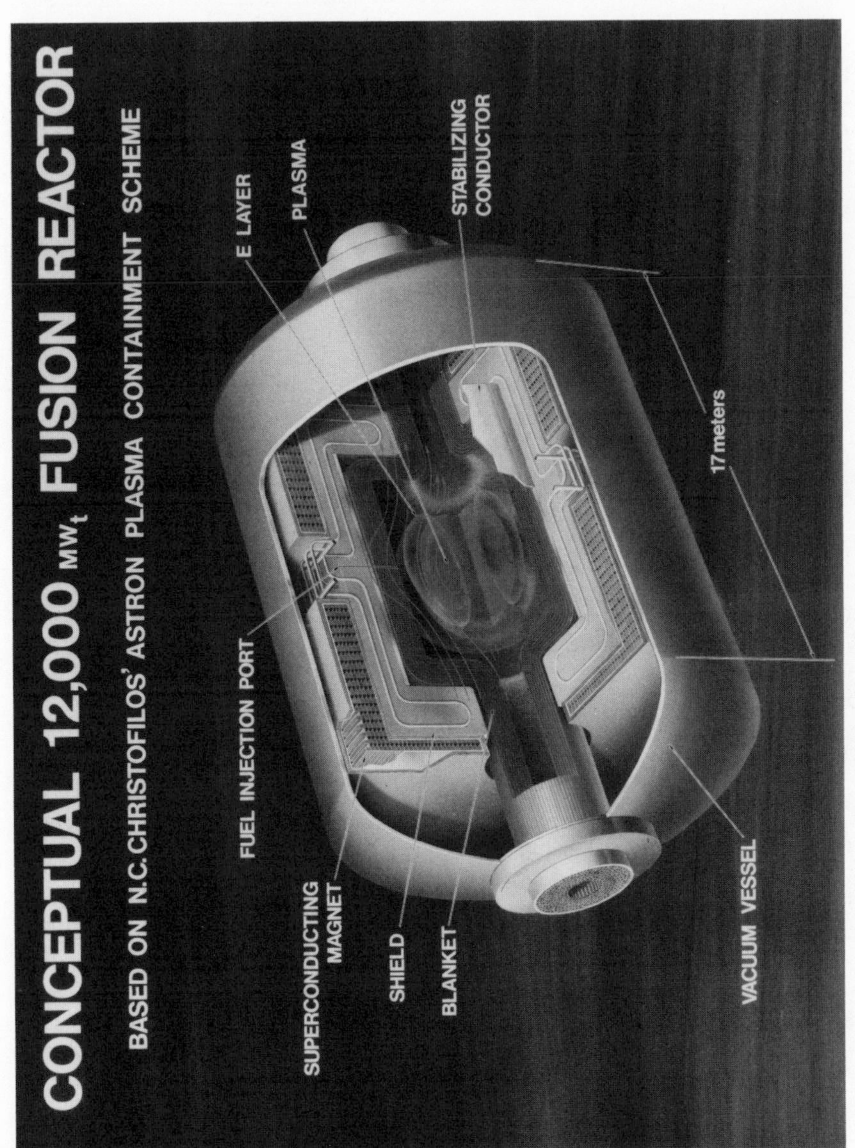

Fig. 1: Schematic showing the essential features of a CTR.

design is laser fusion, which nevertheless will produce most of the same materials problems. Regardless of the confinement geometry, it is generally agreed that the following nuclear reactions are the most popular candidates for a CTR:

	Reaction	ΔE (MeV)	Threshold (keV)
(1a)	$2D \longrightarrow T + H$	3.25	50
(1b)	$2D \longrightarrow {}^{3}He + n^{o}$	4.0	50
(2)	$T + D \longrightarrow {}^{4}He + n^{o}$	17.6	10
(3)	${}^{3}He + D \longrightarrow {}^{4}He + H$	18.3	180

For a variety of reasons, but mainly because of its large cross section (see Fig. 2) low ignition temperature and large energy yield, reaction (2) is now the most highly favored of the lot.

The tritium necessary to sustain the reaction is produced by reacting the neutrons with lithium according to the following reactions:

$${}^{7}Li + n \longrightarrow n + {}^{4}He + T + 4.78\ MeV$$

$${}^{6}Li \quad n \longrightarrow {}^{4}He + T + 2.47\ MeV$$

Forutunately, the naturally occuring relative isotopic abudance, ${}^{6}Li/{}^{7}Li = 0.08$, is very nearly optimum for production of T. ${}^{7}Li$ has its maximum neutron cross section at 2.8 MeV, whereas ${}^{6}Li$ has its maximum at thermal energy. Thus a molten Li moderator serves as the primary heat exchange medium while also serving as a fertile fuel for the production of tritium.

The temperature of the inner wall, which interfaces with the plasma, is conceived to be at a temperature of $\sim 600^{o} \pm 200^{o}C$. Compared to the temperature of the plasma itself, $\sim 10^{8}$°C (one hundred million), this must be regarded as somewhat on the cool side. The reason for sacrificing the obvious increase in Carnot efficiency to be derived from a higher temperature lies not in the physics or engineering complexities of the CTR, but in our inability to use steam economically at significantly higher temperatures. By non-economic, we refer to the shortened lifetimes of generators operating much above 600°C.

The materials which we have elected to investigate are niobium, vanadium and molybdenum metals. Several feasibility studies have indicated that these materials offer the best selection of properties with respect to strength, activation, comptability, etc.

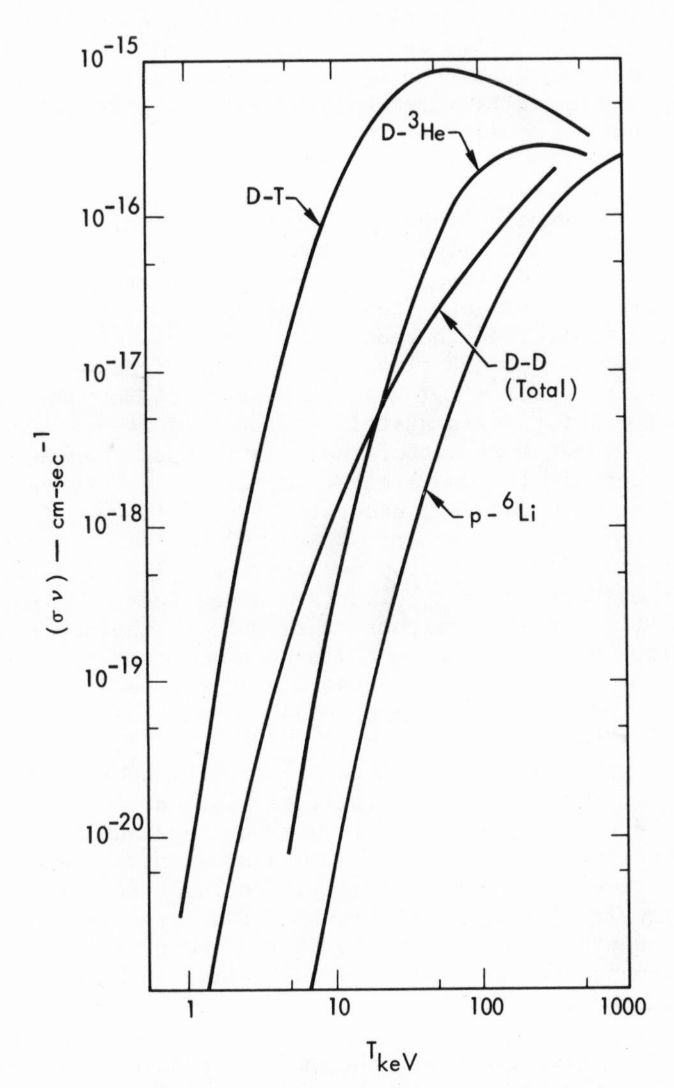

Fig. 2: Plots of fusion reaction rate parameters $<\sigma v>$ for d-t, d-^{3}He, d-d (total), and p-^{6}Li reactions as obtained by averaging over a Maxwellian distribution; 1 keV kinetic temperature = 1.16×10^7 deg Kelvin (taken from Ref. 1, courtesy of R.F. Post).

Unfortunately, from the engineering point of view, these elements and their alloys have received very little attention from those doing basic research in radiation damage. It should also be noted that relatively little scientific research into the nature of neutron-induced radiation damage has been completed thus far. Our ignorance is therefore compounded by our choice of materials as well as our choice of radiation.

Because of this overall lack of information, we have embarked on a fundamental investigation of the interaction of energetic neutrons with metals and alloys as distinct from an engineering study. Another justification for basic research stems from our inability to achieve the very high fluxes of high-energy neutrons anticipated for controlled thermonuclear reactors. Several estimates[2] have been made of CTR fluxes which range from $\sim 3 \times 10^{14}$ to 10^{15} n/cm^2/sec, of which $\sim$ 20% will consist of 14-MeV neutrons. There does not exist today a sustained source of 14-MeV neutrons which can meet even the lowest of these estimates. Consequently, we must rely upon simulation rather than direct measurement, which in turn requires a fundamental understanding of the physics.

Neutron bombardment can displace atoms in a solid matrix by two different mechanisms, viz, elastic and nonelastic. The former is the simplest and most straightforward whereby the neutron collides with the nucleus and the resultant displacement conserves both energy and momentum. The nonelastic process itself can be divided into two mechanisms: one which leaves the nucleus unchanged, e.g., Nb(n')Nb and one which does not, e.g., ^{93}Nb(n,β^-)^{94}Zr. Energetic neutrons can give rise to a complete spectrum of nuclear reactions which include (n, α) and (n,γ) as well as fission. Further atomic displacements are then made possible via the interaction of the various radioactive emanations with the host atoms. The neutron has no interaction with electrons and hence loses its kinetic energy only as the result of nuclear collisions. The cross section for such collisions in nonfissile material is of the order of 1 barn (10^{-24} cm^2), and the mean free path between collisions is of the order of several centimeters.

Even at very high energies, the neutron behaves classically in elastic collisions, and a straightforward consideration of the conservation of momentum and energy permits one to calculate the energy transferred. For a mono-energetic beam of neutrons, of kinetic energy energy E_1, the energy transferred, E_2, is given by

$$E_2 = \frac{4M_1M_2}{(M_1 + M_2)^2} E_1 \sin^2 \frac{\theta}{2} = \frac{4M_2}{(1 + M_2)^2} E_1 \sin^2 \frac{\theta}{2}, \quad (1)$$

where θ is the angle between the direction of the incident neutron and the displacement direction of the recoiling atom. For example, a 14-MeV neutron colliding headon with a niobium atom would transfer as its maximum amount of energy

$$E_2 = (4 \times 93 \times 14)/(94)^2 = 0.6 \text{ MeV} \tag{2}$$

Assuming the scattering to be isotropic, we can estimate an average value of $\overline{E}_2$ = 0.300 MeV. This estimate ignores the effect of non-elastic collisions which will lower somewhat the average energy of the primary knock-on (PKO). Taking $\overline{E}_2$ = 150 keV, we can estimate the total number of knock-ons, n, according to the simple formula of Kitchen and Pease,

$$n = \overline{E}/2E_d = 150/2 \times 0.025 = 3 \times 10^3 \tag{3}$$

This is also somewhat of an overestimation as it neglects that fraction of the energy $\overline{E}_2$ which is dissipated by simple ionization and is thus not available for causing atomic displacements. More sophisticated models include corrections for this as well as the energy dependence of the displacement efficiency. Nevertheless, the elementary theory is sufficient to demonstrate the relationship between the neutron energy and the total number of displacements. What has not been shown is the relationship between the ultimate character of the damage and the energy spectrum of the PKO's. It is intuitively obvious that high-energy PKO's will produce regions of highly concentrated damage in constrast to less energetic events which will create more-or-less isolated Frenkel pairs. The ability of the crystal defects produced by radiation damage to survive the annealing which is concurrent with the irradiation will depend upon their initial geometric disposition; the structural properties of the material will in large measure be governed by the character and concentration of the surviving defects. Hence, we have established the practical need for investigating radiation damage caused by 14-MeV neutrons.

Our present source of 14-MeV neutrons is much weaker than anticipated for a CTR. Nevertheless, when combined with the results of high-flux fission neutrons such as produced by EBR-II (Experimental Breeder Reactor No. II), it will provide an adequate basis for simulation. The source we refer to is the Rotating Target Neutron Source (RTNS) located at the Lawrence Livermore Laboratory (LLL).[3] The RTNS, currently operational, delivers a peak fluence of approximately 3×10^{12} n/sec. Approximately half this amount can be obtained as useful flux on the experimental specimen. These neutrons are produced by impinging a 400-keV, 5-10 mA beam of D^+ upon a titanium tritide target ($\sim TiT_{1.2}$). The target is 3-5 mg/cm^2 thick ($\sim 10\mu$), supported on a copper backing which is cooled by flowing water over the outer surface; the entire assembly is shown in Fig.3.

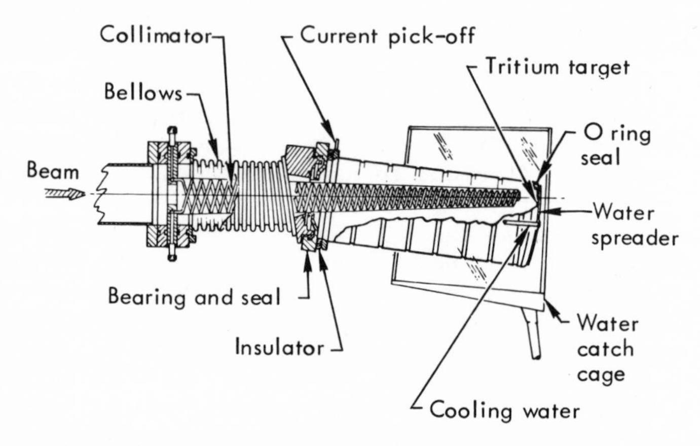

Fig. 3: Schematic of the Rotating Target Neutron Source (taken from Ref. 3, courtesy of R. Booth).

The water-cooled target is rotated at a velocity of 1100 rpm which avoids the problem of local heating while simultaneously maximizing the lifetime of the target. The neutron flux is determined by means of proton recoil telescopes which are in turn calibrated by neutron activation analysis. Reference 3 contains a more detailed description of the RTNS.

Recent calculations by C.M. Logan[4] imply that 16-MeV protons create a displacement cascade in which the energy distribution is very nearly identical to that calculated for 14-MeV neutrons. His results are shown in Fig. 4, and the interested reader is referred to Ref. 4 for complete details of the calculations. The preliminary experiments to test the validity of these results are discussed next. The possibility of using protons to simulate high-energy neutrons is indeed tantalizing, for one can easily obtain particle fluxes comparable to a CTR. The major drawbacks of proton experiments are first, the very low penetration distances ($\sim$ 100 - 200 μ) and second, the experimental difficulty in removing the heat efficiently so as to maintain a reasonable target temperature.

The initial experiments have consisted of comparing the appearance of various metals which have been irradiated with fission neutrons at ambient temperatures with others irradiated by the RTNS. The dosages are comparable, although an accurate knowledge of the total cross section as a function of neutron energy makes it impossible to obtain an accurately normalized comparison. The fission neutrons are obtained in the Livermore Pool-Type Reactor (LPTR) in which the ambient temperature is $\sim 45^{o}C$ and the total flux is $\sim 10^{13}$ n/cm^2/sec. The irradiation of high-purity Cu[5] makes clear the difference between damage caused by 14-MeV neutrons and those of lower energy, as shown in Fig. 5. The visible radiation damage consists of clusters of vacancies or self-interstitials produced by irradiation to $\sim 3 \times 10^{16}$ particles/cm^2 of 16-MeV protons, 14-MeV neutrons and fission neutrons. These clusters lie on the {111} planes and are in effect stacking faults with Burgers vectors a/3<111>. The difference in the concentration and size of the damage clusters between the LPTR specimen and the other two is quite obvious. In the case of both the 14-MeV n^o and p^+ irradiated specimens, the result appears to be qualitatively similar to a specimen having received a much larger dose of fission spectrum neutrons. Figure 6 records a semi-quantitative measurement of the cluster size and density for all three specimens and serves to emphasize the similarity of the 16-MeV p^+ and 14-MeV neutrons.

The most direct method of simulating the PKO spectrum is by self-ion bombardment. We have not actively pursued such a program at LLL; nevertheless, for the sake of completeness we include a photograph (Fig. 7) of a 304 stainless steel foil bombarded by 5-MeV

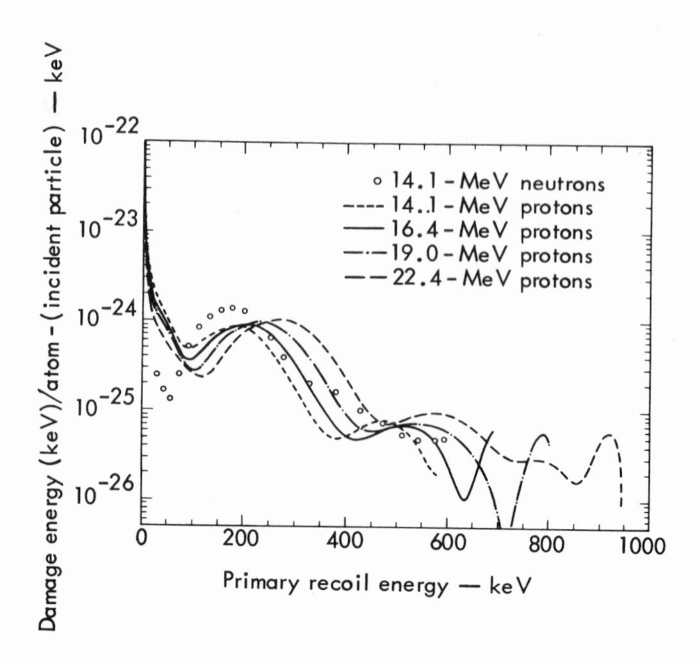

Fig. 4: Calculated total PKO spectra in NB. Energy spectra of recoiling Nb atoms from collisions with 14.1-MeV neutrons and 16.4-MeV protosn (taken from Ref. 4, courtesy of C.M. Logan).

Fig. 5: Transmission electron micrographs of radiation damage structure in high-purity Cu. Kinematic diffraction contrast (taken from Ref. 5, courtesy of J.B. Mitchell). (Figure reduced 31% for reproduction.)

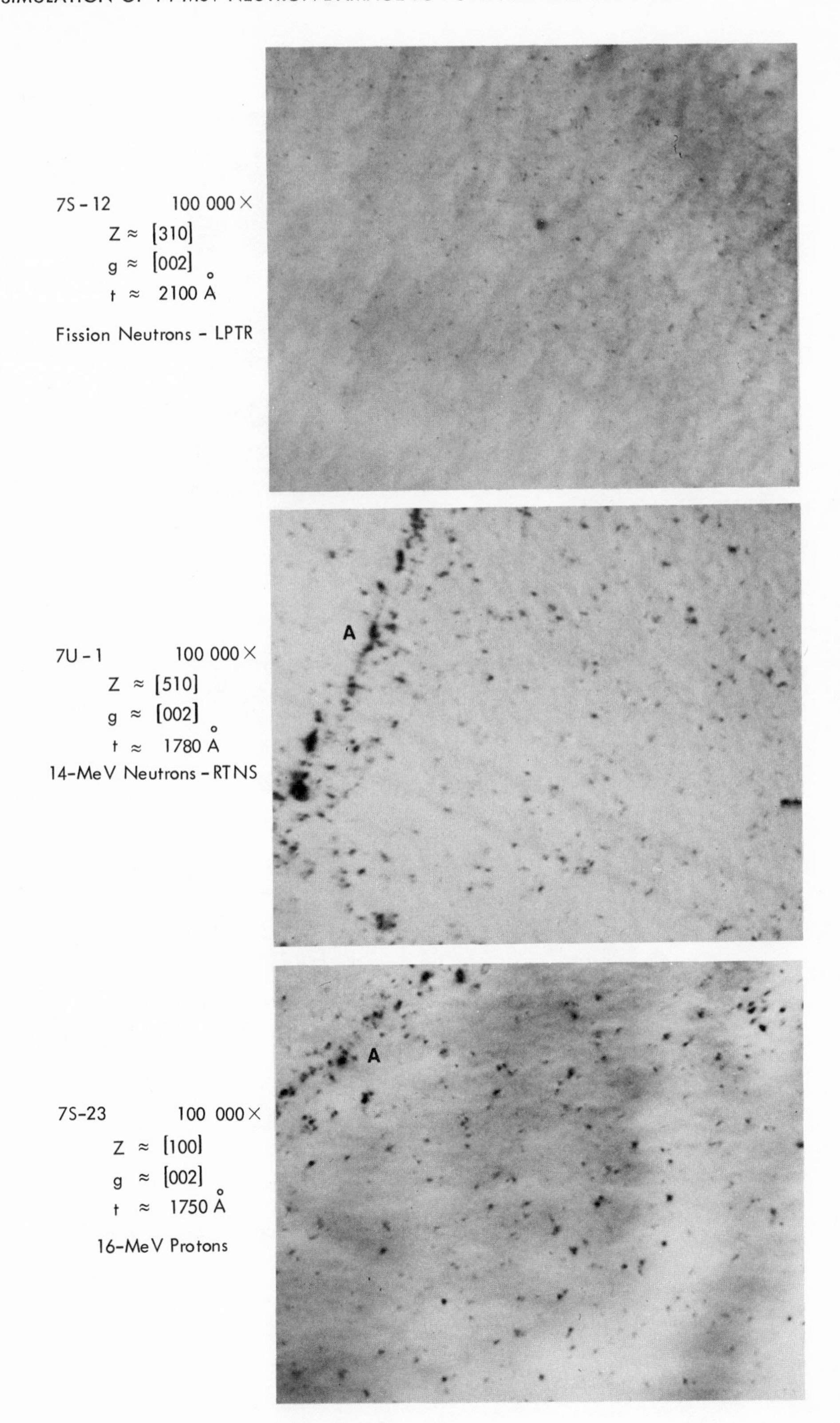
7S - 12 100 000×
Z ≈ [310]
g ≈ [002]
t ≈ 2100 Å
Fission Neutrons - LPTR
A
7U - 1 100 000×
Z ≈ [510]
g ≈ [002]
t ≈ 1780 Å
14-MeV Neutrons - RTNS
A
7S-23 100 000×
Z ≈ [100]
g ≈ [002]
t ≈ 1750 Å
16-MeV Protons

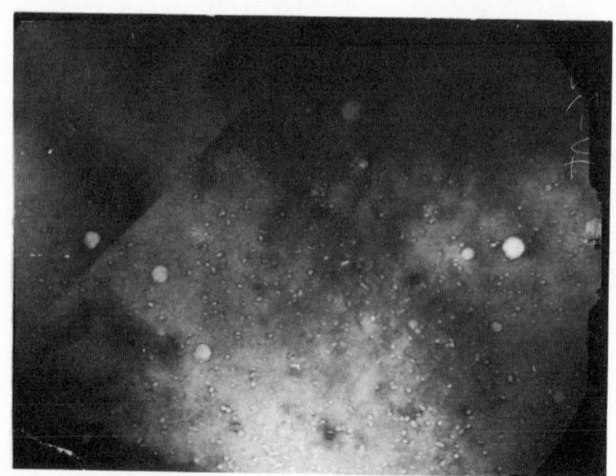

Fig. 6: Voids in 304 stainless steel as a result of bombardment by 5 meV Ni^{++} ions (courtesy of P.B. Mohr).

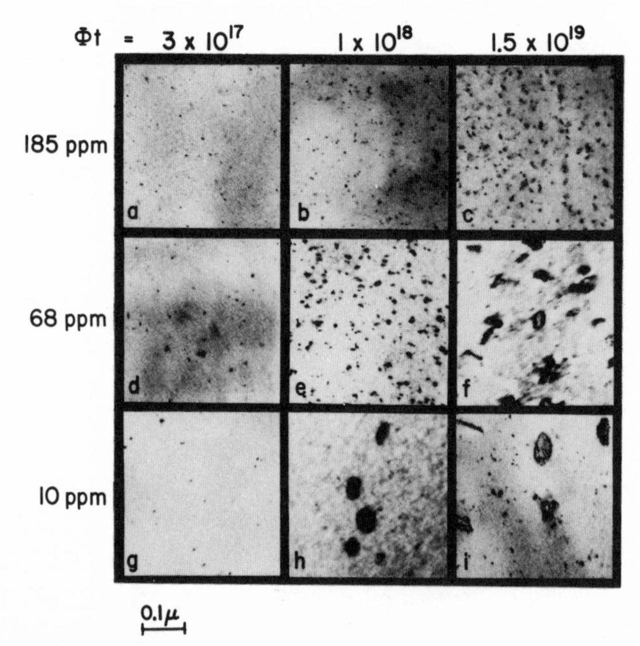

Fig. 7: Radiation damage in Nb doped to various concentrations with O_2 and irradiated with fission spectrum neutrons (courtesy of B.A. Loomis).

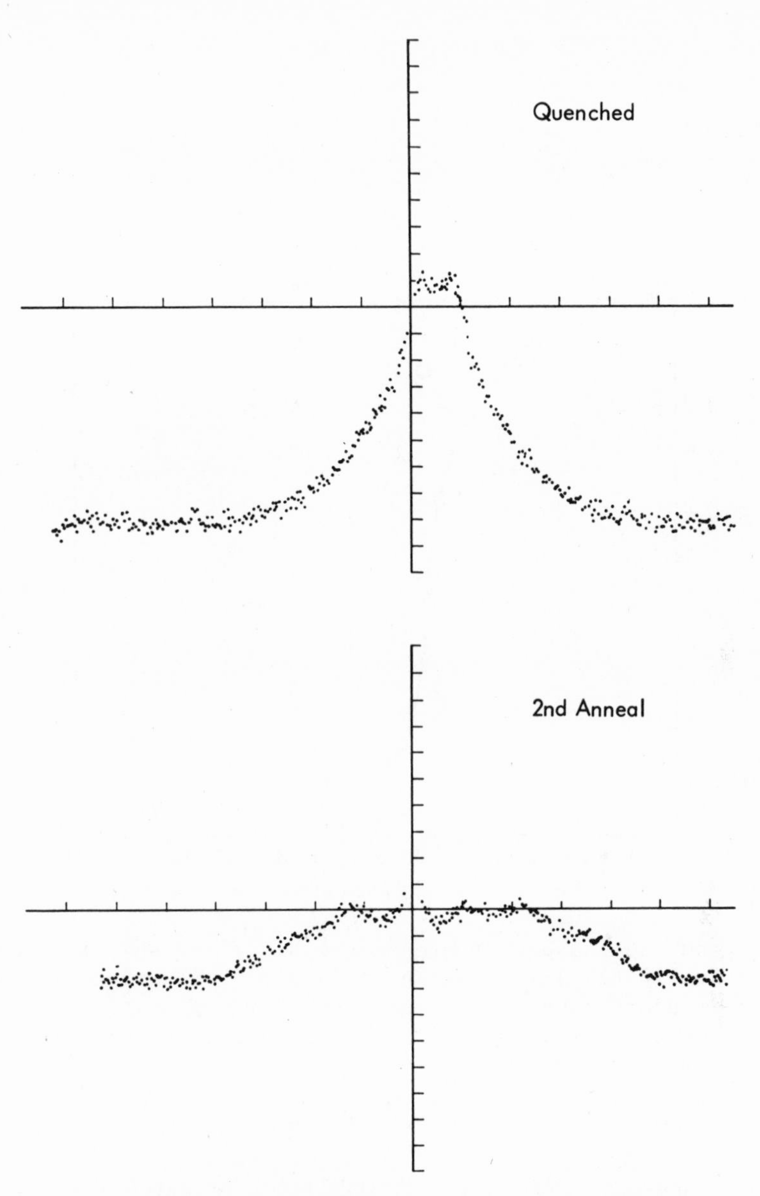

Fig. 8: Comparison of the Mössbauer spectra of a Au-17% Fe alloy which has been (a) quenched from the single-phase region, (b) quenched, then annealled for several days in the two-phase region, and (c) quenched, then irradiated to $\sim 10^{16}$ nvf with reactor spectrum neutrons.

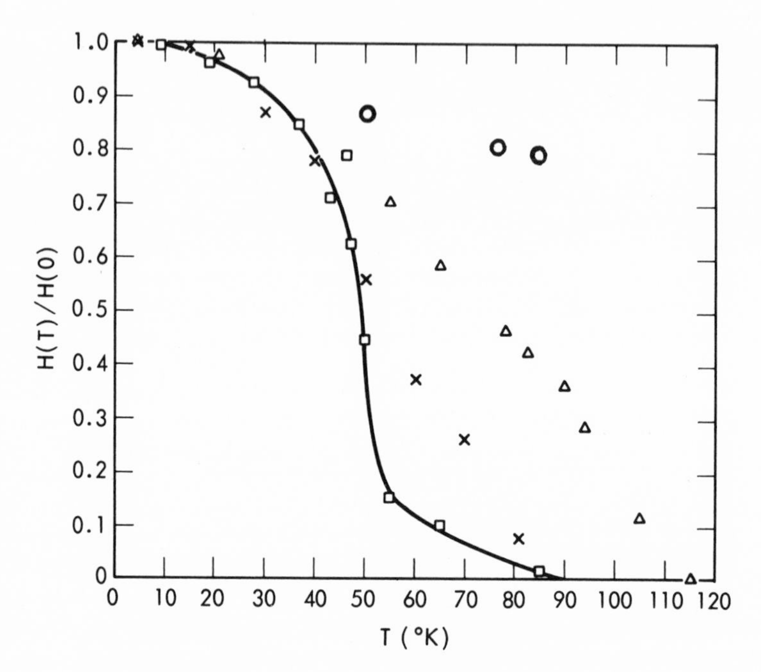

Fig. 9: The magnitude of the magnetic hyperfine splitting of Au-17% Fe alloy shown as a function of various degrees of short range order (see text for explanation). Note the greater values of H at equivalent temperatures for the irradiated specimen.

ions.* One can easily see the high concentration of small voids resulting from the large number of displacements. (Each atom displaced ∿ 50 times.) Similar microstructures have been produced by high neutron fluences, and this phenomenon will be discussed in another paper at this Conference. (See Problem Areas in the First Wall Materials of a Controlled Thermonuclear Reactor by A.C. Damask.)

The profound effect of impurities upon the damage structure can scarcely be overemphasized. For example, the addition to vanadium of 3-20% titanium effectively suppresses the formation of voids.[6,7] Figure 7 shows Nb doped to various levels with oxygen and then irradiated in the Argonne CP-5 Reactor to various total fluence.†

Radiation-enhanced diffusion offers an indirect method for observing the effect of varying dose and neutron energy. It can also yield qualitative information about the stability of phases or compositional, short-range order in radiation fields. We have chosen to observe the change in the magnetic order of Au-Fe alloys by means of the Mossbauer effect as a function of the irradiation variables. A previous investigation[8] established that the temperature at which magnetic order begins and the temperature dependence of the magnetic hyperfine splitting are quite sensitive functions of the short-range compositional order. The order was varied by simply annealing in the two-phase region a single-phase, α-solid solution of Au-17% Fe. The annealing temperature was high enough to allow atomic migration, and the net driving force is in the direction leading to the separation of two phases, one Fe-rich and the other Au-rich. The anneal was interrupted before the actual precipitation of a second phase, but only after considerable clustering into Au-rich and Fe-rich regions had occurred. The increased temperature at which magnetic order starts as a result of such clustering is illustrated in Fig. 8. Note, however, that irradiation by fission neutrons causes an even more pronounced segregation as evidenced by the much better developed hyperfine spectrum. Additional evidence for the efficacy of neutrons in aiding the production of solute clusters is shown in Fig. 9. Here, the measured magnitude of the splitting, H(T), is assumed to be a measure of the departure from random order, and it is obvious that ∿ 10^{16} n/cm^2 at ∿ 45°C induces a greater degree of clustering than annealing for several days in the two-phase region.

---

*We wish to thank Mr. P.B. Mohr of LLL for permission to use these results.
†We wish to thank Dr. B.A. Loomis of ANL for permission to use these results.

REFERENCES

1. R.F. Post, Ann. Rev. Nucl. Sci. 20 509 (1970).

2. J.D. Lee, Lawrence Livermore Laboratory Rept. UCID-15944 (1971).

3. R. Booth, H.H. Barschall, Nucl. Inst. Methods 99 1 (1972).

4. C.M. Logan, Lawrence Livermore Laboratory Rept. UCRL-51224 (1972).

5. J.B. Mitchell, C.M. Logan, C.J. Eicher, Lawrence Livermore Laboratory Rept. UCRL-74518 (1973).

6. F.W. Wiffen, Oak Ridge National Laboratory, USAEC Rept. ORNL-4440, p. 135.

7. J.D. Elen, Septiéme Congrs. Internationale de Microscopie Electronique, Grenoble (1970) Vol. II, 351-352.

8. R.J. Borg, D.Y.F. Lai, C.E. Violet, Phys. Rev. B5 1035 (1972).

DEFORMATION LUMINESCENCE OF IRRADIATED CsI (Tl) CRYSTALS

B.G. Heneish* and L.M. El Nadi

Physics Department, Faculty of Science

Cairo University, Guiza

The luminescence of plastically deformed irradiated CsI (Tl) crystals was studied. Samples of 2x8x5 mm were cleaved from the single crystal and were used without any pretreatment. Aluminium wrapped samples were irradiated using radium, ^{60}Co, ^{137}Cs as gamma sources and Pu-Be as a neutron source. All deformations were carried out in compression along the (100) dimension of the crystal sample by applying successive loads at rates of 0.01 to 0.001 $Kg.cm^{-2}$ sec^{-1}. Detection of the deformation induced luminescence was done using a DuMont 6292 photomultiplier coupled to a D.C. amplifier and luminescence intensities were measured in corresponding millivolts. Pulse rise and decay times of the glow luminescence were recorded using a fast oscilloscope. The absorption spectra of deformed and undeformed irradiated and unirradiated samples were studied using a Bausch & Lomb 500 monochromator.

The following measurements were carried out: (a) radiophoto-luminescence of undeformed samples; (b) load and load rate-luminescence intensity dependance for integral doses of 1R - 10^5R; (c) luminescence intensity dependance on dose rate; (d) dependance of luminescence intensity on the quality and of irradiation sources. A proposed model of the process of deformation luminescence was tried and disscussed. A possibility of application of such studies for dosimetry is presented.

---

* Radiation Protection Dept., Atomic Energy Authority, Cairo.

INTRODUCTION

Certain kinds of solid substances containing small percentages of other elements possess the property of forming a stable luminescent center when exposed to radiation. This center emits fluorescence under excitation either by absorption of ultra-violet or visible rays, or by thermal treatment, or by mechanical deformation. The first type in which the center never vanishes is known as radiophotoluminescence (RPL). Recently, many RPL substances, such as types of phosphate glass or alkali halide crystals doped with Ag, have been applied to dosimetry.[1,2,3] The second type of excitation is thermoluminescence (TL). Some types of crystals and powders doped with manganese or other elements are now used as dosimeters.[4,5,6] Glass thermoluminescent dosimeters are also used on a wide scale.[7,8]

The third type of excitation, deformation induced luminescence (DL), has been investigated for some crystals of alkali halides irradiated to gamma doses higher than 10^5 R. Such methods are not yet applied to dosimetry.

In TL and DL, the centers formed due to irradiation vanish after excitation and several attempts have been made to relate the emission spectra in these processes to specific color center formation.[10,11]

The aim of the present work is to study systematically the RPL and DL processes for CsI crystals activated by 1% thallium and to apply the results to dosimetry.

EXPERIMENTAL

Single crystals of CsI(Tl) provided by the Harshaw Chemical Company were used in this study. The samples, 2 x 8 x 5 mm, were cleaved from the single crystal and were used without any pretreatment.

The gamma ray sources used were 2 mg radium needles, one mC^{60}Co, one mC^{137}Cs, and 4.7 10^4R/h ^{60}Co unit. A Pu-Be source of 0.8 x 10^4 neutrons cm^{-2}sec^{-1} was used as a neutron source.

Samples to be irradiated were wrapped in aluminium foil to exclude daylight effects. The coded samples were irradiated at dose rates of 2 R/h, 5 R/h and 10 R/h, using the above mentioned gamma sources. The integral doses varied from 1-100 R. For high dose irradiation the cobalt unit was used. Upon removal from the irradiation sites, samples were left in their wrapping for one hour to allow the afterglow to decay. Using high dose irradiation, the RPL was tested ten minutes after irradiation. All measurements on the unwrapped samples were made in darkness, in a black-painted, tightly closed compartment.

Detection of RPL. The absorption and emission spectra for unirradiated and irradiated samples were measured using a Bausch & Lomb 500-mm monochromator (range 200-750 mμ). The specimens were placed 5 cm from the entrance slit of the monochromator. The entrance and exit slits used were 5 mm (spectral width 80 A^o). When emission spectra were to be measured two light filters were used, one between the light source of the system (100 watt mercury lamp or hydrogen lamp), and the second after the sample to absorb the rays with λ less than 3900 Å. The light at the output of the monochromator was measured by 6292 DuMont photomultiplier coupled to a DC amplifier. The high voltage applied to the cathode of the phototube was obtained from a Dynatron stabilised power supply. The potential, due to the incident light intensity on the photocathode of the multiplier, was measured by a millivoltmeter.

Detection of DL. The load applying system shown in Fig. 1 was situated in a black painted dark compartment, and consisted simply of two stainless steal jaws, the lower one fixed and the upper one movable in a way suitable for applying successive loads. The samples could be positioned reproducibly at the same distance from the detector by means of a shallow groove at the centre of the lower jaw of the loading system. All deformations were carried out in compression along the long dimension of the sample by applying successive loads at rates of 0.01-0.001 kg mm^{-2}sec^{-1}. During all measurements, sample to detector distance was kept equal to 2 cm. The detector, enclosed in the dark compartment with the sample, was also a magnetically sheilded DuMont 6292 photomultiplier. The DL intensities in millivolts and the glow pulse shape were measured using a fast oscilloscope type 230 Philips. The stability of the system was checked before each measurement by the constant luminescence of a CsI(Tl) crystal mixed with 0.1 uC^{137}Cs source.

All measurements were made of the sample at room temperature.

RESULTS AND DISCUSSION

RPL Measurements. The measured absorption spectra for the unirradiated or the irradiated crystals was found to range from 2300 up to 3150 A^o, with 100% absorption at λ 3000A^o. This is in agreement with absorption peaks at λ 2990, 2690 and 2410 A^o observed by Knoepfel et al.[12]

Excitation of the unirradiated crystal with rays less than 3100 A^o, gave the spectrum in Fig. 2.1. The emission spectrum covers the range 4000-6000 A^o with broad peaks at 4200, 4500, 4900

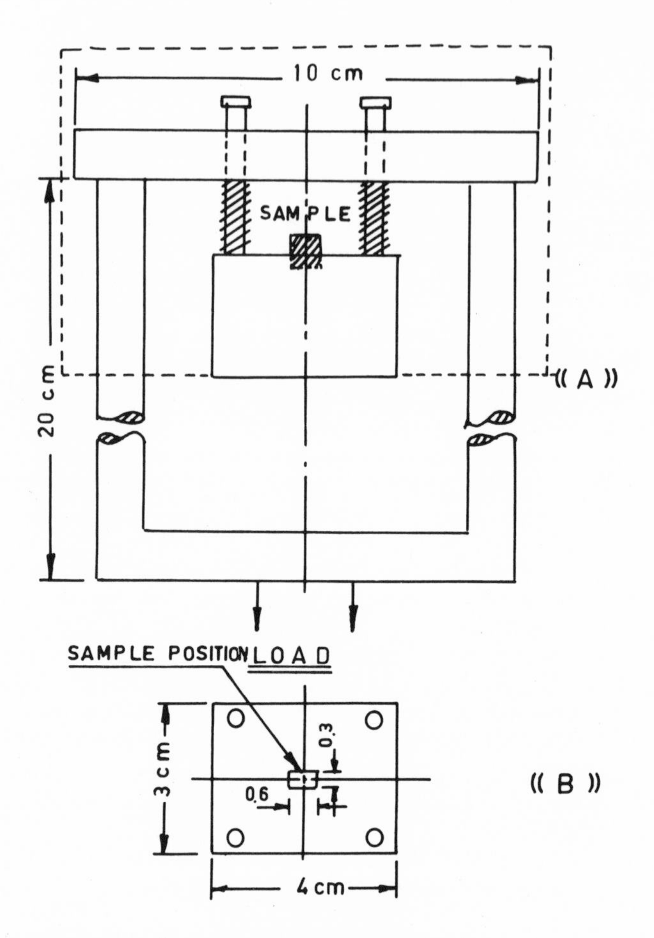

Fig. 1: Schematic diagram of the load applying system, where the sample position is indicated by the arrow. A top view of the fixed jaw is shown at the lower part of the Figure, where the shallow groove in which the sample is reproducibly placed, is indicated at the center. The dashed area indicates the dimensions of the dark compartment.

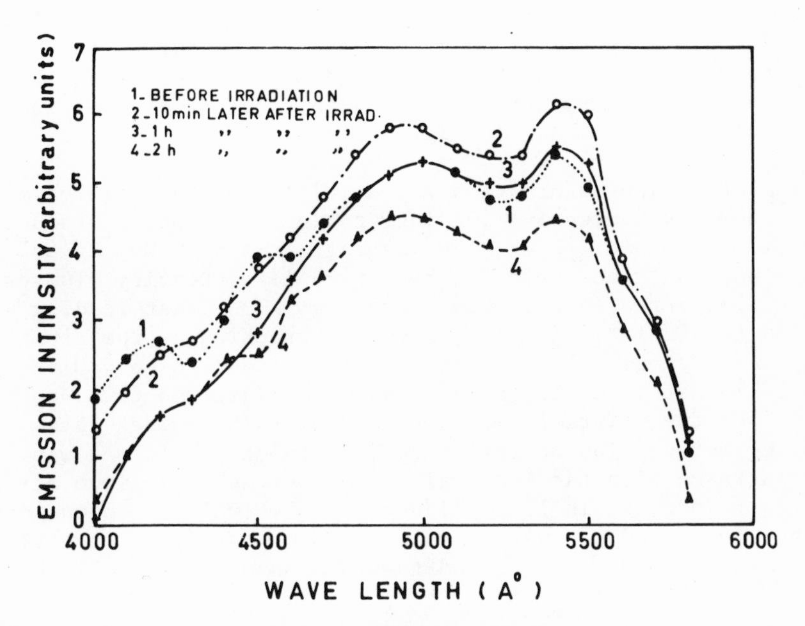

Fig. 2: Comparison of typical emission spectra obtained in RPL measurements under the conditions of (1) non-irradiation, (2) after 10 minutes irradiation to 10^4 R, (3) after one hour irradiation to the same dose, and (4) after two hours irradiation at the same dose.

and 5600 A°. This range is in fair agreement with that measured by Lagu and Thosar[13] for CsI (0.1% Tl).

Crystals irradiated up to 100R at different dose rates showed no changes in the fluorescence intensity nor in the emission spectral distribution. As higher dose 4.7 10^3 was applied by irradiation in the Cobalt for 6 minutes a slight increase in intensity was observed, when the measurement started 10 minutes after the exposure was over. This increase was mainly observed at 4900 and 5600 A° peaks. Then it reached the same intensity as before irradiation after one hour while the peaks at 4200 and 4600 A° completely disappeared. After two hours the emission spectrum became stable but lower than that before irradiation, as shown in Fig. 3.

DL Measurements. The deformation was initiated on the samples as described above by successive loading, taking care not to produce cracking of the sample. When slow loading rates of 0.001 kg mm^{-2} sec^{-1} were applied, the crystal showed very low intensity fluorescence in a flat curve as the load increased. When fast loading was applied the DL intensity increased but cracking of the crystal occurred. An optimum value of load rate was found to be 0.01 Kg $mm^{-2}sec^{-1}$. The DL intensity as a function of load was found to be very small for the unirradiated samples, and triboluminescence could be neglected. The DL intensity as a function of load was found to increase with the load value and to reach a maximum after which further increase in load produced less DL values. When load was further increased a constant value existed until the crystal cracked, a sharp rise occurred and then the sample was completely spoiled.

The effect of dose rate in the range of 2R/h - 5R/h and 10R/h was studied using only radium sources for an integral dose of 100R. No significant change in the DL intensity load was observed, indicating that fading of the formed centers during irradiation was negligible.

The effect of energy dependance of the DL intensity was checked using radiation of different energies from ^{60}Co source (1.17 and 1.33 MeV), ^{137}Cs (0.663 MeV) and radium (range 0.3 - 2.2 MeV average energy 1 MeV), for an integral dose of 100 R at dose rate of 5 R/h. No deviations were observed, which might be attributed to the fact that at such energies the mode of interaction is through the Compton scattering process, which is the main process by which energy is absorbed in the sample. The behaviour in the KeV range was not checked.

When the sample was irradiated with neutrons from the Pu-Be source for approximately 48 hours producing 130 R integral dose, it was found that no DL was observed other than that of the

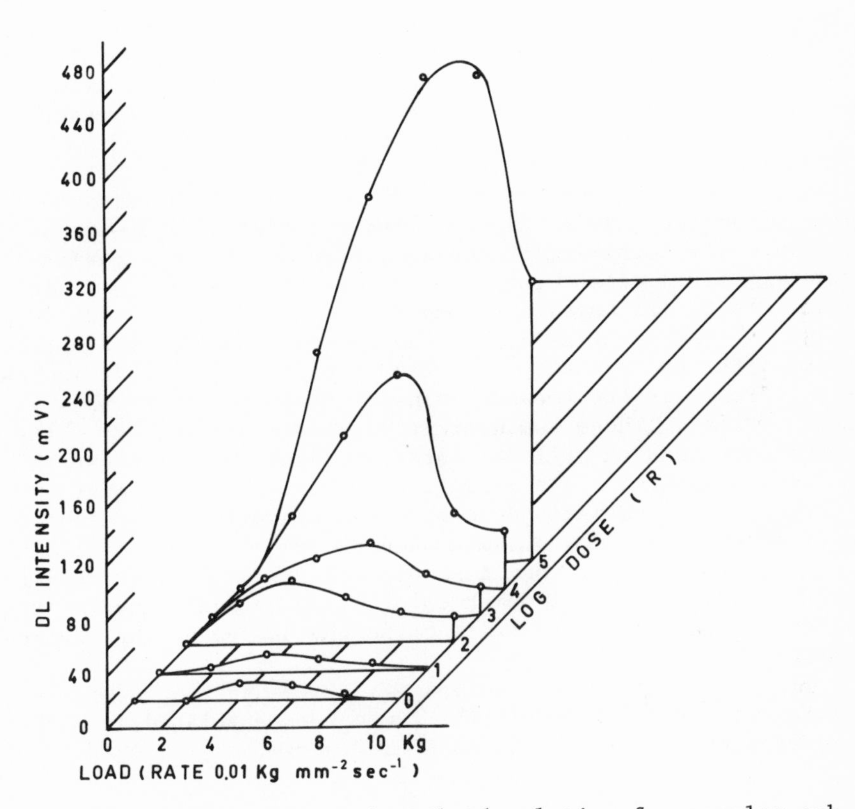

Fig. 3: The DL intensity against load relation for samples subjected to integral doses from 1 to 10^5 R.

background. There were no centers formed during neutron irradiation although one expects neutron absorption by iodine would help to produce iodine vacancies due to the recoil of active iodine.

The effect of increasing the integral dose, while keeping the dose rate constant at a suitable value was checked and the DL intensity-load curves are shown in Fig. 3. As the integral dose increases there is a general increase in the intensity of the DL at all the values of the applied load. It is also observed that the maximum yield occurred at load value of 4 Kg on the sample face of area 10 mm^2; this was true for integral doses 10, 100, and 1000 R respectively. But when the dose increased to 10^4 R the maximum was shifted to 6 Kg, and for 10^5 R the maximum was shifted further to 7.8 Kg.

Calculating the area under the DL intensity-load curve corresponding to different integral dose, the relation in Fig. 4 was obtained for the total luminescence emitted during deformation against the values of the integral dose. There was a linear increase in the DL intensity, the logarithm of the integral dose up to 8.0 10^3 R. Above this value a sudden deflection in the rate of increase of the DL intensity with the dose is clear. This indicates a sharp increase in the formed luminescent centers above this value of integral dose. These luminescent centers could either be the same as those produced at lower integral dose, or they might be attributed to the formation of different sorts of color centers. When the irradiated samples were observed visually one could not see a difference in color between samples irradiated in this range of integral doses, while when a sample was irradiated for an integral dose of 10^{10} R a faint yellowish color could be recognized. This might be due to iodine liberation or to the damage of the chemical bond of the CsI crystal.

In all these measurements the luminescence emitted at each loading step has a pulse duration of 1-2 μ sec.

CONCLUSION

From these results one might conclude that there is no response for RPL at lower doses than 10^4 R. The results of the DL measurements indicate that samples of CsI(Tl) crystals could be used as dosimeters in the range of integral doses from 1 up to 10^3 R. In this range of integral dose the nature of the luminescence center could be attributed to free electrons released in the crystal lattice by the gamma rays' interaction with the sample material and these electrons being trapped, forming the luminescent center.

For integral doses above 8.0 10^3 R the maximum of the DL yield

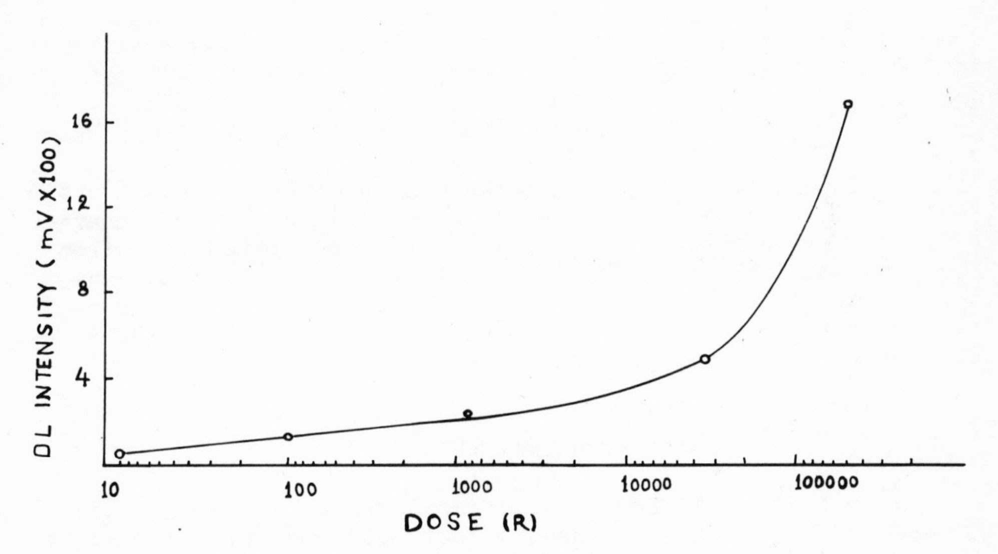

Fig. 4: The total DL intensity as a function of integral dose.

was found to be shifted towards higher load values. It was mentioned in previous work[9,10] that the maximum yield of the DL occurs at load values corresponding to the position of the plastic deformation range on the stress-strain curve of the sample. Considering the same idea valid for the present case, one might conclude that the samples harden when irradiation exceeds 10^4 R.

The emission spectra studied by Knoepfel et al.[12], Morgenstern[14] and Lagu and Thosar[13] suggest that the 4100-4200 Å emission band in pure CsI crystals due to iodine vacancies is partially suppressed and shifted to longer wave length (4750 Å or even more) at high concentrations of Tl. The function of Tl in the crystal is not only to act as an emission center, but also to attract several iodine ions creating iodine lattice vacancies. The emission bands at λ = 5800 to 6000 Å are due to Tl ions and surrounding lattice ions acting increasingly as emitting centers, as Tl concentration increases.

From this information and the observed increase in the emission peaks at 4700 and 5600 Å wavelength in the emission spectrum (Fig. 2-2) as well as the disappearance of the peaks at λ 4200 and 4400 Å, one may propose that at doses higher than 10^4 R the energy dissipated in the sample is sufficient to release iodine ions from their bonding. These ions can be easily attracted to the Tl ions, thus increasing both iodine vacancies and the ions surrounding the Tl ions. This might explain the sudden rise in the intensity of the luminescent centers at such doses.

ACKNOWLEDGMENTS

The authors would like to thank Prof. M. El Nadi, Prof. K.A. Mahmoud and Dr. T. El Khalafawy for their encouragement and assistance.

REFERENCES

1. W.A. Weyl, J.H. Schulman, R.J. Ginther, L.W. Evans, J. Electrochem. Soc. 95 70 (1949).

2. R. Yokota, S. Nakajima, H. Osawa, Health Phys. 11 241 (1965).

3. R. Yokota, Y. Mutoh, K. Fukuda, Toshiba Rev. 52 (1970).

4. R.J. Ginther, R.D. Kirk, J. Electrochem. Soc. 104 365 (1957).

5. J.H. Schulman, F.H. Attix, E.J. West, R.J. Ginther, Rev. Sci. Inst. 31 1263 (1960).

6. D.E. Jones, K. Petrock, D. Denham, TID-4500, UC-41 (1966) 33.

7. W.R. Hendee, Health Phys. 13 (1967).

8. F. Morgan Cox, The Harshaw Chemical Co. Review (1972).

9. A.S. Krugloff, I.A. Elshanshory, J. Phys. Soc. Japan 21 2147 (1966).

10. C.T. Butler, Phys. Rev. 141 750 (1966).

11. I.A. Elshanshory, A.S. Krugloff, UAR J. Phys. 1 1 (1970).

12. H. Knoepfel, E. Loepfe, P. Stoll, Helv. Phys. Acta 30 521 (1957).

13. R.S. Lagu, B.V. Thosar, Proc. Ind. Acad. Sci. 53A 219 (1961).

14. Z.L. Morgenstern, Opt. Spectrosc. 7 146 (1959).

A BOND-TYPE CRITERION TO PREDICT BOMBARDMENT-INDUCED STRUCTURE CHANGES IN NON-METALLIC SOLIDS*

H.M. Naguib, Dept. of Metallurgy, Faculty of Engineering
Cairo University, Egypt

Roger Kelly, Institute for Materials Research
McMaster University, Hamilton, Ontario, Canada

Radiation damage, as found in non-metallic solids bombarded to high doses, often takes the form of structure or stoichiometry changes. For example, many instances are known in which a bombarded specimen is rendered either crystalline or amorphous independent of the initial structure. A previously formulated thermal-spoke model is reviewed and shown to remain a viable means of understanding the crystal structure. In addition, a bond-type criterion, which has apparently not been treated in detail elsewhere, is discussed. It is surprisingly successful, as is shown by the fact that 21 substances which amorphize on ion impact have ionicities of ≤ 0.47 while 23 substances which are stable in crystalline form have values of ≥ 0.60. A further 7 substances have intermediate ionicities (whence varying behavior) while 4 constitute exceptions (SnO_2, $W_{18}O_{49}$, NiO, Ta_2O_5).

The other main class of high-dose damage concerns stoichiometry changes, as when V_2O_5 becomes V_2O_3 or when TiO_2 becomes Ti_2O_3. The mechanism for this effect is still unsettled.

One of the main applications of the present work lies in predicting which minerals ought to contain fission-fragment tracks and therefore be useful for dating. It is argued that minerals consisting wholly or in part of the following may be interesting: moS_2, WO_3, Nb_2O_5, TiO_2, and Ta_2O_5.

*Work supported by the National Research Council of Canada and the Geological Survey of Canada. Communications should be addressed to Prof. Kelly.

INTRODUCTION

The use of ion beams to inject a chosen atomic species into the surface layers of a target is now a well-established method. Until a few years ago the method was limited to such areas as sputter deposition, ion-bombardment cleaning, and isotope-source preparation. However, more recently, it has been realized that the solid-state aspects of ion bombardment are important because of the many properties that are sensitive to small concentrations of impurities. Thus, amongst these properties are the electrical behavior of semi-conductors and transition-metal oxides, as well as the optical behavior of some electroluminescent materials. It is conceivable that chemical, mechanical, magnetic, and superconducting properties may also be modified in a useful way. Severe problems are inherent to the use of ion beams, however, owing to radiation-damage effects. Thus in low-dose work ($\lesssim 1\text{x}10^{13}$ ions/cm^2) one has to contend with the formation of point defects, defect clusters, and amorphous zones, while in high-dose work, as we will be concerned with here, structure and stoichiometry changes become important.

An apparently unrelated body of experimental work concerns the dating of minerals by counting fission-fragment tracks. The tracks, formed by the spontaneous fissioning of U^{238} impurities with a half-life of $1\text{x}10^{16}$ years, are in some respects equivalent to a bombardment-induced structure change. They are difficult to detect, however, and an understanding of the origin of structure changes would thus be of direct importance in pin-pointing which minerals ought to contain tracks.

In previous papers in this series (1-6), we have presented a variety of examples of how non-metallic solids respond to high doses of heavy ions. Briefly, the following groups have come to be recognized:

(a) The first group includes 18 materials which either retain their amorphous structure to a high dose or else undergo a crystalline-to-amorphous transition. Such transformations are the most commonly observed result of high-dose bombardment as far as non-metallic solids are concerned. They are, at the same time, very rare with metals, having been reported apparently only three times (U_6Fe (7), $Pd_{80}Si_{20}$ (8), and U_3Si (9)). A number of experimental techniques have been used to study amorphization, the most powerful of which is probably electron diffraction as exemplified in Fig. 1 (due to Naguib[10]).

(b) The second group includes 29 materials which either retain their crystallinity to high doses or else (as shown explicitly in 9 of the 29 cases) undergo an amorphous-to-crystalline transition. These nine are the following: HfO_2, MoO_2, NbO, SnO_2, Ti_2O_3, UO_2,

V_2O_3, $W_{18}O_{49}$, and ZrO_2. The relevant experimental techniques here include both electron diffraction and transmission microscopy, an example of the latter (due to Wilkes[11]) being shown in Fig. 2.

(c) The third group of 8 materials consists of certain transition-metal oxides which, when bombarded to high doses, have shown first an amorphization (or retention of amorphousness) but then a crystallization to a lower oxide. In effect, a bombardment-induced stoichiometry change has occurred. Despite recent efforts[12,13] the understanding of the mechanism is still in a state of flux. It is still unsettled whether a stoichiometry change is due to internal precipitation, to sputtering, or to vaporization. Transmission microscopy is again an appropriate technique, an example of which (due to Murti[14]) is shown in Fig. 3.

We will be concerned in what follows with trying to understand the origin of crystalline-to-amorphous transitions, as well as the converse, as found in specimens bombarded to high doses. Low-dose damage is already fairly well understood, while bombardment-induced stoichiometry changes, though strictly speaking a high-dose effect, are best left unelaborated until more experimental data are available and general trends become discernible. The 55 examples to be considered are summarized in Table 1.

CRITERIA FOR STRUCTURE CHANGES

Thermal-spike model. In an attempt to formulate a criterion for the response of non-metallic substances to ion impact, we have previously presented a model based on the concept of the thermal spike.[3] The essence of the argument was that an impact can, provided the ion is sufficiently heavy, be regarded as creating a small disordered region equivalent to a liquid. This region cools rapidly (10^{-11} to 10^{-12} sec) and crystallization begins when the temperature falls below the melting point. Assuming a $t^{-3/2}$ cooling law, as for a point source, the problem can be set up as follows. Crystallinity will be preserved provided the following is true:

$$(1/\lambda) \int_{t1}^{t2} D_c dt > \lambda \qquad (1)$$

where λ is the mean atomic spacing, t_1 is the time when $T = T_m$ (the melting point), and t_2 is any subsequent time when T is somewhat greater than T_∞ (the macroscopic target temperature). D_c is the diffusion coefficient for crystallization, having the form

$$D_c \approx (0.3) \exp [-(H_c)/(R\{T_\infty + At^{-3/2}\})]$$

ΔH_c is the activation enthalpy for crystallization, which can be shown by using an equality related to that in eq. (1) to be given

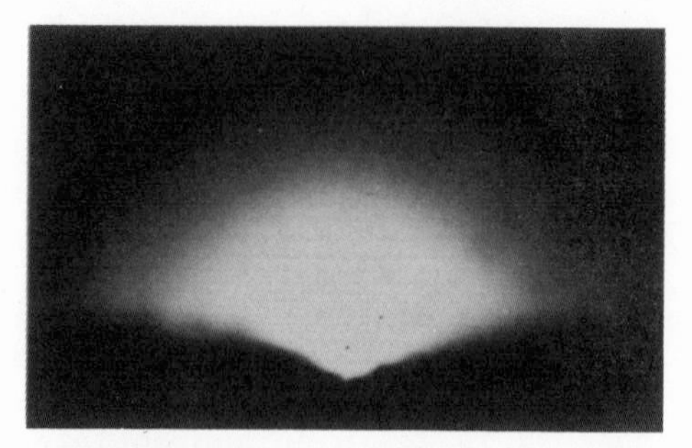

Fig. 1: Left: Reflection electron diffraction pattern of sintered Bi_2O_3 before bombardment. The specimen is evidently well crystallized.

Right: As before, but after bombardment with 40-keV Kr to a dose of $1x10^{16}$ ions/cm^2. The specimen is now amorphous. Due to Naguib (10).

Fig. 2: Transmission electron micrograph of amorphous HfO_2 which has been bombarded through an overlying grid with 20-keV Kr to a dose of $2x10^{14}$ ions/cm^2. Monoclinic crystallites have developed in the exposed region at the left. Mag. 34,000. Due to Wilkes (11).

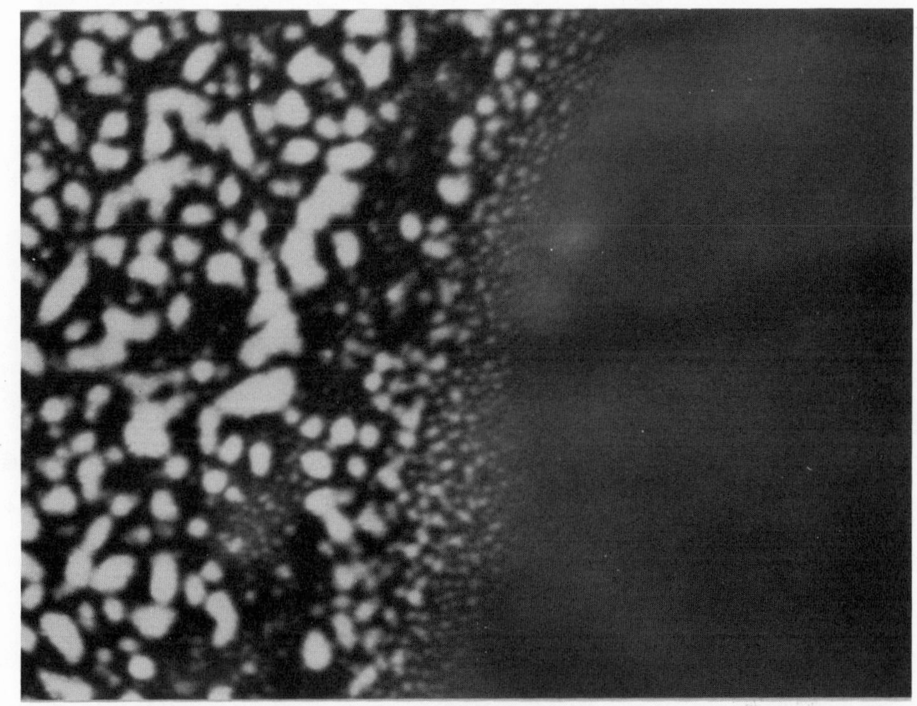

Fig. 3: Transmission electron micrograph of amorphous Nb_2O_5 which has been bombarded through an overlying grid with 35-keV oxygen to a dose of $1x10^{17}$ ions/cm^2. Crystalline NbO has developed in the exposed region at the left. Mag. 132,000. Due to Murti (14).

Table I

Response of Non-Metallic Solids to Ion Impact

Initial state	Structure for medium doses ($<1\times10^{16}$ ions/cm^2)	Structure for high doeses ($>1\times10^{17}$ ions/cm^2)	Examples
Cryst. or amorphous	Cryst.	Cryst.	BaF_2 (15), BeO (16), CaF_2(15), CaO (15), CdS (17), CdTe (18), Cu_2O (19), Fe_3O_4 (20), HfO_2 (11), KBr (15), KCl (15), KI (15), LiF (21), MgO (15), MoO_2 (22), NaCl (15), NbO (14), NiO (15), SnO_2 (23), ThO_2 (15), TiO (24), Ti_2O_3 (13), UO_2 (19), VO (14), V_2O_3 (22), $W_{18}O_{49}$ (24), ZnS (25), ZnSe (25), ZrO_2 (26)
Cryst. or amorphous	Amorphous	Amorphous	Al_2O_3 (15), As_2Se_3 (27[a]), Bi_2O_3 (10), C (2), Cr_2O_3 (21), Fe_2O_3 (21[c]), GaAs (28,29), GaP (30), GaSb(30A,31,32[a]), Ge (1), GeO_2 (33), InAs (31), InSb (30A,31,32[a]), MoS_2 (27[a]), Si (1), SiC (34), SiO_2 (35), TeO_2 (10)
Cryst. or amorphous	Amorphous	Lower oxide	CuO (19[b]), MoO_3 (10,22), Nb_2O_5 (14), Ta_2O_5 (14), TiO_2 (13), U_3O_8 (19), V_2O_5 (10,22), WO_3(6,24)

[a]The evidence for amorphization is here somewhat weak in that low-temperature bombardment was used.

[b]The existence of an amorphous intermediary was not established directly though may be assumed by analogy with other substances which change stoichiometry.

[c]Oxygen loss is here in principle possible.

by

$$\Delta H_c \approx (69 \pm 5) T_c \quad \text{cal/mole},$$

T_c being the crystallization temperature for a 1-10 min time scale. $\underline{A}$ is the term governing the decrease of temperature and is given in general by $\underline{A} = 2(E/c\ \rho\kappa^{3/2})$ or more specifically (for the ion energy E in keV) by

$$A = 2(E_{keV}/c\ \rho\kappa^{3/2})(8.59\times10^{19}).$$

The quantities c, ρ, and κ are the heat capacity, density, and thermal diffusivity.

Eq. (1) is readily integrated by parts to yield an asymptotic series, the leading term of which can be expressed as $T_c/T_m < 0.23 \pm 0.07$. This enables one to formulate the following rules:

$$\left.\begin{array}{l}\text{a substance remains crystalline if } T_c/T_m < 0.23 \pm 0.07 \\ \text{a substance amorphizes if } T_c/T_m > 0.23 \pm 0.07\end{array}\right\} \quad (2)$$

Applied to some 19 different non-metallic substances, the criterion of eq. (2) accounted for the response to ion impact correctly except for two cases where either T_c or T_m was unknown (C, U_3O_8). The criterion also served to predict bombardment-induced amorphization for Bi_2O_3, Nb_2O_5, SiC, and TeO_2, and would have made a similar prediction for WO_3 had a better value of T_c been available. Stability was anticipated for BeO. Recent studies (Table 1) have shown each of these materials to behave as expected.

More extensive examples of the application of the thermal-spike model are given in columns 3 and 4 of Tables II, III, and IV. Provided the critical value of T_c/T_m is raised from 0.23 to 0.27, the theory works in all 33 cases for which information is available.

Crystal-structure criterion. A crystal-structure criterion has been proposed by a number of investigators[15,46] to explain response to ion impact. It was generally formulated in the form that anisotropic substances tend to amorphize under ion impact whereas cubic substances tend to remain crystalline. Provided the term "anisotropic" is replaced by "non-cubic", the criterion can be readily tested. As shown in column 2 of Tables II, III and IV, the results are only in part satisfactory, there being 16 exceptions amongst the 55 cases for which information is available.

As far as a rationalization for a crystal-structure criterion is concerned, the usual one, namely that non-cubic substances become unstable due to non-isotropic strains, is unsatisfactory by

Table II

Substances which are predicted to amorphize under ion impact

Substance	Crystal	T_c (°C)	T_c/T_m (°K/°K)	Ionicity (Phillips)	Ionicity (Phillips)	Observed structure[e]
C	cubic	650 (2)	...	0	0	Am
Si	cubic	720 (1)	0.59	0	0	Am
Ge	cubic	470 (1)	0.61	0	0	Am
As_2Se_3	monoclinic	$\gtrsim$25[b] (27)	$\gtrsim$0.47	$\gtrsim$0.04		Am
SiC	hex./cubic	~1600 (34)	~0.61	0.12	0.18	Am
MoS_2	hexagonal	~300 (27)	~0.39	0.18		Am
GaSb	cubic	350 (30A)	0.64		0.26	Am
MoO_3	orthorh.	250 (10)	0.49	0.30		Am
TeO_2	tetragonal	260 (10)	0.53	$\gtrsim$0.30		Am
GaAs	cubic	300[a]	0.38		0.31	Am
InSb	cubic	350 (30A)	0.78		0.32	Am
WO_3	monoclinic	400 (6)	0.39	0.34		Am
InAs	cubic	~ 500 (31)	~0.64		0.36	Am
GaP	cubic	500 (30)	0.44		0.37	Am
Bi_2O_3	monoclinic	260 (10)	0.49	0.43		Am
CuO	monoclinic	...	....	0.43		Am(?)
GeO_2	tetragonal	627 (36)	0.65	0.43		Am
SnO_2	tetragonal	<200[b] (37)	<0.21	0.43		Cr
$W_{18}O_{49}$	monoclinic	...	....	~0.43		Cr
Fe_2O_3	hexagonal	535 (1)	<0.43[c]	0.47		Am
NiO	~cubic	...	...	0.47		Cr
SiO_2	hexagonal	675[b] (35)	~0.56[d]	0.47		Am
U_3O_8	orthorh.	...	...	0.47		Am
V_2O_5	orthorh.	330 (10)	0.63	0.47		Am

[a]This temperature, for a high bombardment dose (28), is to be preferred to 150°C as obtained with a medium dose (29).

[b]This temperature applies to the amorphous phase on a crystalline substrate. The value is considerably higher when the amorphous phase is isolated.

[c]Upper limit since T_m is known only as a lower limit.

[d]Calculated using the value 1400-1450°C for T_m (38).

[e]"Am" = remains or becomes amorphous under ion impact; "Cr" = remains or becomes crystalline under ion impact.

Table III

Substances predicted to have varying structural sensitivity

Substance	Crystal structure	T_c (^oC)	T_c/T_m ($^oK/^oK$)	Ionicity (Pauling)	Ionicity (Phillips)	Observed structure
Fe_3O_4	cubic	...	...	0.49		Cr
Cu_2O	cubic	...	...	0.51		Cr
MoO_2	monoclinic	...	...	0.55		Cr
Al_2O_3	hexagonal	730 (1)	0.43	0.59		Am
Cr_2O_3	hexagonal	445 (1)	0.27	0.59		Am
Nb_2O_5	monoclinic	585 (14)	0.49	0.59		Am
TiO_2	tetragonal	480 (1)	0.35	0.59		Am

Table IV

Substances which are predicted to crystallize under ion impact

Substance	Crystal structure	T_c (°C)	T_c/T_m (°K/°K)	Ionicity (Pauling)	Ionicity (Phillips)	Observed structure
BeO	hexagonal	300 (39)	0.21	0.59	0.60	Cr
NbO	cubic	...	...	$>$0.59		Cr
ZnS	hex./cubic	...	...		0.62	Cr
Ta_2O_5	tetragonal	740(14)	0.47	0.63		Am
ThO_2	cubic	$\sim$425(40)	$\sim$0.20	0.63		Cr
Ti_2O_3	hexagonal	...	...	0.63		Cr
V_2O_3	monoclinic	...	...	0.63		Cr
ZrO_2	cubic[a]	530(26)	0.27	0.63		Cr
VO	cubic	...	...	$>$0.63		Cr
HfO_2	cubic[a]	$\sim$ 500(41)	$\sim$0.25	0.67		Cr
TiO	cubic	...	...	0.67		Cr
UO_2	cubic	$\sim$ 400(42)	$\sim$ 0.22	0.67		Cr
CdTe	cubic	...	...		0.68	Cr
ZnSe	cubic	...	...		0.68	Cr
CdS	hex./cubic	...	...		0.69	Cr
MgO	cubic	$\sim$200(43)	$<$0.16[b]	0.73	0.84	Cr
CaF_2	cubic	$<$25 (44)	$<$0.18	0.89		Cr
BaF_2	cubic	$<$25 (44)	$<$0.19	0.91		Cr
CaO	cubic	$\sim$350(45)	$\sim$0.22	0.79	0.91	Cr
alkali halides[c]	cubic	...	...		high	Cr

[a]This is the crystal structure assumed under ion impact.

[b]Upper limit since T_m is known only as a lower limit.

[c]LiF, KBr, KCl, KI, NaCl.

virtue of being difficult to test. More satisfactory is to note that the distinction between cubic and non-cubic structure is to some extent equivalent to one of bond type. This will be considered next and we will see that when the emphasis is placed on bond type rather than structure, there are only 4 exceptions amongst the 55 cases enumerated in Tables II to IV.

Bond-type Criterion. Let us now consider to what extent one can understand response to ion impact on the basis of bond type. For example, we would propose that a straightforward argument can be developed based on the extent to which a substance tolerates substitutional disorder such as that represented schematically as follows:

```
   |      |      |      |
 - Mg  -  O  -  Mg  -  O  -
   |      |      |      |
 - O   -  O  -  Mg  -  Mg -
   |      |      |      |
 - Mg  -  O  -  Mg  -  O  -
   |      |      |      |
 - O   -  Mg -  O   -  Mg -
   |      |      |      |
```

The point is that substitutional disorder is related to amorphization in that both will to a certain extent cause like atoms to approach too near each other. One infers that covalent solids (which can readily accommodate substitutional disorder) will be more easily amorphized than ionic solids (which cannot tolerate substitutional disorder even to a small extent). Alternative arguments can perhaps be made based on such features as (a) the directed nature of covalent bonds, (b) the fact that covalent bonds involve short-range interactions, and (c) the difference in the lattice strain due to a defect depending on whether the atoms neighboring the defect are ionic or not (cf. Ref. 30).

A bond-type approach has, in fact, already been considered in early work by geologists[47,48] as an empirical guide to whether minerals will be found in a metamict (amorphous) state. Crawford and Wittels[49] have also pointed out the importance of bond type on the extent of radiation sensitivity. They studied the effect of fast neutrons on certain minerals and proposed that the order of increasing radiation sensitivity of the relevant bonds, namely Be-O, Al-O, Zr-O, Si-O, was also the order of increasing covalency. This is not objectionable provided more recent information is used, in which case the order both of increasing sensitivity and of increasing covalency is Zr-O, Be-O, Al-O, Si-O. (The additional details, unknown to Crawford and Wittels[49], are that ZrO_2 is crystallized by ion impact[26] whereas Al_2O_3 is amorphized[15]; Al_2O_3 is, however, not as readily amorphized as SiO_2[35].)

In order to put the correlation between the response to ion impact and bond type on a quantitative basis, use will be made where possible of Pauling's equation[50]. This has the form

$$\text{amount of ionic character} = 1 - \exp\{-1/4(X_A - X_B)^2\}, \quad (3)$$

where X_A and X_B are the electronegativities of atoms A and B as tabulated, for example by Batsanov[51]. The results are given in column 5 of Tables II, III and IV.

According to eq. (3) the bond type is determined only by the difference $(X_A - X_B)$. The III-V compounds have $(X_A - X_B)$ lying between 0.3 and 0.5, while with the II-VI compounds the values are 0.6 to 1.0. Both groups should apparently be classed as covalent, thence subject to amorphization, inspite of experimental evidence that only the III-V's are readily amorphized (Table I). The problem of the ionicity of compound semiconductors has been discussed by Phillips[52], who showed that a more satisfactory result follows from the equation:

$$\text{amount of ionic character} = C^2/(E_h{}^2 + C^2) = C^2/E_g{}^2 \quad (4)$$

where E_g is the average energy gap, E_h is the covalent contribution to E_g, and C is the ionic contribution. The results applying eq. (4) to the compound semi-conductors as well as to certain other compounds are given in column 6 of Tables II, III, and IV.

A total of 21 substances which amorphize on ion impact are seen to have ionicities of ≤ 0.47, 23 substances which are stable in crystalline form have values of ≥ 0.60, while 7 substances with varying behavior have intermediate ionicities. A bond-type criterion can evidently be taken as a reliable guide for the response of a substance to ion impact, there being only 4 exceptions amongst the 55 substances considered. The exceptions are SnO_2, $W_{18}O_{49}$, NiO, and Ta_2O_5.

CONCLUSIONS

We have shown that the response of a number of substances to ion impact can be understood in terms both of a thermal-spike model and of a bond-type criterion. The thermal-spike approach was introduced by the authors in an earlier publications[3] though is here extended to cover a greater number of examples; the bond-type approach has apparently not been treated in detail elsewhere. A crystal-structure approach is shown to be less satisfactory and is in any case largely equivalent to considering the bond type -- i.e. non-cubic structures often have significant covalency.

An unanswered question is whether the thermal-spike and bond-type approaches are equivalent. We are not yet in a position to explore this question in detail though would note the following. The thermal-spike model attributes structural evolution to a competition between the disordering tendency of an ion impact and the self-healing tendency of a crystal lattice. The mathematical result is that amorphization should occur whenever T_c/T_m is large (>0.27). But a large value of T_c/T_m is characteristic of a covalent substance and in this respect the two approaches can perhaps be regarded as different aspects of a single, more inclusive model.

We will in addition have to refrain from exploring to any great extent the question as to whether the present work serves as a guide as to which minerals offer a reasonable prospect of containing fission-fragment tracks useful for dating purposes. The reason is that the occurrence of tracks is dependent not only on the possibility of their formation as related to amorphizability but also (a) on the possibility of formation as related to uranium content, (b) on the possibility of retention over a geologically meaningful time scale (favored by T_c being high), and (c) on the possibility of detection (favored by the ill-defined concept of whether the amorphous form has a high chemical reactivity). Nevertheless, even the limited information contained in Tables II to IV is sufficient to suggest that minerals consisting in whole _or in part_ of the following may be useful for dating purposes:

> MoS_2 (molybdenite); WO_3 (mixed with FeO in wolframite); Nb_2O_5 (mixed with FeO in niobite); TiO_2 (rutile); Ta_2O_5 (mixed with FeO in tantalite).

ACKNOWLEDGMENT

Work supported by grants from the National Research Council of Canada and the Geological Survey of Canada.

REFERENCES

1. C. Jech, R. Kelley, J. Phys. Chem. Sol. 30 465 (1969).
2. C. Jech, R. Kelley, J. Phys. Chem. Sol. 31 41 (1970).
3. R. Kelly, H.M. Naguib, Proc. Intern. Conf. on Atomic Collision Phenomena in Solids, North-Holland, Amsterdam (1970) 172.
4. R. Kelly, Rad. Effects 2 281 (1970).
5. R. Kelly, N.Q. Lam, Rad. Effects 10 247 (1971).

6. N.Q. Lam, R. Kelly, Can. J. Phys. 50 1887 (1972).

7. J. Bloch, J. Nucl. Mat. 6 203 (1962).

8. D. Lesueur, C. R. Acad. Sci. Paris 266 1038 (1968).

9. D.G. Walker, J. Nucl. Mat. 37 48 (1970).

10. H.M. Naguib, R. Kelly (to be published).

11. J.C. Wilkes, unpublished work at McMaster University, 1972.

12. R. Kelly, N.Q. Lam, Rad. Effects (in press).

13. T. Parker, R. Kelly, Proc. 3rd Intern. Conf. on Ion Implantation in Semiconductors and Other Materials (in press).

14. D.K. Murti, unpublished work at McMaster University, 1973.

15. Hj. Matzke, J.L. Whitton, Can. J. Phys. 44 995 (1966).

16. W. Yeniscavich, M.L. Bleiberg, Westinghouse Report WAPD-BT-20 (1960).

17. P.K. Govind, F.J. Fraikor, J. Appl. Phys. 42 2476 (1971).

18. G. Langguth, E. Lang, O. Meyer, Proc. 2nd Intern. Conf. on Ion Implantation in Semiconductors, Springer-Verlag, Berlin (1971) 228.

19. E. Kuczma, unpublished work at McMaster University, 1973.

20. J.J. Trillat, J. Chim. Phys. 53 570 (1956).

21. Hj. Matzke, Can. J. Phys. 46 621 (1968).

22. H.M. Naguib, R. Kelly, J. Phys. Chem. Sol. 33 1751 (1972).

23. E. Giani, unpublished work at McMaster University, 1973.

24. T. Parker, unpublished work at McMaster University, 1973.

25. J.A. Olley, P.M. Williams, A.D. Yoffe, Proc. Europ. Conf. on Ion Implantation, P. Peregrinus, Stevenage (UK) (1970) 148.

26. H.M. Naguib, R. Kelly, J. Nucl. Mat. 35 293 (1970).

27. J.A. Olley, A.D. Yoffe, Proc. 2nd Intern. Conf. on Ion Implantation in Semiconductors, Springer-Verlag, Berlin (1971) 248.

28. D.J. Mazey, R.S. Nelson, Rad. Effects 1 229 (1969).

29. R. Bicknell, P.L.F. Hemment, E.C. Bell, J.E. Tansey, Phys. Stat. Sol. a12 K9 (1972).

30. H.M. Naguib, W.A. Grant, G. Carter, Rad. Effects (in press).

30A. D. Haneman, Phys. Rev. 121 1093 (1961).

31. A.U. MacRae, G.W. Gobeli, J. Appl. Phys. 35 1629 (1964).

32. B.L. Crowder, J.E. Smith, M.H. Brodsky, M.I. Nathan, Proc. 2nd Intern. Conf. on Ion Implantation in Semiconductors, Springer-Verlag, Berlin (1971) 255.

33. H.M. Naguib, unpublished work at McMaster University, 1971.

34. O.J. Marsh, H.L. Dunlap, Rad. Effects 6 301 (1970).

35. Hj. Matzke, Phys. Stat. Sol. 18 285 (1966).

36. J. Drowart, F. Degrève, G. Verhaegen, R. Colin, Trans. Faraday Soc. 61 1072 (1965).

37. E. Giani, R. Kelly, submitted to J. Electrochem. Soc., 1973.

38. J.D. Mackenzie, J. Am. Ceram. Soc. 43 615 (1960).

39. I.S. Kerr, Acta Cryst. 9 879 (1956).

40. V. Balek, J. Mat. Sci. 4 919 (1969).

41. I.A. El-Shanshoury, V.A. Rudenko, I.A. Ibrahim, J. Am. Ceram. Soc. 53 264 (1970).

42. O. Gautsch, C. Mustacchi, A. Schürenkämper, H. Wahl, Euratom Report EUR 3267.e (1967).

43. L. de Brouckère, J. Inst. Metals 71 131 (1945).

44. W.D. Kingery, J.M. Woulbroun, R.L.Coble, in Advances in Glass Technology, Part 2, Plenum Press, New York (1963) 24.

45. Hj. Matzke, unpublished results at Euratom, Karlsruhe.

46. R.M. Berman, M.L. Bleiberg, W. Yeniscavich, J. Nucl. Mat. 2 129 (1960).

47. H.D. Holland, J.L. Kulp, Science 111 312 (1950).

48. A. Pabst, Am. Mineralogist 37 137 (1952).

49. J.H. Crawford, M.C. Wittels, Proc. 2nd Intern. Conf. on Peaceful Uses of Atomic Energy, United Nations, Geneva (1958) 1679.

50. L. Pauling, The Nature of the Chemical Bond, Cornell University Press, Ithaca (1960) 98.

51. S.S. Batsanov, Russ. Chem. Rev. 37 332 (1968).

52. J.C. Phillips, Rev. Modern Phys. 42 317 (1970).

SURFACE AREA AND PORE STRUCTURE OF LOW-POROSITY CEMENT PASTES

R.Sh. Mikhail, S.A. Abo-El-Enein, and G.A. Oweimreen

Department of Chemistry, Faculty of Science, Ain Shams University and Department of Materials Engineering and Physical Sciences, The American University in Cairo, Egypt

Low-porosity portland cement pastes were prepared by two methods: by reducing the water/cement ratio to 0.2, and by pre-hydration compression at 5, 10, 30 and 70 atmospheres, followed by hydration in liquid water at 25°C for about two years. The thickness of the produced tobermorite sheets was found to differ in the two methods of preparation, and for the pressed samples to depend on the pressure used in the pre-hydration compression. Changes in the degrees of hydration, specific surface areas, and pore structures are reflected in the compressive strength values measured for the various pastes, and possible interpretations are offered.

INTRODUCTION

It is generally recognized that the strength of fully compacted cement-based products is determined largely by the cement content of the cement-water paste present in the raw mix and by the degree of hydration of the cement in the cured materials. The cement content of the paste can be increased by alternative techniques, each of which offers potential advantages. One technique, developed largely by Brunauer and co-workers,[1,2] might be called a physico-chemical method, and is based on reducing the water/cement ratio to a minimum while maintaining workability by chemical additives and maintaining the rate of hydration by increasing the surface area of the unhydrated clinker.

The other technique is based, by analogy with other compared powder systems, on the idea that the cement content of the paste can be increased by applying mechanical pressure, and accordingly might be classified as a mechanical method. Proposals to utilize this principle industrially to form high-strength materials have appeared in several patents.[3,5] Several speculative laboratory investigations of pressure-compacted cement paste have been published.[6,7] A survey of the results obtained is given in a research reported by Lawrence.[8]

Powers and Brownyard[9] adopted the term "gel-space ratio," analogous in concept to the Feret cement-space ratio. Since the cement gel produced from a given cement has a characteristic specific surface area proportional to V_m/V_g, where V_m is the monolayer coverage obtained from the BET equation, and V_g is the volume of cement gel, the quantity of gel in a unit volume of specimen is properly represented by V_m/V_p, where V_p is the volume of the paste. The volume of the paste is the gel volume plus the pore volume; the first is related to the solid and can be represented by the surface area, and the other is related to the "space", represented by total porosity of the paste.

Surface area and pore structure, as well as the strength development for low-porosity pastes obtained by the Brunauer physico-chemical method, have been the subjects of numerous studies, by Brunauer and co-workers[10] and by Mikhail and co-workers.[11] Similar studies, as far as the authors know, have not been carried out for pressure-compacted pastes produced by the mechanical method. These studies are the subject of the present investigation, and the two methods are critically compared.

Specific surface areas, pore structure analysis, and compressive strength results are presented and discussed.

EXPERIMENTAL AND MATERIALS

Cement Pastes. The composition of the compacted low proposity cement samples* used was C_3S - 49%; C_2S - 28%; C_3A - 8%; C_4AF - 5%. It has a Blaine area of 4260 cm^2/g. Four samples were prepared by pressing the dry cement under 5, 10, 30 and 70 atmospheres and immersing them in water for two years at 25°C. These four samples were designated 11F5, 10F10, 10F30 and 12F70R, respectively.

The other sample, obtained by the physico-chemical method, was prepared with a water-to-clinker ratio equal to 0.2 by weight.

---

* Donated by the Research and Development Division of the Cement and Concrete Association, England.

Experimental details were given in a previous publication.[11] This sample received the notation IA_o.

Free-line determination was performed using a modification of Franke's solvent variation method.[12] Evaporable water was determined by drying at 105°C and non-evaporable water was determined by further ignition at 1000°C to constant weight. Results of the analyses are shown in Table 1.

The complete degree of hydration was determined either by pulverizing and continuous hydration in water in a steel ball mill, or was computed theoretically using an equation given by Copeland et al.[13] The degrees of hydration of the samples 11F5, 10F10, 10F30, 12F70R and IA_o were 64.17%, 53.82%, 44.03%, 32.08%, and 71.50%, respectively.

The adsorption of water vapour was measured using silica spring balances of the McBain-Baker type,[14] with sensitivities about 35 cm/g^{-1}, thermostated at 30±0.1°C. Prior to the adsorption runs the pastes were dried in situ, by equilibrating them at the vapour pressure of ice at -78°C, 5×10^{-4} mm Hg.

The paste samples were cut into 12 mm cubes, and compressive strength measurements were carried out on the Ravenstein Compressor Model 11/2612 (max. load 1000 kg).

RESULTS AND DISCUSSION

I - Adsorption Isotherms and the BET-Surface Areas

Adsoprtion-desorption isotherms were measured for each of the five samples. The equilibration was slow, and the shortest time needed for an experimental point was about 8 days, whereas for most of the points much longer periods of time were required. The adsorption was followed up to the saturation vapour pressure of water at 30°C; the desorption was measured down to extremely low relative pressures. Typical isotherms are shown in Figs. 1 and 2 for samples 11F5 and 12F70R, which were pre-compressed at the lowest and highest pressures.

The adsorption-desorption cycles obtained for the samples 11F5, 10F10, 10F30 and IA_o exhibit gigantic hysteresis loops which remain wide open till the lowest pressures. For the pre-compressed samples, the size of the loops decrease with increase in the prehydration pressure, and actually for the sample 12F70R

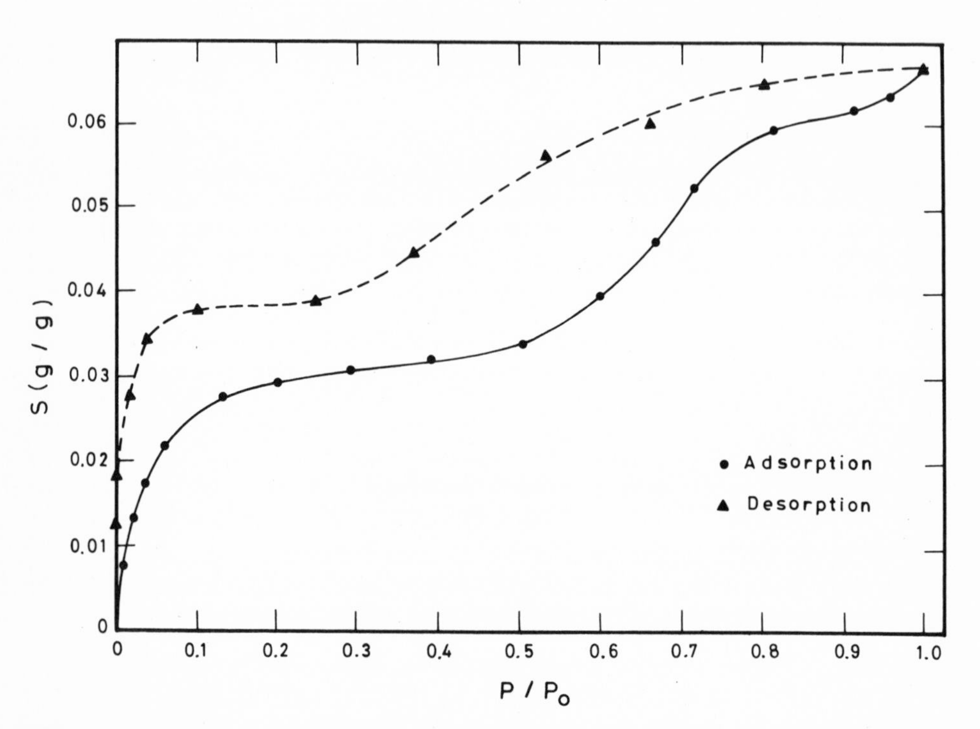

Fig. 1: Adsorption-desorption isotherms of water on Sample 11F5.

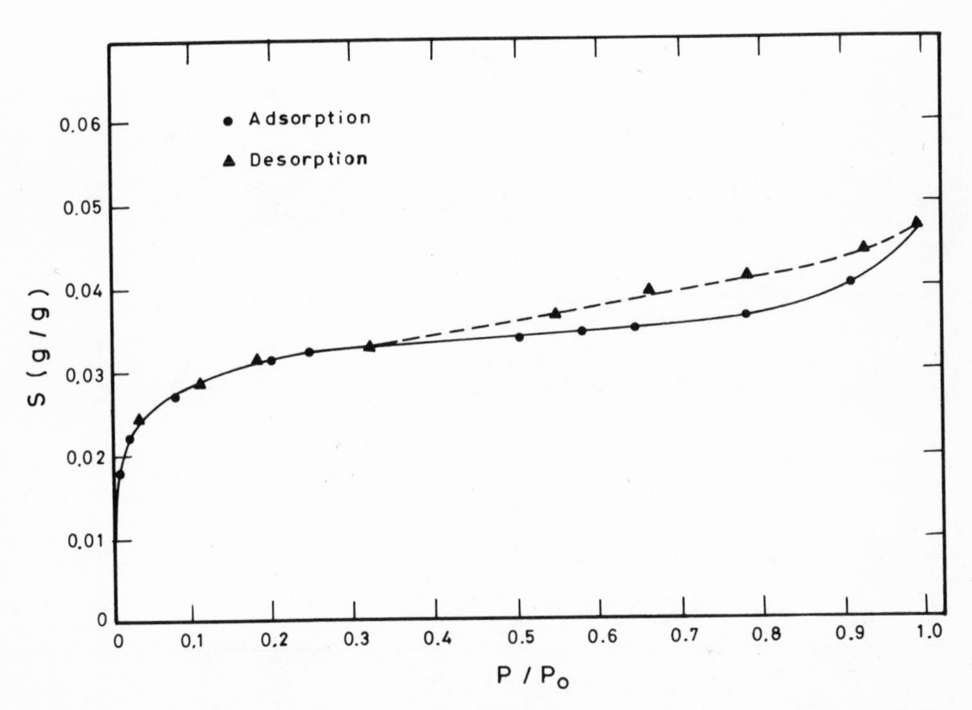

Fig. 2: Adsorption-desorption isotherms of water on Sample 12F70R.

Table (1)

Free-lime and Water Analysis of the Low-porosity Pastes

Sample	% Evaporable water	% Non-evaporable water	% Free-line
	Ignited weight basis		
IA_o	20.31	14.01	5.96
11F5	15.41	12.46	5.21
10F10	9.71	10.95	4.50
10F30	8.20	8.55	3.41
12F30	8.00	6.23	2.81

Table (2)

Specific Surface Areas of the Cement Paste Samples Calculated from Water Vapor Adsorption

1	2	3	4	5
Sample	BET-C Constant	BET S_{ign} (m^2/g)	S_{corr} (m^2/g)	S_{tob} (m^2/g)
IA_o	50	105.9	147.0	220.0
11F5	50	96.0	149.0	224.3
10F10	72	122.7	228.0	342.0
10F30	120	93.0	211.3	316.9
12F70R	125	102.8	320.2	480.0

a much smaller hysteresis loop was obtained which closes at some intermediate pressure. These gigantic hysteresis loops might be associated with the existence of a special pore shape, namely "ink-bottle" pores; i.e. pores with constricted entrances. This has been suggested in several previous publications dealing with either low porosity or normal pastes (10, 11, 15). The process by which adsorbate molecules enter such pores is activated diffusion.[10] The gigantic hysteresis loops obtained in the present investigation might indicate a big difference in size between the "neck" and the "body" of the "ink-bottle". The results obtained also indicate that as the pre-hydration pressure increases this difference in width between the "neck" and "body" decreases. This discussion would imply that along the desorption branch much longer time is needed to reach equilibrium than along the adsorption branch. In view of these considerations, the adsorption branches were considered to represent true equilibrium or close to it, and were accordingly used in computing some surface characteristics of these pastes.

Specific surface areas could be evaluated through the application of the BET equation, adopting a molecular area value of 11.4 $Å^2$.[20] Column 3 of Table 2 shows the BET surface areas of the various samples, calculated on ignited weight basis (S_{ign}^{BET}).

Since the samples were not completely hydrated, the specific surface areas represented in column 3 of Table 2 would be more meaningful if some correction were made to calculate the specific surface area of the hydration products themselves. In making this correction, the main assumption is that the area development is exclusively due to hydration. The corrected surface areas are shown in column 4 of Table 2.

If we assume that all the surface is in tobermorite gel, then one-third of the hydration products by weight do not contribute anything to the surface. The surface of the tobermorite gel is shown in column 5 of Table 2. For both IA_o, prepared with low water/cement ratio, and 11F5, pre-compressed at the lowest pressure, then hydrated in water, the surface of the tobermorite gel is about 220 m^2/g, which is close to a three-layer sheet (252 m^2/g), with some four layer sheets. On the other hand for samples 10F10 and 10F30 the tobermorite areas indicate 2-3 layer sheets, while for the sample 12F70R, pre-compressed at the highest pressure, the tobermorite area indicates 1-2 layers thick sheets. Brunauer et al.[16] calculated the area for tobermorite sheets composed of one, two, three and four layers to be 755, 377, 252 and 189 m^2/g. respectively.

The result obtained in this investigation is of particular interest. It indicates that pre-hydration compression could affect the hydration process at least in two ways. It leaves no space for complete hydration, and simultaneously leaves insufficient space for sheet thickening to take place. The results obtained for sample IA_o indicate that for low-porosity pastes obtained by the physico-chemical method,[3,4] layer sheets are produced, because these are thermodynamically more stable than thinner sheets.[10,11] For prehydration pressed samples and depending on the applied pressure, steric effects seem to counteract the formation of thick sheets.

II. Total Pore Structure Analysis

Total pore structure analysis gives the totality of surfaces and volumes associated with both micro and mesopores. These were analyzed by two recently developed methods by Brunauer and coworkers. One of them is the "MP-method"[17] suited for the analysis of micropores, and the other is the "corrected modelless method"[18] suitable for the analysis of mesopores. Results of the analysis are shown in Table 3.

In Table 3, S_n and V_n represent the surfaces and volumes associated with micropores, while S_w^{cp} and V_w^{cp} represent the surfaces and volumes associated with mesopores. For micropores, the shape of the pore has no effect on the results of analysis, while for mesopores the cylindrical pore idealization seems to fulfill the criteria for correct analysis.[17] These results indicate that for the pre-compressed samples the fraction of narrow pores present in each sample increases as the prehydration compression increases. This is indicated by the increase in the fraction of surface located in micropores relative to the BET surface area; these fractions are 0.79, 0.82, 0.83 and 0.93 for samples 11F5, 10F10, 10F30 and 12F70R, respectively. On the other hand, the sample IA_o prepared with low water/cement ratio shows a comparatively high ratio of 0.92.

Pore volume distribution curves could be constructed. These are shown in Fig. 3. For the pre-compressed samples the dashed curves represent the distribution of micropores, while the solid curves represent those of mesopores. The interesting conclusion which might be drawn from these distribution curves is that upon the increase of pre-hydration compression, the most probable size of the mesopores gradually decreases and merges into the range of size characteristic for micropores. This ultimately leads to an increase in the number of micropores at the expense of mesopores. The sample 12F70R might be considered totally microporous with dual distribution of pore sizes.

The distribution curve for the low water/cement ratio sample IA_o, is widely different from the distribution curves of the pre-compressed samples. Although samples IA_o and 11F5 are expected to

Table (3)

1	2	3	4	5	6	7	8
Sample	S_n (m^2/g)	V_n (ml/g)	S_w^{cp} (m^2/g)	V_w^{cp} (ml/g)	$S_n+S_w^{cp}$ (m^2/g)	$V_n+V_w^{cp}$ (ml/g)	V_p (ml/g)
IA_o	97.0	0.0614	8.3	0.0105	105.3	0.0809	0.0803
11F5	75.0	0.0245	23.6	0.0443	98.6	0.0688	0.0670
10F10	100.5	0.0272	22.0	0.0465	122.5	0.0737	0.0720
10F30	77.5	0.0221	17.6	0.0462	95.1	0.0683	0.0680
12F70R	96.0	0.0314	9.6	0.0170	105.6	0.0484	0.0469

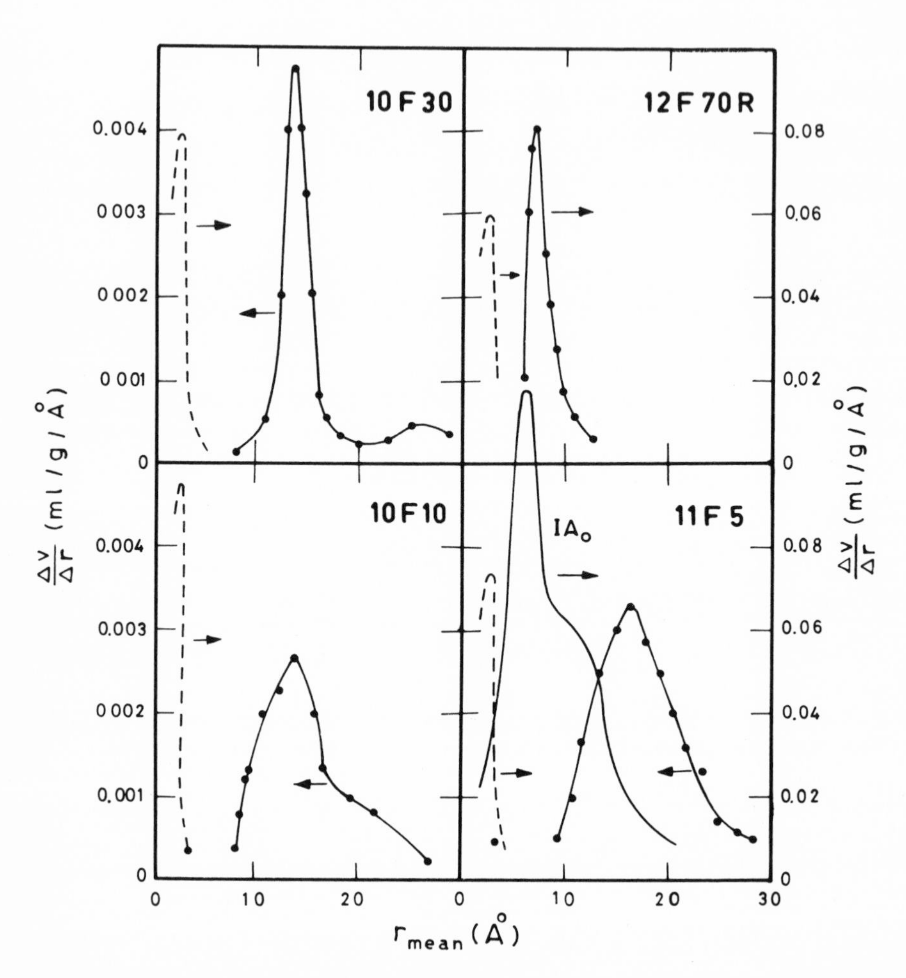

Fig. 3: Pore volume distribution curves obtained from water vapour adsorption.

show the closest similarities in their properties, their distribution curves are very different from each other. The most obvious differences are the absence of any bimodal distribution of pore sizes in the iA_o sample with the result that the distribution is more even, and that the sizes of pores are also different for the two different methods of preparation. For the low water/cement sample IA_o, the most frequent micro-pore size is much larger than the micro-pore size of the pre-compressed samples.

III - Compressive Strength

Compressive strength measurements were taken for the five samples IA_o, 11F5, 10F10, 10F30 and 12F70R, and these were 30,500; 17,300; 17.550; 19.450 and 19,450 psi, respectively. Obviously, the compressive strength of the sample IA_o is appreciably higher than the rest of samples prepared by pre-hydration compression. The compressive strength values for all the pre-compressed samples can be considered to be approximately the same, although somewhat lower values were obtained for the two samples pre-compressed at the lowest pressures. This result is unexpected, particularly in light of the fact that the degree of hydration of the various samples differs widely.

It has been stated by Powers et al.[9,19] that the compressive strength is developed upon hydration, and is mainly due to the formation of the gel (tobermorite). This effect might partly explain the higher value obtained for the IA_o over the rest of samples, but cannot by itself explain the behaviour of the pre-compressed samples.

An increase in the specific surface area is expected also to lead to an increase in compressive strength. Taking into consideration that the gel surface of both the IA_o sample and the 11F5 sample are nearly the same, this factor seems unable to explain the much higher strength value of IA_o sample, although it might partly contribute to the results obtained for the pre-compressed samples.

A major factor governing the compressive strength of hardened cement pastes is the total porosity. The significance of this factor will be first discussed in relation to the pre-compressed samples. Thus the sample 12F70R has a higher surface (Table 2) and considerably lower total porosity (Table 3) than 10F30. It should have a considerably higher compressive strength while the results reported in this investigation show equal strength values. The compensating factor might be the lower degree of hydration as mentioned earlier, but more significant is the larger volume in the micropores (Column 3, in Table 3). Apparently micropores have a greater effect on reducing strength than the wide pores.[20]

In comparing 11F5 and 10F10, the latter has a larger surface and a larger pore volume, which two effects are mainly responsible for the compensation. A part in the compensation may also be played by the larger micropore volume of the latter.

The above comparisons were made within the two sets of strength results separately. A cross comparison of the two sets with each other can be made using, for example 11F5 and 10F30, and this would eventually lead to the same result, although this situation is not as clear as in the comparison of 10F30 and 12F70R, because in the latter case the micropore volume difference is much greater.

The fact that micropores play a significant role in reducing the compressive strength seems to be the most interesting conclusion drawn from the present results, and it could also explain IA_o over the rest of samples prepared by pre-hydration compression. Thus, it is evident from the distribution curves (Fig. 3), that the average size of micropores in the IA_o sample in much higher than the average size of micropores in the pre-compressed samples, and are therefore much less effective in reducing the strength of the IA_o sample. Extremely fine pores may be regarded as cracks, and the number and the legnth of cracks may be important factors in causing failure under load. This agrees with Griffith's theory of the structural failure of brittle materials, based on the formation and propagation of cracks in solids.[21]

REFERENCES

1. M. Yudenfruend, I. Odler, S. Brunauer, Cement and Concrete Research 2 313 (1972), and refs. cited there.

2. M. Yudenfruend, I. Skalny, R.Sh. Mikhail, S. Brunauer, Cement and Concrete Res. 2 331 (1972).

3. E.W. Roberts, L.F.W. Leese, Patent Spec. No. 181, 745, London Patent Office (21 March 1921).

4. E. Freyssinet, Concrete and Constructional Engineering 31 209 (1936); Patent Spec. No. 431 (9 July 1935), and Patent Spec. No. 453-555 (14 Sept. 1936), London Patent Office.

5. Svenska Enterprenad Aktiebologet Sentab-London, Patent Spec. No. 906, 573 (26 Sept. 1962) London Patent Office.

6. T. Kluz, F. Leconner, Insynieria i Budownictwo 17 204 (1960).

7. G. Wischers, Schriftenreihe der Zementindustrie 28 117 (1961).

8. C.D. Larence, Cement and Concrete Association Research Report 19, p. 21 (1969).

9. T.C. Powers, T.L. Brownyard, Res. Lab. Portland Cement Assoc. Bull. 22 (1948).

10. I. Odler, J. Hagymassy, M. Yudenfruend, K.M. Hanna, S. Brunauer, J. Colloid Interface Sci. 38 265 (1972).

11. R. Sh. Mikhail, S.A. Abo-El-Enein, Cement and Concrete Res. 2 401 (1972).

12. E.E. Pressler, S. Brunauer, D.L. Kantro, C. Weise, Anal. Chem. 33 877 (1961).

13. L.E. Copeland, D.L. Kantro, G. Verbeck, Proc. of the 4th Int. Symp. on Chem. of Cement, Wash. 1960-NBS Monog. 43 I 429.

14. I.W. McBain, A.M. Baker, J. Am. Chem. Soc. 48 690 (1926).

15. R. Sh. Mikhail, L.E. Copeland, S. Brunauer, Canadian J. Chem. 42 426 (1964).

16. S. Brunauer, S.A. Greenberg, 4th Int. Symp. on Chem. of Cement, Wash. 1960, NRS Monog. 43 I 149-50.

17. R. Sh. Mikhail, S. Brunauer, E.E. Bodor, J. Colloid Interface Sci. 26 45 (1968).

18. S. Brunauer, R. Sh. Mikhail, E.E. Bodor, J. Colloid Interface Sci. 24 451 (1967).

19. T.C. Powers, Proc. 4th Int. Symp. on Chem. of Cement, Wash. 1960 NRS Monog. 43 II 601.

20. S. Brunauer, private communication.

21. A.A. Griffith, Phil. Trans. R. Soc. A221 163 (1920).

ADSORPTIVE BEHAVIOUR OF ALKALI AND ALKALINE-EARTH CATIONS ONTO QUARTZ

Selim F. Estefan and Mounir A. Malati*

Metallurgy Department, National Research Center

Cairo

Adsorption isotherms of the alkali and alkaline-earth cations onto quartz surfaces were determined employing radioactive tracer techniques at different temperatures and pH values. For each temperature, the adsorption affinity follows the lyotropic sequence, i.e., for cations of a given charge, the affinity decreases with the increase in the radius of the hydrated cation. The results of the adsorption and zeta-potential measurements indicate that the hydrated alkaline-earth cations are adsorbed in the quartz/electrolyte double layer and may be considered to be held electrostatically in the outer Helmholtz layer. The cation/quartz surface attraction seems to play an important role in fixing the collector to the mineral particles in flotation.

INTRODUCTION

The cation-water interaction is important in understanding the mechanism of cation adsorption from aqueous solution. The Gouy-Chapman theory treats the ions as point charges in the double layer and neglects specific interactions between the ion and the particles, or the ion and the water molecules. Jyo[1] has shown that variation of the zeta-potential of mineral particles in a solution of univalent ions agrees with the Stern theory of the electrical double layer, but this is not true for bivalent and trivalent ions.

The hydration energy of the alkali metal cations decreases down the group, and the same applies for the alkaline-earth cations. The

*Medway and Maidstone College of Technology, Chatham, Kent, England.

hydrated radii can be estimated by different methods.[2,3] It has been suggested[4] that adsorption of Ba^{2+} ions onto a quartz surface is accompanied by partial dehydration. Onoda and Fuerstenau[5] have assumed that dehydrated Ba^{2+} ions are adsorbed by the quartz surface. Ahmed and Van Cleave[6] and Tadros and Lyklema[7] concluded that below pH 7.5, alkaline-earth cations are physically adsorbed whereas at higher pH values, they are chemisorbed. Chemisorption is presumably accompanied by dehydration. On the other hand, in previous work, we have shown that the hydrated cations are electrostatically held to the quartz surface.[8,9]

EXPERIMENTAL

Selected Aswan quartz was finely ground and wet screened. The -74 + 37, and -37 + 10 μm fractions were leached several times with hot concentrated hydrochloric acid, washed repeatedly with bi-distilled water and dried at 110°C. Chemical and spectrographic analyses of the samples were carried out and revealed that they assayed 99.5% SiO_2. Surface area measurements on the fraction -37 + 10 μm by the krypton adsorption method gave a value of 1400 ± 200 cm^2/g.

Zeta-Potential Studies: Batches of 5 g of quartz (-37 + 10 μm) were conditioned for one hour with 100 ml of 0.001 M sodium nitrate solution as a supporting electrolyte and in presence of varying concentrations of the alkaline-earth chlorides at pH 10.5. Electrophoretic mobility measurements were carried out at 21°C on a portion of each suspension after readjusting the pH to 10.5 using a cylindrical microelectrophoretic cell. The zeta-potential was calculated using the Helmholtz-Smoluchowski equation* in the form:

$$\xi = \eta U/\varepsilon$$

where η is the viscosity and ε is the permittivity of the medium and U is the electrophoretic mobility defined as:

$$U = V/E$$

where V is the velocity of the particle and E is the potential gradient across the cell. The potential gradient was calculated from the specific conductance, measured by a Metrohm E 182 conductometer. The initial value of the zeta-potential of quartz in absence of the alkaline-earth cations was determined for a sample suspended in 0.001 M sodium nitrate solution under the same conditions of pH and temperature.

---

* This is permissible since the ratio of the particle radius to the thickness of the double layer was greater than 300.

Adsorption Studies: Batches of 5 g of quartz (-37 + 10 μm) were conditioned for one hour at pH = 7 or 10.5 at 10°, 30° or 45°C with varying concentrations of radioactively tagged solutions of the chlorides of the alkali and alkaline-earth metals at a solid/liquid ratio of 5%. The radionuclides used were ^{22}Na, ^{86}Rb, ^{134}Cs, ^{45}Ca, ^{89}Sr and ^{140}Ba. Aliquots of 1 ml of each filtrate were counted by a scintillation or a Geiger Muller counter depending on the type of radiation of the radionuclide used. Details of the technique have been published elsewhere.[9]

For the determination of the co-adsorption of Ca^{2+} and Na^+ ions, 5 g of quartz were conditioned for 30 minutes at 30°C and pH 11.8 with 100 ml of the $NaCl-CaCl_2$ solution, using ^{22}Na or ^{45}Ca-labeled solution. Aliquots of the filtrate were evaporated on a planchet and counted in the scintillation or Geiger-Muller counters respectively. Deionised water was used for preparing the solutions.

RESULTS AND DISCUSSIONS

The adsorption isotherms of Na^+, Rb^+ and Cs^+ ions onto quartz at pH = 7 and at different temperatures are shown in Fig.1. It is clear that at each temperature, the adsorbability of Cs^+ ions is higher than that of Rb^+ and Na^+ ions. At higher equilibrium concentrations, the adsorbability of Rb^+ ions is higher than that of Na^+ ions.* The adsorbability seems to decrease in the series: $Cs^+ > Rb^+ > Na^+$, following the lyotropic or Hofmeister series. This is the order of the decrease in $1/a^o$, where a^o is the Debye and Huckel distance of closest approach and this is also the order of decrease in $1/r'$, where r' is the radius of the hydrated ion.[10] This order of affinity between the alkali metal ions and the surface of silica has been reported by a number of authors, using different techniques.[11,12] This is also the order of preference of silica gel for monovalent cations as found by Tien.[13]

Recently, attention has been focused on the state of hydration of cations at the oxide/electrolyte double layer, in particular the silica/electrolyte system. Wiese et al[14] have reported the zeta-potential for vitreous silica in electrolyte solution as a function of the concentration of potassium, barium and lanthanum nitrates. For satisfactory agreement between theory and experiment, they added a term, $\Delta G_{solv.}$, representing the change in free energy as the cation adsorbs at the silica/water interface. However, Levine[15] has pointed out that the treatment of Wiese et al is oversimplified, and has discussed some limiting cases for calculating $\Delta G_{solv.}$.

* By repeating the Na^+ adsorption isotherm at 10°C using plastic containers for the adsorption tests and for preparing the solutions, the isotherm was found to be clearly lower than that of Rb^+ ions (J. Hamsley and P.J. Mason, G.R.I.C. Research Project, Medway and Maidstone College of Technology, England, 1972).

The sequence of the affinity of adsorbed cations for the oxide surface has been recently discussed by a number of authors. Dumont and Watillon[16] have considered the role of ions and of the solid in promoting or breaking the structure of water. In the light of the flickering cluster theory of water, hydration of ions can be discussed in terms of their function as cluster-promoting or structure-breaking.[17] Large ions of low charge, e.g. Cs^+, ions are net "structure breakers" whereas small highly charged ions are net "structure promoters". Dumont and Watillon[16] emphasised the distinction between surfaces of oxides that behave as structure breakers, and those that behave as structure promoters such as quartz. Although no theory of a "structured double layer" has been developed, Stumm et al.[18] have assumed that a strongly structured, extensively hydrogen-bonded layer of water is formed adjacent to the hydrophilic oxide surface. They have considered the association of the oxide surface with H^+ ions and other cations. The affinity sequence $Cs^+ > Na^+$ which they observed is taken to indicate a weak field strength of the adsorbent. However, the validity of estimating the field strength of the solid from its crystal structure is questionable, since a surface layer of an oxide may be quite different from the bulk of the solid.[19] The higher affinity of bentonite for Cs^+ ions compared to Na^+ ions has been correlated with the polarisation energy of the two ions.[20]

The exothermic character of the adsorption of the alkali metal ions shown in Fig. 1, is expected for physical adsorption. Such monovalent ions are not considered to be specifically adsorbed and do not reverse the sign of the ξ-potential of quartz.[4] Monovalent cations do not activate the flotation of quartz with anionic collectors and may be described as indifferent ions. The isotherms in Fig. 1 have a Langmuir shape and the applicability of the low coverage form of the Langmuir isotherm may be assumed for the initial parts.[21]

The initial adsorption isotherms of the alkaline-earth cations on quartz at 30°C and at pH 7 and 10.5 are shown in Fig. 2. The adsorbability increases in the order $Ca^{2+} < Sr^{2+} < Ba^{2+}$ and this may be explained by the decrease of the hydrated radii, r', in the order: $Ca^{2+} > Sr^{2+} > Ba^{2+}$. When the adsorbability of the cations, calculated from Fig. 2 or Fig. 3, is plotted against $1/r'$, straight lines are observed (Fig. 4). This may be taken as an indication that the hydrated cations are held in the outer Helmholtz layer by electrostatic forces.[9] It has been suggested[6,7] that the alkaline-earth cations are physically adsorbed on the quartz surface below pH 7.5, whereas at higher pH values, they are chemisorbed and presumably dehydrated. Assuming the unhydrated cations are chemisorbed forming surface silicates, the strength of the alkaline-earth silicate bonds formed would decrease[2] in the order: $Ca^{2+} > Sr^{2+} > Ba^{2+}$, which is the reverse of the observed order of adsorbability. Moreover, had there been a change in the mechanism of adsorption of these cations at pH

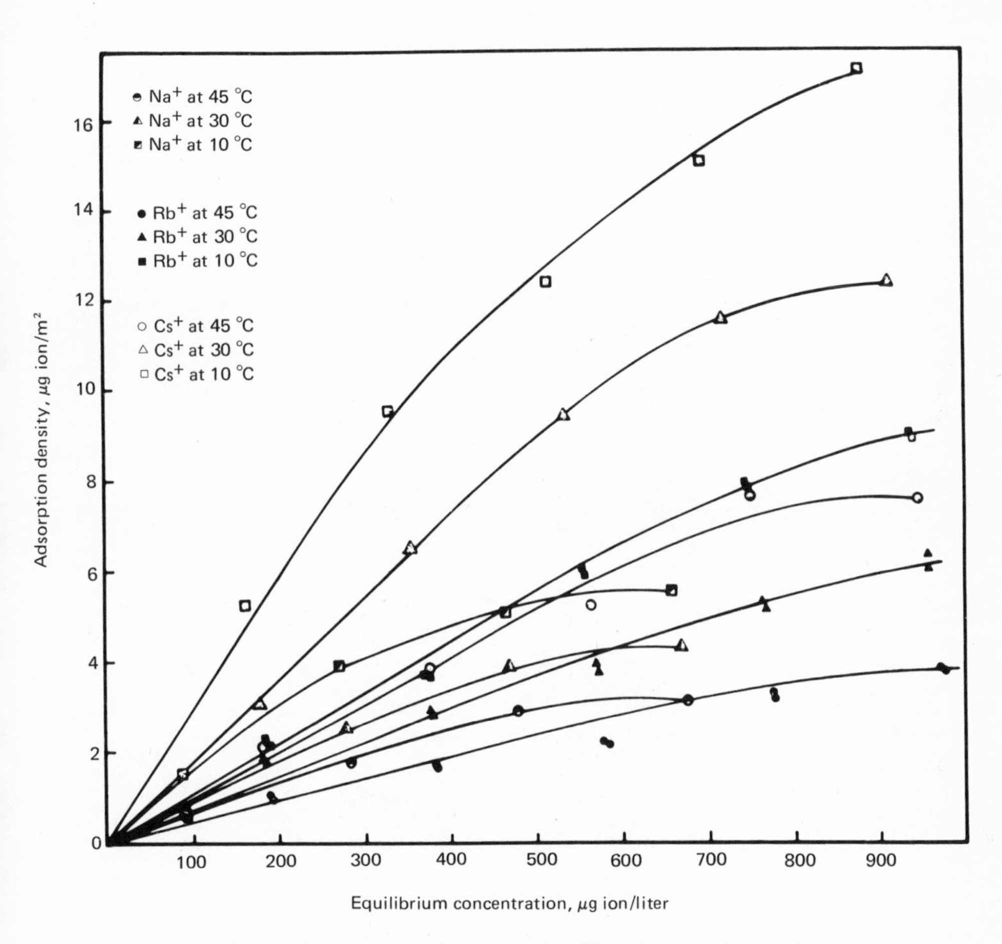

Fig. 1: Adsorption isotherms of alkali metal cations onto quartz at pH 7.

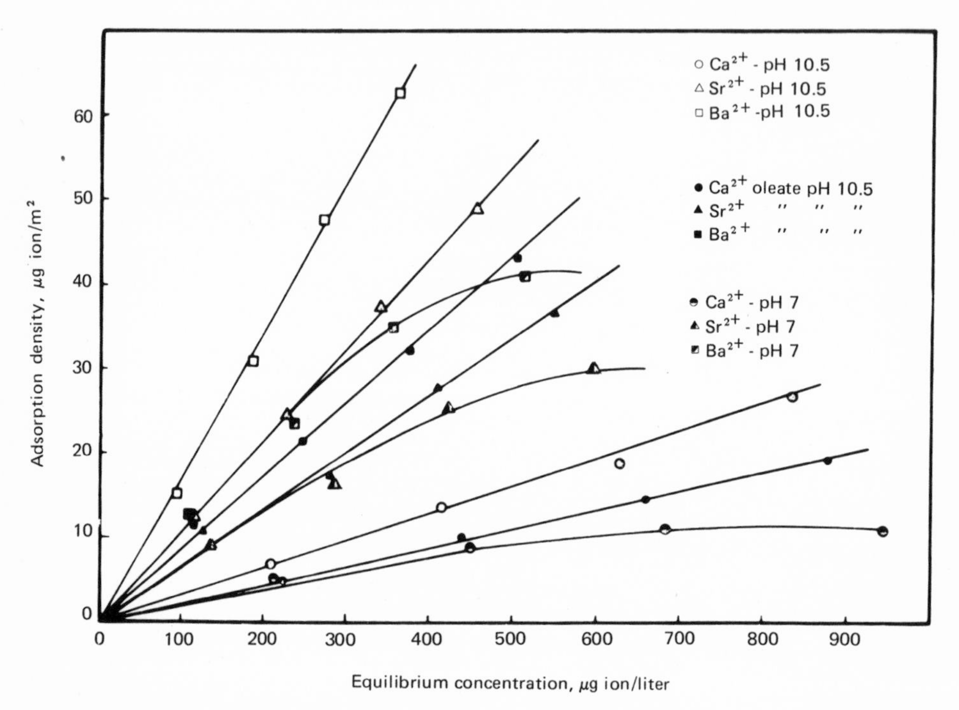

Fig. 2: Adsorption isotherms of alkaline-earth cations onto quartz at 30°C and at pH 7 and 10.5.

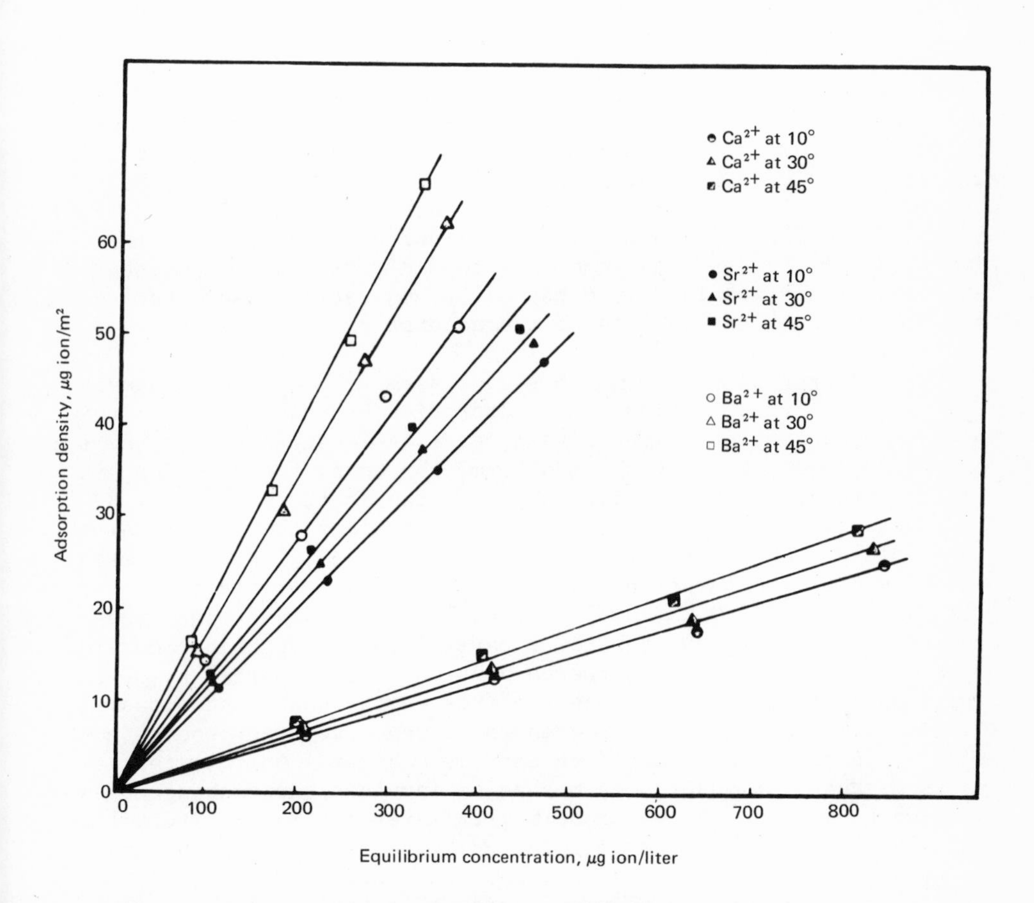

Fig. 3: Adsorption isotherms of alkaline-earth cations onto quartz at pH 10.5 and at 10, 30, and 45°C.

values higher than 7.5, the order of adsorbability would have been different at pH values 7 and 10.5. Figure 4 shows that this is not the case.

The decrease in adsorbability of the alkaline-earth cations in presence of a constant concentration of oleate (2×10^{-5} M dm^{-1}), shown in Fig. 2 is approximately proportional to 1/r'. This behaviour has been tentatively explained by a cation-oleate attraction affecting the quartz-cation attraction.[9]

Figure 5 shows that the ξ-potential of quartz particles in solutions of the alkaline-earth cations becomes less negative from Ca^{2+} to Sr^{2+} and from Sr^{2+} to Ba^{2+} ions. Assuming that the hydrated cations are held in the outer Helmholtz layer, the outer Helmholtz plane would be further from the quartz surface in the order: $Ba^{2+} < Sr^{2+} < Ca^{2+}$. The ξ-potential is then expected to become less negative from Ca^{2+} to Sr^{2+} and from Sr^{2+} to Ba^{2+}, as found experimentally. Thus the results obtained from the electrophoretic measurements are in agreement with the results of the adsorption measurements.

The adsorption isotherms in Fig. 2 seem to be of the Langmuir type. For the three cations, the adsorption process is endothermic, unlike the adsorption of the alkali metal cations. Assuming the low coverage form of the Langmuir isotherm, the heat of adsorption may be calculated from the initial slopes of the isotherms in Fig. 3. The plots of the logarithms of the slopes against the reciprocal of the absolute temperature are seen to be linear (Fig. 6). The estimated[22] heats of adsorption from the slopes of the lines in Fig. 6 are: 3.8, 2.7 and 2.0 kJ $mole^{-1}$ for Ca^{2+}, Sr^{2+} and Ba^{2+} ions respectively. However, one has to be careful in interpreting measured heats of adsorption, unless the conclusions are supported by other experimental evidence.[23] Since the free energy change accompanying the adsorption process is expected to be negative, the process must be accompanied by large positive entropy changes. One may tentatively assume that the adsorption mechanism involves the exchange between hydrated H^+ ions in the outer Helmholtz layer (or in the Stern layer) and alkaline-earth cations in solution. The decrease in the endothermic value of the heat of adsorption in the series: $Ca^{2+} > Sr^{2+} > Ba^{2+}$ is in agreement with the observed order of affinity of the quartz surface for these cations.

The endothermic character of adsorption of the alkaline-earth cations, compared with the exothermic character of adsorption of the alkali metal cations, emphasises the role of the entropy of adsorption, since the adsorbability of the divalent cations is greater than that of the monovalent cations. Table 1 indicates that Ca^{2+} ions are preferentially adsorbed compared to Na^+ ions. Only at relatively high Na^+ ion concentration would the Na^+ ions compete with Ca^{2+} ions for adsorption sites.

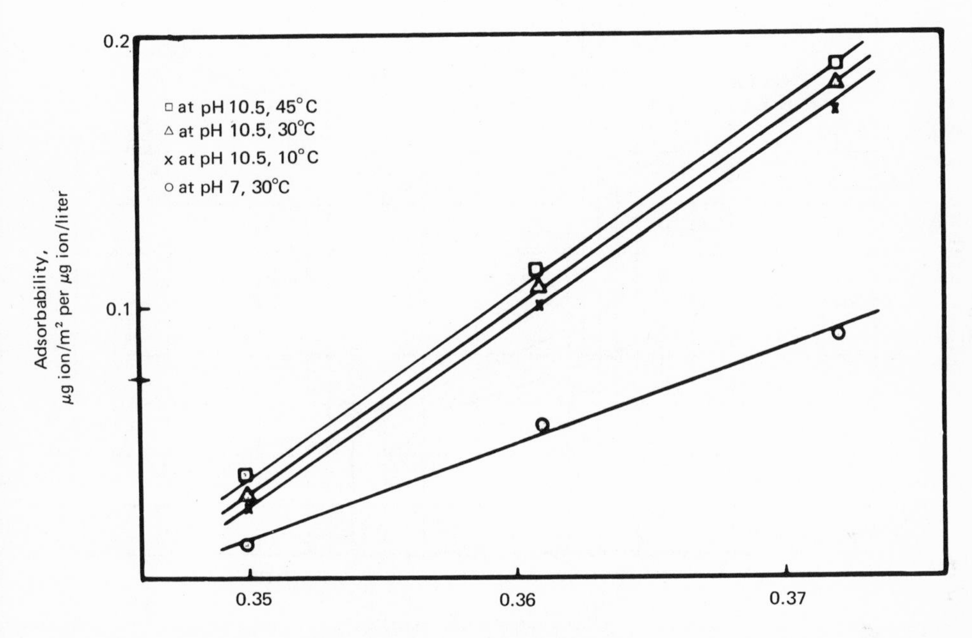

Fig. 4: Relationship between adsorbability of alkaline-earth cations onto quartz and 1/r'.

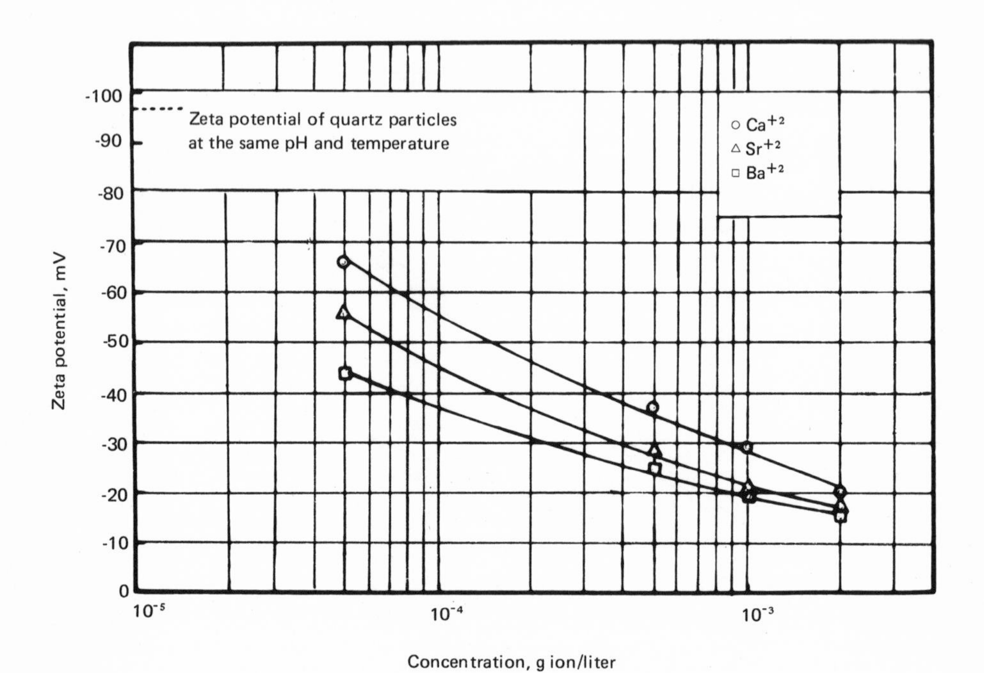

Fig. 5: Zeta-potential of quartz particles in alkaline-earth cations solution.

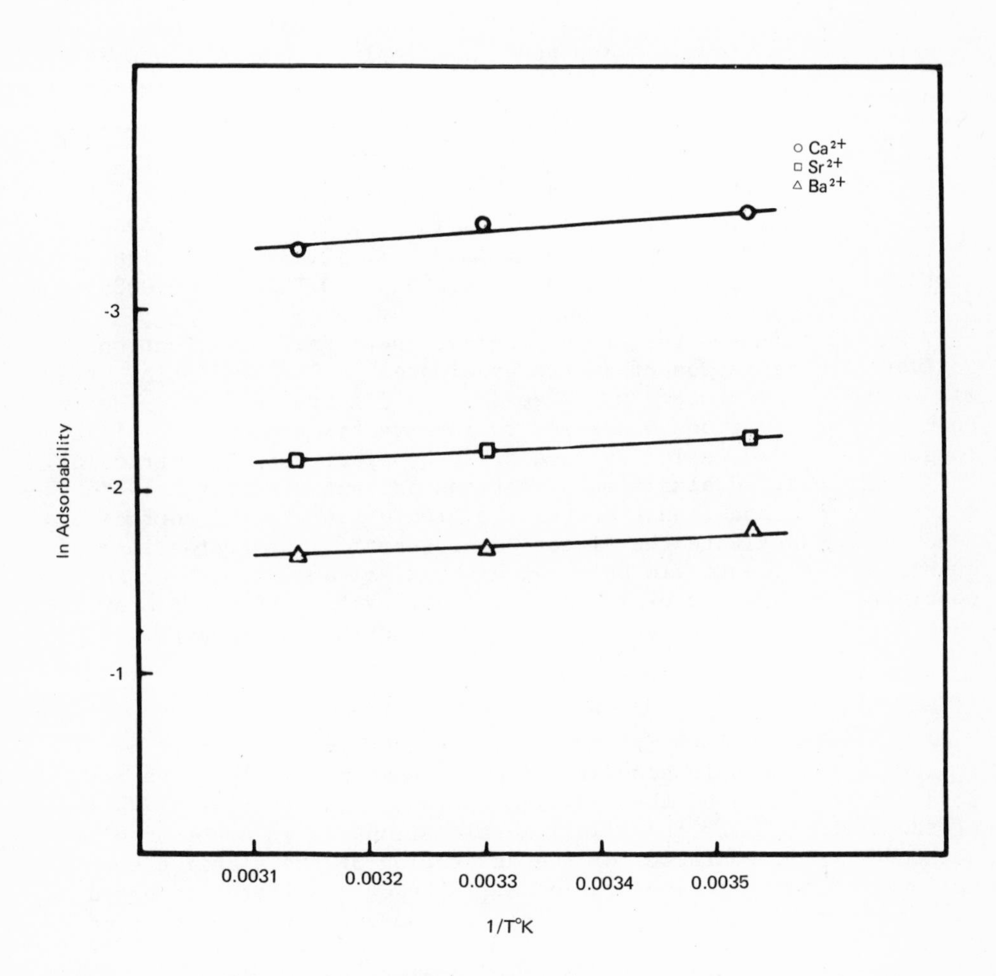

Fig. 6: Relationship between adsorbability of alkaline-earth cations onto quartz and $1/T^{o}$.

Table I

Co-adsorption of Na^+ and Ca^{2+} Ions on Quartz Surface

^{22}Na			^{45}Ca		
Na^+ ion conc. m mole/l	Equilib. conc. m mole/l	Adsorption density m mole/m^2	Ca^{2+} ion m mole/l	Equilib. conc. m mole/l	Adsorption density m mole/m^2
11.5	10.57	0.1330	0.000	0.0000	0.0000
11.5	11.11	0.0557	5.104	4.4215	0.0975
11.5	11.07	0.0615	5.104	4.4310	0.0962
10.5	10.50	0.0000	13.015	12.1320	0.1263
13.5	13.07	0.0615	0.255	0.1958	0.0085

The alkaline-earth cations, unlike the alkali metal cations, activate the flotation of quartz by oleates.[8] The activating effect was found[24] to increase in the series: $Ca^{2+} < Sr^{2+} < Ba^{2+}$. The mechanism of activation is assumed to involve fixation of the oleate ions to the M^{2+}-occupied surface sites by electrostatic attraction. The singly charged alkali metal cations are not strongly held to the quartz surface and cannot anchor the oleate ions to the surface and hence do not activate the flotation of quartz. The depression* of quartz in the oleate flotation at high pH values,[8] i.e. high Na^+ ion concentration, may be explained, at least partly, by the preferential adsorption of Na^+ ions at expense of the activating cations.

CONCLUSION

The adsorbability of the alkali or alkaline-earth cations follows the lyotropic sequence, i.e. it runs parallel to the reciprocal of the radii of the hydrated cations. The exothermic nature of the adsorption of the alkali metal cations is in contrast with the endothermic character of the adsorption of the alkaline-earth cations. The greater adsorption affinity of the latter compared with the former emphasises the role of the entropy of adsorption.

The zeta-potential of quartz particles becomes less negative in the series Ca^{2+} , Sr^{2+} , Ba^{2+} , which confirms the mechanism of adsorption suggested earlier.

The activation of quartz flotation by oleate in presence of the alkaline-earth ions can be explained by electrostatic attraction

* Depression is a term used in ore flotation. A depressed species sinks rather than being incorporated into the surface froth.

between the hydrated ions in the outer Helmholtz layer and the oleate ions.

ACKNOWLEDGMENTS

The authors are grateful to Professor Dr. M.K. Hussein for his kind help and for the facilities he offered, to Professor Dr. M.Y. Farah for his cooperation, and to Professor K.S.W. Sing and Mr. D.A. Payne for the krypton adsorption measurements.

REFERENCES

1. Onzo Jyo, Bul. Res. Inst. Mineral Dressing and Metal. Tohoku Univ. Japan 9:1 (1953).

2. M.A. Malati, S.F. Estefan, J. Coll. & Interface Sci. 22 306 (1966).

3. M.A. Malati, S.F. Estefan, Disc. Faraday Soc. 52 377 (1971).

4. A.M. Gaudin, D.W. Fuerstenau, Trans. AIME 202 66 (1955).

5. G.Y. Onoda, D.W. Fuerstenau, 7th Int. Mineral Proc. Congress, New York (1964) 301.

6. S.M. Ahmed, A.B. Van Cleave, Can. J. Chem. Eng. Feb. (1965) 23-26; 27-29.

7. Th. F. Tadros, J. Lyklema, J. Electroanal. Chem. 22 1 (1969).

8. M.A. Malati, S.F. Estefan, J. Appl. Chem. 17 209 (1967).

9. M.A. Malati, A.A. Youssef, S.F. Estefan, Chimie et Industrie 103 1347 (1970).

10. T.R.E. Kressman, J.A. Kitchener, J. Chem. Soc. 259 1190 (1949).

11. Th. F. Tadros, J. Lyklema, J. Electroanal. Chem. 17 267 (1968).

12. D.L. Dugger, J.H. Stanton, B.N. Irby, B.L. McConnell, W.W. Cummings, R.W. Maatman, J. Phys. Chem. 68 757 (1964).

13. H.T. Tien, J. Phys. Chem. 69 350 (1965).

14. G.R. Wiese, R.O. James, T.W. Healey, Disc. Faraday Soc. 52 302 (1971).

15. S. Levine, Disc. Faraday Soc. 52 320 (1971).

16. F. Dumont, A. Watillon, Disc. Faraday Soc. 52 352 (1971).

17. H.S. Frank, W.Y. Wen, Disc. Faraday Soc. 24 133 (1957).

18. W. Stumm, C.P. Huang, S.R. Jankins, Croatica Chemica Acta 42 223 (1970).

19. Y.G. Bérubé, P.L. De Bruya, J. Coll. & Interface Sci. 27 305 (1968).

20. I. Shainberg, W.D. Kemper, Soil Sci. Soc. of Amer. Proc. 30 700 (1966).

21. A.W. Adamson, Physical Chemistry of Surfaces, 2nd ed., John Wiley & Sons, London (1967).

22. A.A. Youssef, M.A. Arafa, M.A. Malati, J. Appl. Chem. Biotechnol. 21 200 (1971).

23. C. Mellgren, R.J. Gochin, H.L. Shergold, J.A. Kitchener, Intern. Mineral Processing Congr., London (April, 1973), Preprint 21.

24. S.F. Estefan, M.A. Malati, submitted to Institution of Mining & Metallurgy.

ADSORPTION OF SOME STARCHES ON PARTICLES OF SPAR MINERALS

H. S. Hanna

Metallurgical Department, National Research Center

Cairo, Egypt

The adsorption mechanism of starch at mineral-solution interface was investigated for the system, water soluble potato starch-spar minerals. Adsorption isotherms of the Langmuir type and the high occupation area of starch molecules suggested relatively flat orientation of starch molecules on the mineral surface. Mineral solubility, crystal lattice structure, associated mineral impurities and the ionic composition of the pulp were shown to control the adsorption process.

Adsorption-desorption data pointed to rather weak adsorption forces of the hydrogen bonding and electrostatic interaction types. Direct or indirect bonding between starch molecules and the divalent mineral cations is also important in describing some specific adsorption properties. A mechanism proposed for starch adsorption based on the above interactions is developed.

A critical starch concentration for optimum depression of these minerals exists. A coadsorption of both starch and oleic acid (collector) on the mineral surface was proved by I.R. analysis. This supports the idea of a starch-collector clathrate complex, and the wrapping of the collector inside the starch helixes can explain why the resulting surface is not hydrophobic.

INTRODUCTION

Selective dispersion of certain solid particles by some organic colloids is of vital importance in the flotation separation of minerals having similar cations or anions in their cyrstal lattices e.g. sulfides or calcium containing minerals fluorite, calcite, apatite, etc. Such separation is more complicated in the use of non-selective collectors such as fatty acids and their soaps. The depressant property of starches and their derivatives in flotation[1] and their flocculation characteristics on a wide variety of ores[2] have long been recognized. However, research on the mode of interaction of starches with mineral surfaces, in presence or absence of surfactant molecules, has not been extensively evaluated.[3,7] Neither hydrogen bonding nor electrostatic mechanisms have explained why starches are selectively adsorbed on some oxide and salt-type minerals.

The aim of the present work is to examine the effects of crystal lattice structure and ionomolecular composition on the adsorption characteristics of starch on the spar minerals fluorite (CaF_2), calcite ($CaCO_3$) and barite ($BaSO_4$). Correlating this information with flotation results will elucidate the adsorption mechanism of starches on mineral-solution interfaces.

EXPERIMENTAL

Samples of barite, calcite and fluorite (97.0% purity) were crushed, then dry ground in a screen ball mill. The size fraction -0.16 + 0.04 mm (sample I) was collected after scrubbing, desliming and drying and kept in air-tight glass bottles for use in normal adsorption and flotation tests. Extra pure optical grade calcite and fluorite samples (Carl Zeiss Jena Company) and twice recrystallized $BaSO_4$ (99.9% purity) were specially prepared for I.R-studies (<5μ) and adsorption tests (sample II: -0.10 + 0.032 mm). The measured surface areas of the above described samples were:

	Sample I	Sample II	Sample III
Barite	0.04 ± 0.01	0.07 ± 0.01	2.6 ± 0.02 m^2/gm
Calcite	0.05 ± 0.01	0.09 ± 0.01	4.0 ± 0.02 "
Flourite	0.05 ± 0.01	0.10 ± 0.01	2.0 ± 0.02 "

The water soluble starch (S-starch) used was a product of VEB Laborchemie Apolda (pro analyse). The oleic acid and all electrolytes applied in this investigation were of analytical grade. S-starch solutions were prepared according to Gorloviskij's method[8] as 0.2% solutions. To minimize chemical as well as biological degradation freshly prepared solutions were always used.

The adsorption isotherms of S-starch were determined at 20°C and 20% solids. Adsorption was virtually complete in a few minutes but analyses were not carried out before 1 hour of shaking. The amount adsorbed was calculated from the change of starch concentration before and after adsorption was complete. Determination of the starch concentration was based on the starch-iodine method.[9] Desorption tests were conducted, under the same conditions of adsorption, by the weight dilution technique.[10] I.R-spectral analyses were done by means of a UR 20 VEB Carl Zeiss Jena apparatus on mineral samples pressed in KBr.

Flotation tests were performed in a 250-ml. FIA-type flotation cell. Fifty-gram mineral samples in 200 ml. reagent solution were conditioned at a speed of 1100 r.p.m. and floated at 1.6 l/min air.

RESULTS AND DISCUSSIONS

The adsorption of S-starch on spar minerals is rapid and reversible. The adsorption isotherms obtained are shown in Fig. 1 and were found to fit the Langmuir adsorption equation (Fig. 2). The affinity of these minerals for starch is observed to increase from barite to calcite, to fluorite, following more or less the order of mineral solubility, 2.4, 14.3, to 9.1 mg./liter respectively. In other words, it seems that starch adsorption increases with the increase of lattice ionic species in solution. The addition of such ions in the form of saturated mineral solutions was shown to improve starch adsorption on the corresponding mineral.

The adsorption of starch molecules on solid-solution interfaces creates considerable interest in their conformation. It is generally accepted that polymer molecules are attached with short sequences of segments on the surface, the unattached segments extending in the solution as loops and bridges.[11] The relative number and arrangement of the attached and the unattached segments define the conformation of the adsorbed molecule. It can be inferred from the Langmuir-type adsorption isotherms of starch that there is preferential attachment of the polar groups of the starch on the mineral which results in a configuration more extended from the surface into solution. If the polymer is adsorbed in a coiled configuration, then the Langmuir isotherm is not obeyed, due to compressibility.[11,12]

An area of about 6×10^4 Å^2 per starch molecule, calculated from adsorption data, accords well with the value of 4×10^4 Å^2/mol. obtained from polymer thin film data[13] and the dimensions of a flat oriented starch molecule (helix). For many cases, investigators suggest a relatively flat (unfolded, uncoiled) polymer adsorption similar to polymer orientation at water-air interface. Thicker layers may then build up on subsequent addition of polymer.[14]

The relation between starch adsorption and mineral solubility may suggest that starch, like many organic polymers, could be adsorbed preferentially in the inner part of the electrical double layer, i.e., attached directly to a solid surface by any physical

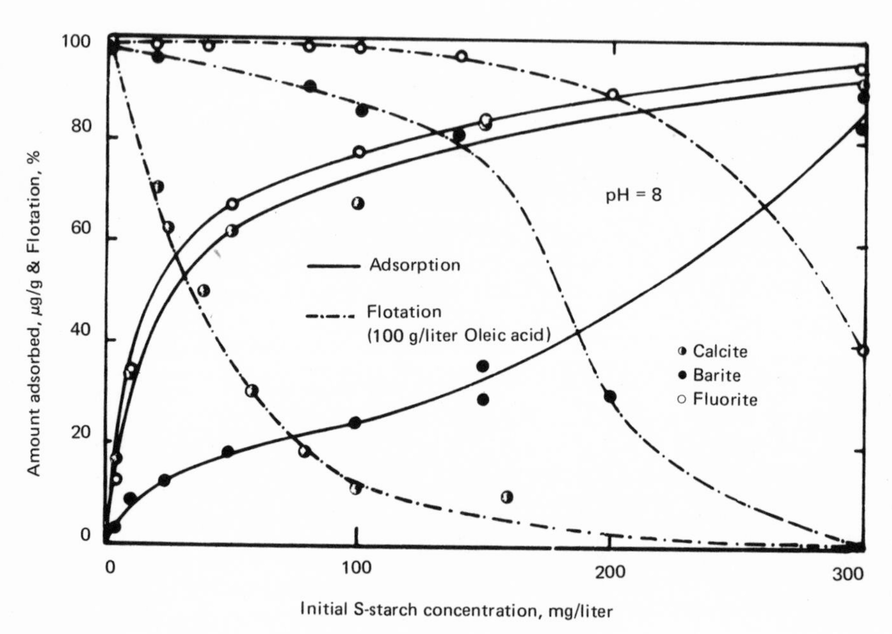

Fig. 1: Starch adsorption and depression on spar minerals.

or chemical bonding (depending on the nature of the solid and polymer). The character of the atomic-molecular interaction between ions and molecules in this region of the double layer and the solid surface is the same as in the bulk of the crystal lattice.[1] Therefore, the energy of adsorption of such ions and molecules is very close to the energy of dissolution of lattice constituents. The latter is known to depend on the hydration energy of the ionic species and crystal lattice forces. In a previous work[15] it was shown that there is a preferential release of mineral cations in aqueous solutions. Such cations are, therefore, involved in the starch adsorption process, as will be discussed later on.

Barite, calcite and fluorite are three salt-type low-solubility minerals with orthorhombic, hexagonal, and cubic crystal lattice respectively. The possible cleavage planes of barite and calcite are characterized by the coexistance of both cations and anions, while fluorite has only one plane between the fluoride ions. The electrostatic field built on the uniform fluorite surface is, thereby, much weaker than the non-uniform calcite and barite surfaces.[6] Therefore, we expect stronger adhesion of the adsorbed water and starch molecules to calcite and barite surfaces than to the fluorite one. This may explain why starch is a more effective depressant for calcite than for fluorite, in spite of the higher starch density on fluorite particles (Fig. 1).

Thus, there exists a direct relation between starch adsorption and crystal lattice structure, solubility and floatability of barite, calcite and fluorite particles. To understand the functional mechanism, the structure and conformation of starch molecules must be taken into consideration.

Starch is a highly polymeric carbohydrate built up of D-(+)-glucose units arranged either in straight- (amylose) or branched-chain (amylopectine) form. In amylose molecules the primary alcohol groups at C-6 form perhaps the strongest OH....O bonds, while the secondary OH-groups at C-2 and C-3 are more available in amylopectin. Thus, the adsorption of S-starch (20% amylose, 80% amylopectin) may take place via the huge number of OH-groups at C-2, C-3, and C-6 and the electronegative elements of the mineral surface, O_2 and F_2 (hydrogen bonding). The blocking of these groups either with iodine[7] or esterfication[10] has been proved to decrease the flocculating and depressing activities of starch sharply and hence its adsorption. Ionization of starch hydroxyls (pK $\simeq$ 12), especially those at C-2 and C-6, together with the ionization of other groups present as impurities (fatty acids, phosphates, etc.), would produce a negative charge on its micells.[15] Electrostatic adsorption of starch on positive sites of barite, calcite and barite surface (isoelectric point at pH 3.0, 6.8 and 6.3 respectively) is then possible. This is supported by the significant shifts of zeta-potential values of spar minerals towards negative values when brought in contact with starch solutions.[5,6,16]

Adsorption-desorption results shown in Fig. 2 indicate that the process is completely reversible, i.e., starch molecules are easily washable from the mineral surface. Consequently the forces attaching them to the surface must be rather weak and of the hydrogen bonding type and electrostatic interaction. I.R.-examination of the adsorbed starch layers supports this view. No new adsorption bands or even serious shifts in the location of the characteristic bands of starch were identified. Nevertheless, direct bonding between starch and mineral cationic species cannot be ruled out, since starches have been reported to form complexes with calcium, barium and other multivalent metal cations.[17] Such a complex formation of starch with mineral cations would contribute selectivity in the adsorption and depression process. A reduction in the adsorption of starch at high pH-values (Fig. 3) is then expected, since the concentration of the metal cationic species at the mineral surface decreases at these pH's. Addition of these cations in the form of saturated mineral solutions or as solutions of $CaCl_2$ and $BaCl_2$ proved to increase starch adsorption (Fig. 4).

From the above discussion it is obvious that the adsorption mechanism of starch on spar minerals involves much more than simple ionic and molecular dynamics. Neither the nonspecific adsorption forces (hydrogen bonding and electrostatic interaction) nor the formation of starch-metal complexes describe starch adsorption satisfactorily. A mechanism is therefore proposed to relate these interactions, based upon:

a) Direct or indirect chemical bonding, between starch hydroxyls with mineral cations, of the type M- 0 or 0 -M- 0,

b) Hydrogen bonding,

c) Electrostatic interaction,

d) Crystal lattice structure and free surface energy.

Application of oleic acid for the flotation of the starch-covered mineral particles is of special importance in the selective separation of spar minerals. The results shown in Fig. 1 indicate a good correlation between starch adsorption and the depressing activity on barite, calcite and fluorite. There exists a critical starch concentration, corresponding to the point of almost complete surface coverage.

Unlike the depressants normally used in froth flotation,

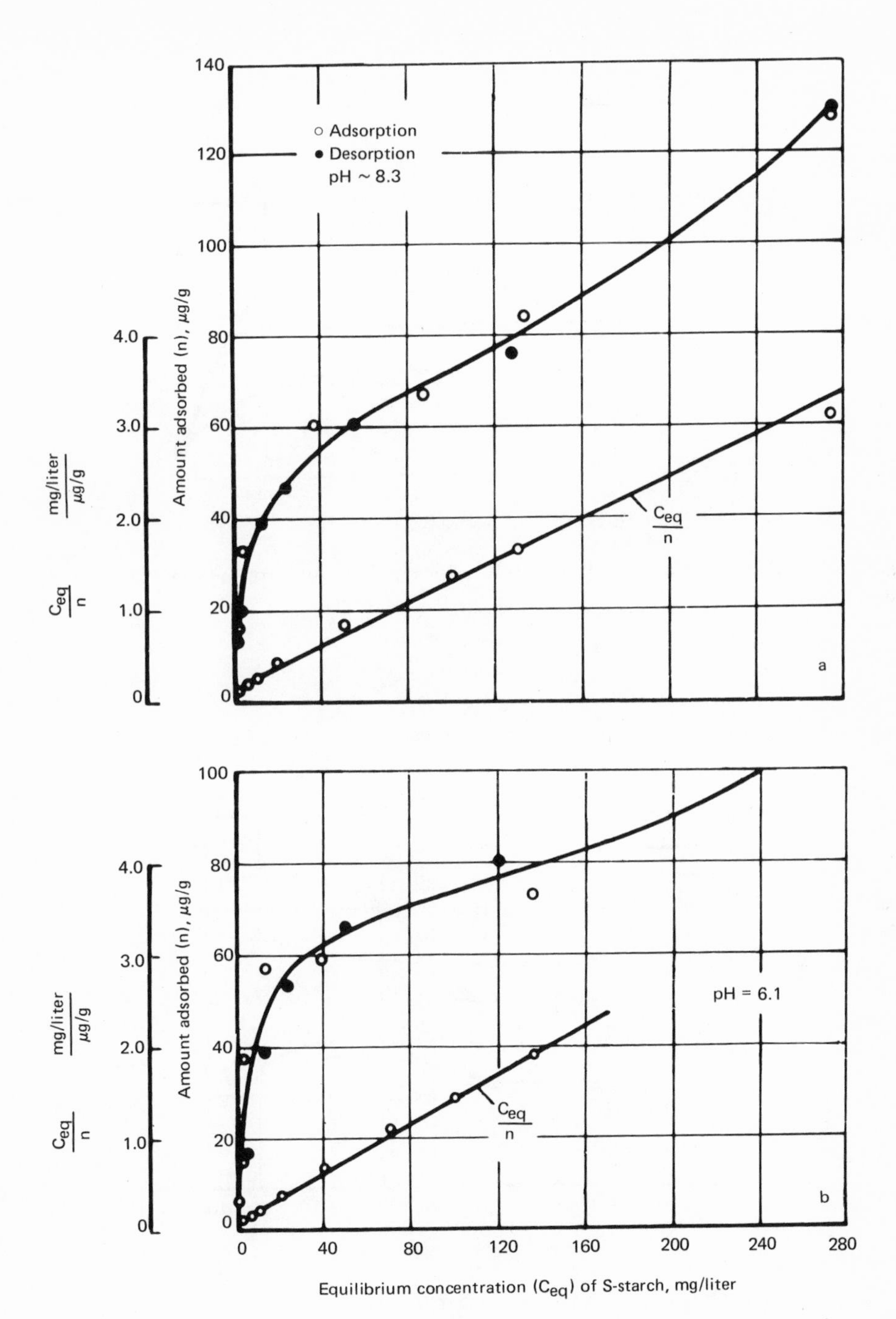

Fig. 2: Adsorption-desorption and Langmuir-isotherms of S-starch on (a) calcite and (b) fluorite.

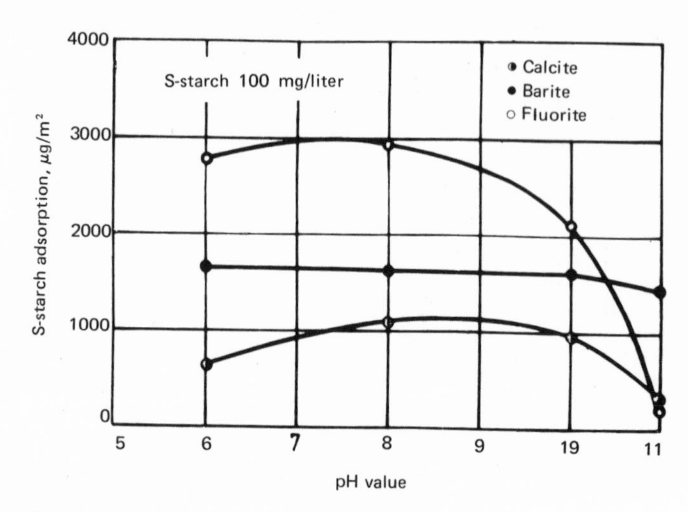

Fig. 3: Effect of pH on S-starch adsorption on extra-pure minerals.

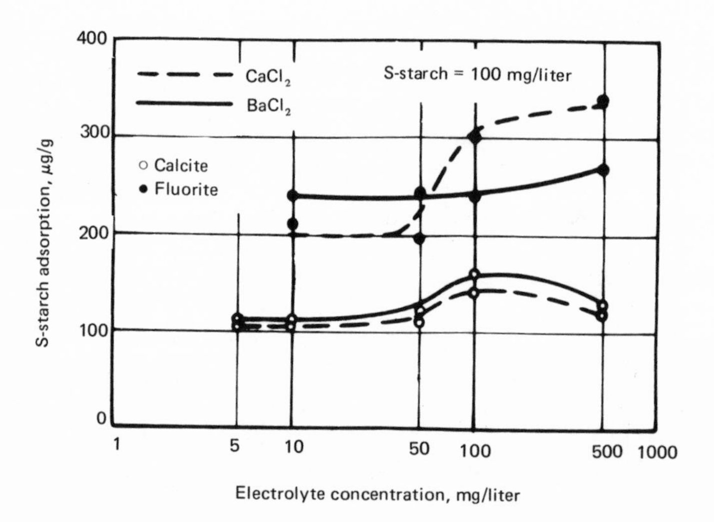

Fig. 4: Starch adsorption on fluorite and calcite in presence of $CaCl_2$ and $BaCl_2$

starch does not act by preventing the adsorption of collector on the mineral particles. A coadsorption of both starch and oleic acid (collector) on the mineral surface was proved by I.R. analysis.[10] This supports the idea of starch-collector clathrate complex and the wrapping of the collector inside the starch helixes, and hence the hydrophilic nature of the resulting mineral surface (depression).

CONCLUSIONS

1. The adsorption of S-starch on spar minerals decreases in the order fluorite > calcite > barite. Mineral solubility, crystal lattice structure, associated mineral impurities and the ionic composition of the pulp are adsorption determining factors. The Langmuir-type adsorption isotherms and the high occupation area of starch molecules suggest flat orientation.

2. A proposed mechanism for starch adsorption, based on: hydrogen bonding, electrostatic interaction and complex formation of starch-metal compounds, is developed.

3. A coadsorption of both starch and oleic acid (collector) on the mineral surface is proved. There is good correlation between starch adsorption and floatability of spar minerals.

REFERENCES

1. Klassen, Mokrousov, An Introduction to the Theory of Flotation, Butterworths, London (1965) 345.

2. G. Schulz, Dr.-Ing. Diss., Weimar Univ. 1971.

3. S.R. Balajee, I. Iwasaki, Trans. AIME 244 40 (1969).

4. P. Somasundaran, J. Colloid Interf. Sci. 31 557 (1969).

5. N.F. Schulz, S.R.B. Cooke, Ind. Enging. Chem. 45 2767 (1953).

6. H.J. Steiner, Radex-Rdsch., Radenthein 6 733 (1965).

7. A. Yazan, Dr.-Ing. Diss. T.H.S. Aachen, 1966.

8. S.I. Gorlovskij, Obogaŝĉenie Rud, Leningrad 6 (1956); 2 (1957).

9. S.R.B. Cooke, N.F. Schulz, E.W. Lindroos, Trans. AIME 193 67 (1952).

10. H.S. Hanna, Dr.-Ing. Diss. Bergakademie Freiberg, 1971 .

11. T.W. Healy, V.K. La Mer, Inter. Mineral Proces. Congr., New York, 1964.

12. Fontana, Thomas, J. Phys. Chem. 57 480 (1961).

13. H.E. Ries Jr., B.L. Meyers, Science 160 1449 (1968).

14. H.E. Ries Jr., Nature 226 72 (1970).

15. J.F. Foster, Starch: Chemistry and Technology, Academic Press, New York (1965) 354,375.

16. P. Blazy R. Houot, J. Cases, 7th Intern. Mineral Proces. Congr., Vol. 1, Gordon and Breach, New York (1964) 209.

17. J.N. Bemiller, in Starch: Chemistry and Technology, Academic Press, New York (1965) 318.

SINTERING OF COBALT-MOLYBDENUM OXIDE FILMS ON ALUMINA SUPPORTS

D. Dollimore and G. Rickett

University of Salford

Lancashire, England

The changes in the surface area during the heat treatment of cobalt-molybdenum oxide films on alumina supports have been related to two processes, surface diffusion occurring below 1073K and volume diffusion occurring above this temperature. The activation energy below 1073K is 35 kcal mole^{-1} (1.46 x 10^{5} joules mole^{-1}) and above this temperature is 85 kcal mole^{-1} (3.55 x 10^{5} joules mole^{-1}).

These results are compared with the sintering trends of single oxides typified by the behaviour of zinc oxide. General relationships are made relating the diffusion processes occurring on sintering with the melting point of the oxide. There is evidence that the presence of the metal oxides at the surface effectively lowers the melting point of the alumina support.

Two measurements may be used to note the progress of sintering: the measurement of density and that of surface area. Here the surface area changes against time are noted. The results for the single oxide and the film of oxide show similar trends. However, the difference lies in the results for the film of oxide spanning a temperature range where two diffusion processes are to be found. The diffusion process is stated by Hüttig to depend on the ratio of the sintering temperature to the melting point of the oxide (T_m in K). When this ratio exceeds 0.33 then sintering occurs via processes of surface diffusion whilst above 0.50 the operative mechanism is that of volume diffusion.

It was not thought that the sintering of zinc oxide resulted in a porous material, nor was the original material porous.

However, a complete low temperature nitrogen adsorption isotherm on the original cobalt-molybdenum oxide film support showed it to be porous. This porosity was retained on sintering at 1088K but the surface area was reduced and micropores were eliminated. It is concluded that the process of pore widening leads to the collapse of the micropores.

INTRODUCTION

The process of sintering is usually studied with respect to the behaviour of single or mixed oxide systems and by reference to properties such as density. In this study the sintering process is followed by reference to the changes noted in surface area, and the system consists of a cobalt-molybdenum oxide film on an alumina support. The sintering of an oxide appears to be dependent upon its initial condition. In many cases the oxide is prepared by the thermal decomposition of oxysalts. In such instances the resultant surface area is a complex function of two competing processes. The first is the increase in surface area accompanying the decomposition and the second is the reduction in the area brought about by the sintering that occurs at the temperature of decomposition. Thus zinc oxalate decomposes to yield a high surface area zinc oxide which is rapidly reduced by sintering.[1] Some aspects of the sintering of active materials have been outlined by Roberts[2] who selected a variety of systems for his review. The extent of sintering depends on the mechanism of diffusion. Hüttig[3] reports that this depends on the ratio of the sintering temperature (in K) to the melting point of the oxide (T_m in K). He considers three processes: (a) adhesion; (b) surface diffusion, and (c) lattice diffusion. Adhesion is the first sintering process to be observed, whereby the separate particles become "fritted" together at points of contact. Surface diffusion, which is caused by movement of ions along the surface of micelles, occurs at $^T/T_m$ = 0.33-0.45. Above the Tamman temperature, where $^T/T_m$ = 0.50, lattice fiffusion is stated to occur. The majority of studies on sintering in these terms are based on studies of the behaviour of metals or similar systems where the reference to T_m is clearly seen. In the case of oxides the melting point may not exist as the oxide may dissociate before such an event occurs. Nevertheless, the occurrence of three different processes of sintering in order of increasing temperature as mèntioned above seems reasonable.

Kuczynski[4] states that volume diffusion is an important factor in the rate of sintering and quotes properties such as non-stoichiometry and dislocations as being of prime importance. An expression for the rate of sintering based on the macroscopic flow of solid materials has been derived by Mackenzie and Shuttleworth[5] and tested by Williams and Murray.[6] Similar work is reported by Clark and White.[7] Sintering of metal oxide systems is usually studied in terms of density changes. Studies on the sintering of zinc oxide include following the change in surface area as determined by gaseous adsorption methods.[8,9,10,11] This method is applicable and especially suitable for catalyst materials, but in many preparations the catalyst itself includes a layer of deposited materials - the catalyst - on a catalyst support.[12] In the case reported here the sintering of a cobalt-molybdenum oxide layer on a γ alumina substrate is reported. This is a typical catalyst used for the removal of organic sulphur compounds from petroleum feedstocks by hydrogenalysis to hydrogen sulphide and hydrocarbons.

EXPERIMENTAL AND MATERIALS

The catalyst used was the standard commercial product Comox manufactured by Laporte Industries Ltd., which consists of cobalt (3.5% calculated as CoO) and molybdenum (12.5% calculated as MoO_3), supported on γ alumina.

For the sintering studies a horizontal muffle tube furnace was used with a temperature control to within ± 2°C. The samples, approximately 2.5 g, were spread as thin layers in a silica boat and could be introduced or removed from the furnace very quickly. Dry air was passed through the furnace at 500 ml/min. Before the samples were sintered they were preheated to 400°C for 2 hrs to eliminate adsorbed water. It was determined by experiment that this pretreatment did not effect any change in the surface area of the samples. The samples were sintered for fixed times at preselected temperatures. When the time had elapsed the sample under investigation was subjected to a surface area determination evaluated from the nitrogen adsorption isotherm at 77 K. The sintered catalyst was outgassed for one hour at 200°C at a pressure less than 10^{-4} mm of Hg (10^{-2} Nm^{-2}). A volumetric unit was used as described by Lippen,[6] Linsen and De Boer.[13]

RESULTS AND DISCUSSION

The variation of the surface area of the catalyst after sintering it at a fixed temperature for various times is given in Fig. 1 in which specific surface area is plotted against time, the original specific surface area being 230 m^2 g^{-1}. The temperatures at which the sintering process was studied were 1013 K, 1038 K, 1068 K, 1088 K and 1123 K. In order to obtain the equilibrium specific surface area it was necessary to extrapolate the data in Fig. 1. This resulted in estimates for the final equilibrium surface area at the various temperatures as follows: 1013 K, 150.6 m^2 g^{-1}; 1038 K, 111.5 m^2 g^{-1}; 1068 K, 95 m^2 g^{-1}; 1088 K, 61.5 m^2 g^{-1}; 1123 K, 14.2 m^2 g^{-1}. Samples heated for one week at the above temperatures were found to reach

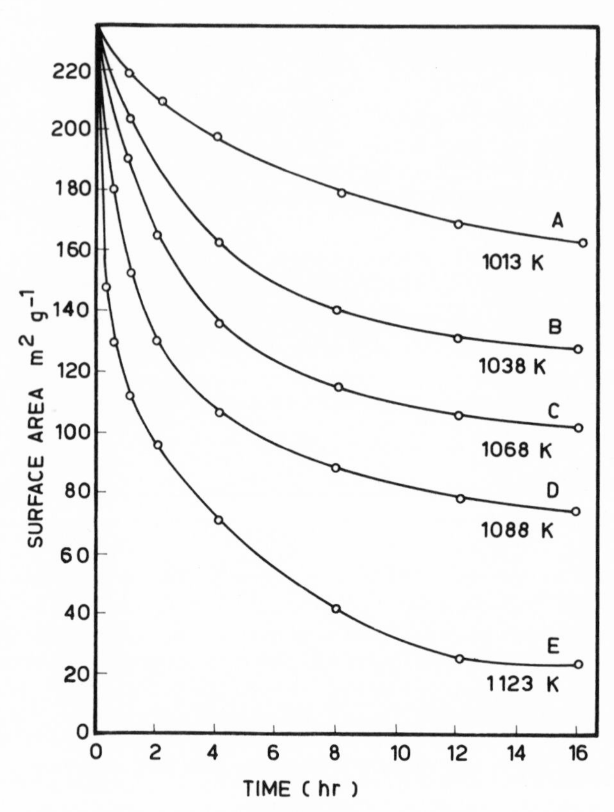

Fig. 1: Variation of the specific surface area of cobalt-molybdenum oxide film on γ alumina during sintering

(A) 1013K (B) 1038K (E) 1123K

(C) 1068K (D) 1088K

these values, but it is preferred to use the extrapolated values because of the difficulty experienced in holding the temperature of the furnace stready over a very long period of time.

The kinetics of the sintering process have been determined by the use of reduced time plots[14] where the fractional reduction in specific surface area (α) is given by:-

$$\alpha = (\underline{S}i - \underline{S}t)/(\underline{S}i - \underline{S}f)$$

where $\underline{S}i$ = Initial Specific Surface Area; $\underline{S}t$ = Specific Surface Area at time t; and $\underline{S}f$ = Specific Surface area at time infinity.

This is plotted against $t/_{t0.5}$, where t is the time and t0.5 is the time taken to reach α equal to 0.5. The plots of α against t/t0.5 are given in Fig. 2. From this it is possible to select two distinct processes, one occurring below 1073 K and one above that temperature.

Activation energies for the sintering of the catalyst have been calculated by the method of Hüttig used by Gray[9] for the sintering of zinc oxide. Hüttig assumes that the velocity constant for a given ratio of sintering at a constant temperature is proportional to the reciprocal of the time taken to reach a given degree of sintering.

That is:

$$K = \text{velocity constant} = k/t$$

where t is the time taken to reach a certain percentage of sintering. When this is applied in the Arrhenius equation;

$$\log t = E/(2.303\ RT) + C$$

where: E = Activation Energy Cal mole^{-1} (joules mole^{-1}); T = Temperature K; R = Universal Gas Constant; and C = Integration Constant. From the variation of log t with 1/T a value for the apparent activation energy can be derived. The results are obtained from Fig. 3. These plots show a change in gradient of the data at 1073 K. The activation energy for the sintering process occurring below 1073 K is 35 cal mole^{-1} (1.46×10^5 joules mole^{-1})and above that temperature is 85 kcal mole^{-1} (3.55×10^5 joules mole^{-1}). It can be concluded from Fig. 2, where at temperatures 1013 K, 1038 K and 1068 K the data fall essentially all on one line, that the results can be represented by a first order kinetic law:-

$$\ln (1 - \alpha) = -Kt + C$$

At these temperatures the above equation describes the behaviour until

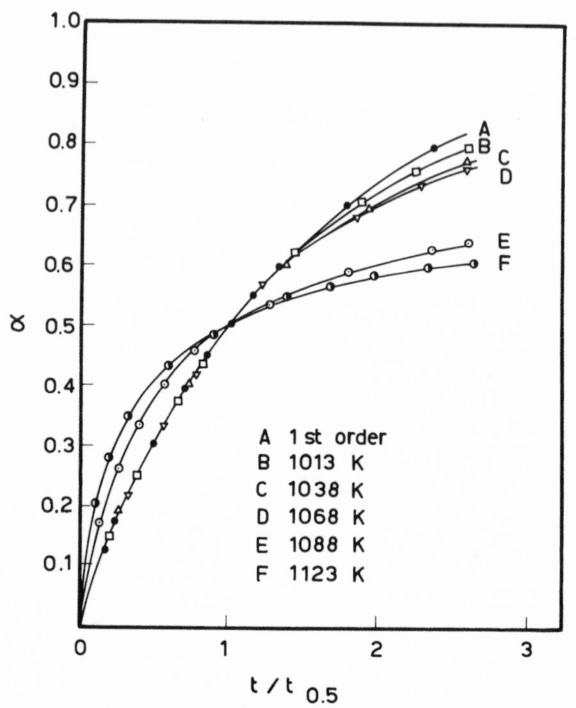

Fig. 2: Relationship between the fractional reduction in specific surface area α and reduced time $(t/_{t0.5})$.

(A) Theoretical 1st order plot

(B) Experimental data at 1013k

(C) Experimental data at 1038K

(D) Experimental data at 1068K

(E) Experimental data at 1088K

(F) Experimental data at 1123K

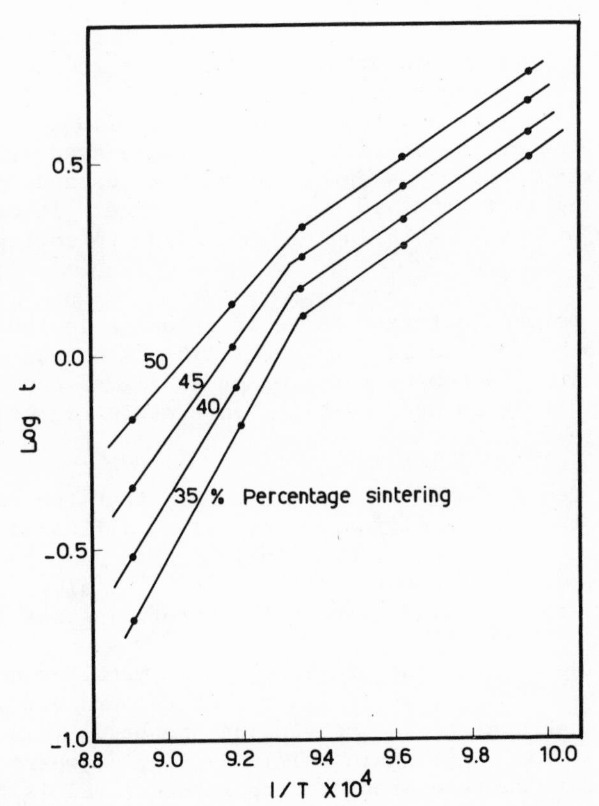

Fig. 3: Activation energy for sintering of the cobalt-molybdenum oxide film on γ alumina.

70 per cent of the surface area of the catalyst is lost. The process below 1073 K deviates from the first order rate law at successively higher values of α as the temperature is reduced. This can be expected as the porous structure of the catalyst is less distorted at lower temperatures. At temperatures 1088 K and 1123 K the plot of α against $t/t0.5$ (Fig. 2) suggests a volume diffusion process is occurring because the ratio of α to $t/t0.5$ is very high when t is quite small.

Figure 4 shows a nitrogen adsorption isotherm before sintering and Fig. 5 shows a nitrogen adsorption isotherm on the catalyst sintered at 1088 K. The pores present in the unsintered catalyst are micropores 0.5 nm - 2.0 nm and transitional pores 2.0 nm - 3.0 nm. They account for the great majority of the surface area of the catalyst. The sintered catalyst is devoid of micropores and the porous structure remaining is in the 4.0 nm - 7.0 nm range. It can be concluded that pore widening is taking place leading to collapse of the micropores.

It has been pointed out that there is a change in the sintering behaviour at 1073 K and this is connected with the nature of the sintering process. The low temperature region in Fig. 3 can be attributed to surface diffusion and the high temperature region of Fig. 3 to volume diffusion.

It has been demonstrated by Smoluchowski[15] that the energy of activation for both surface diffusion and volume diffusion is high, but that of surface diffusion is usually less than half that of volume diffusion. This is borne out also with the values for the activation energy for the sintering of the dyrofining catalyst.

There is evidence that the presence of the metal oxides contained in the catalyst effectively lowers the temperature at which volume diffusion takes place in the alumina support. This can be shown by a plot of log Sf/Si against T/Tm in Fig. 6, where Sf is the minimum specific surface area at temperature T, Si is the initial specific surface area, and Tm the melting point of alumina (from quoted literature values).

There is a distinct break in the plot occurring at T/Tm = 0.47. The change in gradient is to be expected, however, at T/Tm = 0.5, because at this point the process of volume diffusion is reported to commence.[3] The introduction of the metal oxides in the catalyst has thus lowered the T/Tm value of the alumina support where surface diffusion gives way to volume diffusion. It is as though the presence of metal oxides on the alumina makes the surface behave as a solution of the oxide systems with a consequent lowering

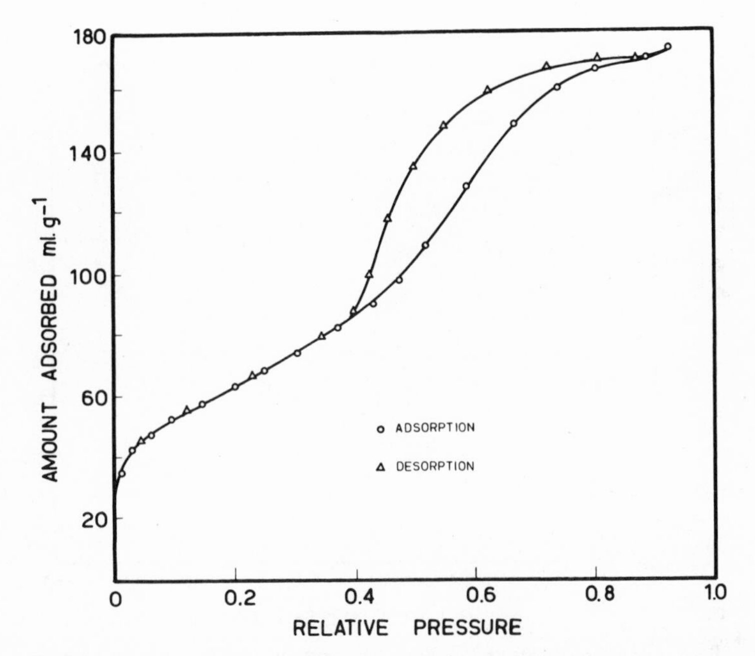

Fig. 4: Low temperature (77K) nitrogen adsorption isotherm of the cobalt-molybdenum oxide film on γ alumina prior to sintering

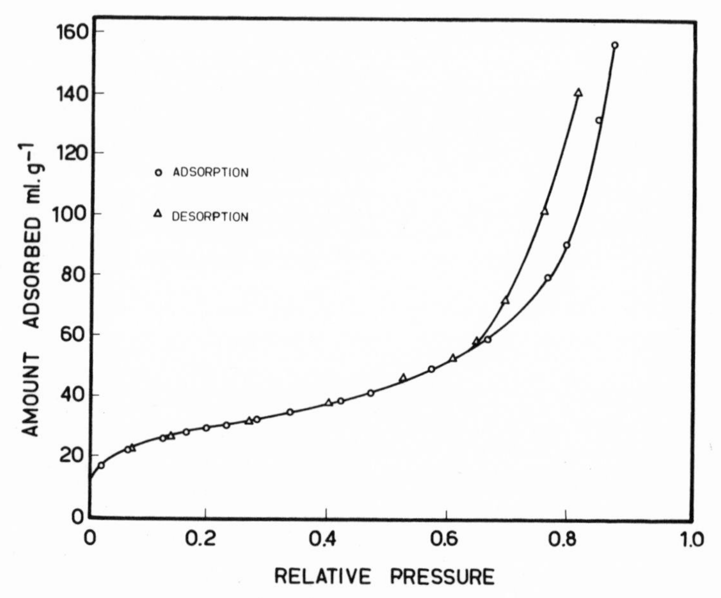

Fig. 5: Low temperature (77K) nitrogen adsorption isotherm of the cobalt-molybdenum oxide film on γ alumina after sintering at 1088K.

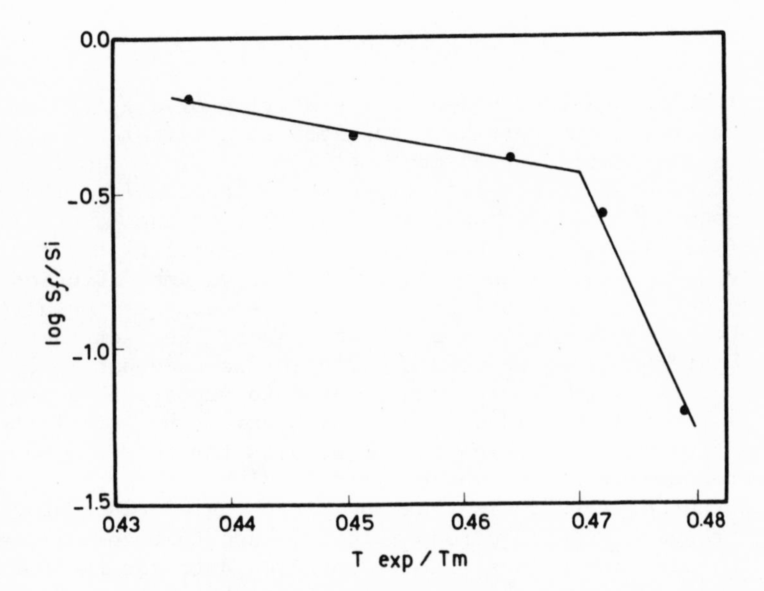

Fig. 6: Logarithmic relationship between surface area and temperature of treatment for the sintering of the cobalt-molybdenum oxide film on γ alumina.

of the melting point of the mixture. A simple calculation based on the temperature of 1073 K at the onset of bulk diffusion in the system under investigation would lead to the prediction that the melting point of the oxide mixture is 2146 K. This is substantially lower than the melting point of alumina. It is not suggested that this melting point could be achieved in practice, as other processes would interfere, but it does serve to indicate that the action of small additions of other metal oxides to the alumina surface could cause that surface to behave as though it was a solution of the oxides with a consequent readjustment of the parameters relating to sintering.

CONCLUSION

Quantitative measurements on the sintering of a hydrofining catalyst on an alumina support have resulted in a calculated value for the energy of activation of 85 kcal $mole^{-1}$ (3.35×10^5 joules $mole^{-1}$) above 1073 K, and 35 kcal $mole^{-1}$ (1.46×10^5 joule $mole^{-1}$) below that temperature. These values establish that at the lower temperature a surface diffusion process predominates, whilst at the higher temperature the mechanism is probably that of volume diffusion. It appears as though at the surface, the alumina support - metal oxide system is behaving as a solution of the oxides; the fact that the temperature of the onset of volume diffusion is somewhat lower than that expected for pure alumina can be used to support this argument. The results for single oxides such as zinc oxide show similar general trends but the difference lies in the results for the film of oxides-panning a temperature range where two diffusion processes are to be found. Another difference is that the sintering of zinc oxide[11] is thought to result in a non-porous material, and that the original material was also non-porous. The adsorption data for the cobalt-molybdenum oxide film on the alumina support showed it to be porous. This porosity was retained on sintering at 1088 K but the surface area was reduced and micropores were eliminated. It is concluded that the process of pore widening causes the collapse of the micropore structure.

ACKNOWLEDGMENTS

G. Rickett wishes to acknowledge the award of a research grant from Esso Petroleum Co. Ltd. The authors also acknowledge Laporte Industries Ltd. for their kind supply of the catalyst.

REFERENCES

1. D. Dollimore, D. Nicholson, J. Chem. Soc. 179 961 (1962).

2. D. Roberts, Metallurgia, August, p. 1 (1950).

3. G.F. Hüttig, Kolloid 2 98 6, 263 (1942); 99 262 (1942).

4. G.C. Kuczynski, J. Appl. Phys. 20 1160 (1949).

5. J.K. Mackenzie, R. Shuttleworth, Proc. Phys. Soc. 62 833 (1949).

6. J. Williams, P. Murray, Technical Report M/TN8 (1952).

7. P.W. Clark, J. White, Trans. Brit. Ceram. Soc. 49 305 (1950).

8. V.J. Lee, G. Paravano, J. Appl. Phys. 30 1735 (1959).

9. T.J. Gray, J. Amer. Ceram. Soc. 37 434 (1954).

10. J.P. Roberts, J. Hutchings, Trans. Farad. Soc. 55 1394 (1959).

11. D. Dollimore, P. Spooner, Trans. Farad. Soc. 67 2750 (1971).

12. D. Dollimore, T.E. Jones, J. Appl. Chem. Biotech. 23 29 (1973).

13. B.C. Lippens, B.G. Linsen, J.H. de Boer, J. Catalysis 3 32 (1964).

14. J.H. Sharp, G.W. Brindley, B.N. Achor, J. Amer. Ceram. Soc. 49 379 (1966).

15. R. Smoluchowski, On Imperfection in Nearly Perfect Crystals, W. Shockley, ed., 467, Chapman and Hall (1952).

PROPERTIES OF ANODIZED ZINC IN ALKALINE SOLUTIONS

S.M. El Raghy, M. M. Ibrahim, and

A. E. El-Mehairy

University of Cairo, Egypt*

The photoeffects, impedance and reflectivity of anodized zinc in alkaline solutions were measured in order to characterize the electronic nature of the oxide. The effect of the forming electrolyte and the rate of growth of the anodic film on its properties were evaluated. The response to illumination increased with the alkalinity of the electrolyte. It was possible to relate the photocurrent to the applied potential mathematically with constants characteristic of the electrolyte.

The specific conductivity of the anodic product formed in sodium hydroxide electrolytes was twice that formed in sodium borate electrolyte. The porosity of the film deposited in a high pH electrolyte was indicated by a higher capacitance and confirmed by a low specular reflectivity.

A theory based upon the changes in the semiconducting properties of the solid surface film of the anodic product is advanced to explain the observed phenomena.

INTRODUCTION

The anodic behaviour of zinc in alkaline solutions has been studied utilizing several electrochemical techniques.[1-5] The steps

*Dr. S.M. El Raghy, Lecturer; Mr. M. M. Ibrahim, Demonstrator; Dr. A. E. El-Mehairy, Professor of Metallurgical Engineering, Faculty of Engineering, University of Cairo, Giza, Egypt.

leading to the formation of an oxide (or hydroxide) film have been summarized by Dirkse.[6] It is generally stated that the passivating film controls the discharge performance of alkaline batteries with zinc anodes. However, the nature of the film cannot be defined from electrochemical measurements alone.

Solid state studies have proved the very sensitive nature of semiconducting zinc oxide.[7-11] Electronic processes in a zinc oxide single crystal were shown to be different from those in a polycrystalline specimen. Also, noncrystalline zinc oxide differs from the crystalline material in several physical properties: it has less density, a higher energy gap and higher electrical resistivity. The incorporation of impurities is very effective in shifting the optical and electrical properties of the oxide.

Huber[12] was the first to attribute the changes in the colour of the anodic oxide film formed on zinc in different electrolytes to the nonstoichiometry of the film. He suggested that the film was zinc oxide with a varying excess of metallic zinc giving rise to the black colour. However, in a study on the nature of the anodic film formed on cadmium in an alkaline electrolyte[13] reference was made to the effect of OH^- incorporated in the oxide.

The present study was therefore undertaken to gain more insight into the properties of the film formed on zinc in alkaline solutions (buffered and unbuffered). This paper represents a progress report on the work carried out to date. Photoeffects, impedance and reflectivity were measured and used to indicate the electronic nature of the oxide.

EXPERIMENTAL

Zinc of 99.9975% purity, the main impurity being lead, was used to fabricate electrodes which were subsequently mechanically polished and etched in nitric acid. The electrolyte was prepared by dissolving analar grade chemicals in distilled water of conductivity 10^{-6} $ohm^{-1}cm^{-1}$.

A Hanovia U.V.S. 220 medium pressure mercury arc in a quartz tube was used with a stabilizing unit to produce a constant intensity of light. Potentiostatic control was effected with a Wittons Electronics potentiostat. A transistorized regulated power supply (Metimpex TR 9/60) supplied the anodizing constant current.

Figure 1 shows the system used for impedance measurements, and an A.C. bridge was used for measuring resistance alone. Figure 2 shows the apparatus used to measure the reflectivity of the dry anode. The main instrument was an 11-stage photomultiplier.

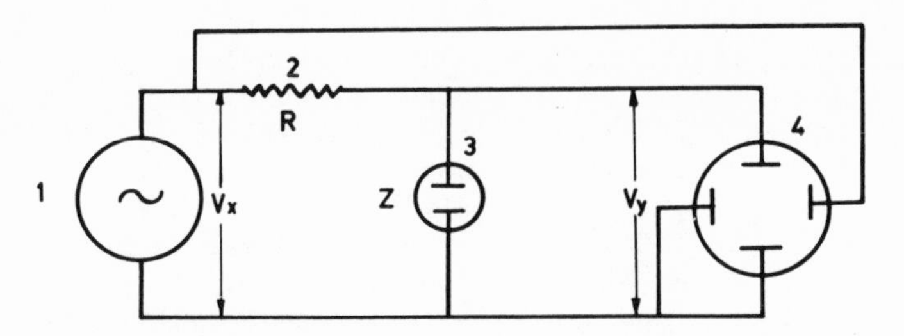

Fig. 1: Schematic layout of the impedance measuring system ($V_y/V_x = Z/R$). (1) A.C. power supply, type E 7128; (2) constant resistance 30 K Ω; (3) cell impedance; (4) cathode ray oscillograph (C.R.O.) type B 256 (England).

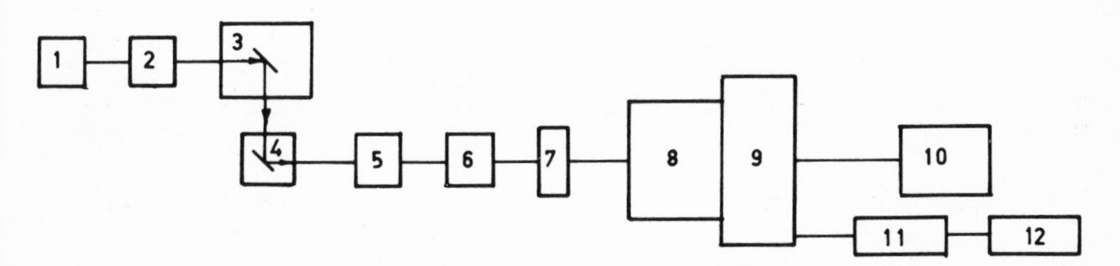

Fig. 2: Schematic layout of the reflectivity measuring system. (1) Transistorized 6V D.C. supply; (2) lamp housing; (3) Reichert inverted-stage microscope; (4) mikes housing; (5) auxiliary magnifying system; (6) aperture slide; (7) filter; (8) photomultiplier; (9) resistance chain; (10) transistorized E.H.T. supply; (11) D.C. amplifier; (12) pen-recorder.

RESULTS AND DISCUSSIONS

Electrochemical photoeffects. The potentiostatic sweep over the -1 to +3 volt anodic potential range was effected at a rate of 50 mV/min and 20 mV/min. Figure 3 shows the E-i behaviour of zinc anodized in sodium hydroxide and in sodium borate electrolyte in darkness. Three stages are indicated in the figure, dissolution, passivation and oxygen evolution. The rate of the anodic processes was much slower in the borate solution. Passivity was consistent with an anodic current of about 10μ Amp over the potential range -0.5 to +1.0 volt. This is taken as a proof of the stability of the passivating film at this pH, and agrees with the solubility measurements[14] which indicate a minimum for zinc oxide in an aqueous electrolyte of pH ∿10.

With sodium hydroxide, the passivity passed through two steps, namely deposition and competititon between the dissolution and deposition of the solid product (oxide or hydroxide).

Flashing the electrode with light at different potentials resulted in an increase of the anodic current at potentials above passivation. The instantaneous increase in current, photocurrent i_p, is shown plotted against the electrode potential for different electrolytes in Fig. 4. The following direct relationship log i_p and E is obvious up to a certain potential. Above this potential "saturation" would control the reaction to light:

$$E = a + b \log i_p$$

where a is a constant related to the passivating potential (Flade potential); b is a constant depending on the nature of the semiconductor, determined by the forming solution; i_p is equal to about 800 mV/each factor of ten in current for films formed in borate, and about 300 mV/each factor of ten in current for films formed in hydroxide.

The photocurrent may be considered as a measure of the absorbed light by the anodic film and of the efficiency of converting these quanta into electrical energy.[10,15] This energy conversion process is a function of the amount and nature of the solid product. However, the amount of deposited film does not vary appreciably with the electrolyte pH in the range studied. In fact the thickness of the anodic film was maximum at a pH around 12,[(16)] whereas the photocurrent increased with the alkalinity of the forming electrolyte. This may be taken as evidence of the changing nature of the anodic product on zinc depending upon the solution in which it is formed.

Reflectivity measurements. The reflectivity of any surface is a measure of its roughness (diffuse reflectivity) and its absorption

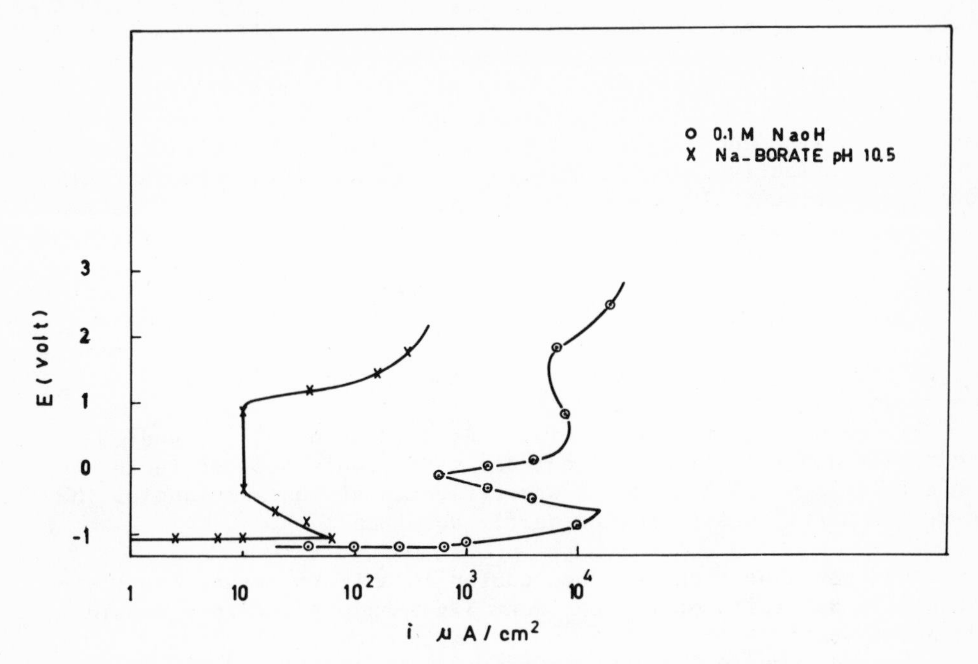

Fig. 3: Current/potential curve at 50 mV/min.

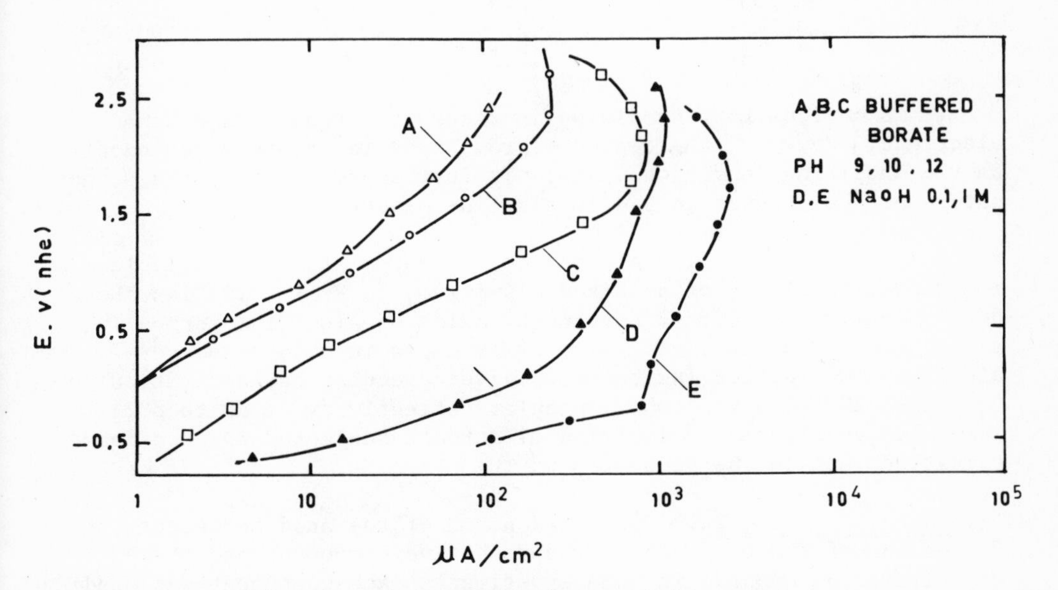

Fig. 4: Photocurrent/potential curves.

spectra (specular reflectivity). A rough surface will show how high scattering of incident light, whereas a mirror-like surface will reflect wave lengths longer than its absorbing limit.

Etched zinc does not absorb light of wave length > 3600 Å. Its specular reflectivity for light of λ = 4900 Å is 40%; the remaining 60% is scattered (diffuse reflectivity). Anodic oxidation changes the surface properties. Figure 5 gives the specular reflectivity of zinc measured as a function of the anodizing current density applied for 30 minutes. It should be noted that in 0.2M sodium borate solution, 28% of the incident light was reflected at the same angle after anodizing at 200μA/cm^2. This percentage decreased to 2% for a surface anodized in 0.2M NaOH at 4mA/cm^2.

Figure 6 shows the reflectivity as a function of the sodium hydroxide concentration. In extreme cases, only 0.5% of incident polarized light of λ = 4900 Å was reflected at the same angle. The rest was partly scattered and partly absorbed.

Assuming that each absorbed photon is able to create one electron, the partition of the incident light on anodized zinc would be as follows:

	Incident light		
Anodizing conditions	absorbed	reflected	scattered
clean zinc	0	40%	60%
borate, 0.2mA/cm^2	0.5%	26%	73.5%
4.0mA/cm^2	8.5%	16%	75.5%
hydroxide, 0.2mA/cm^2	2%	8%	90.0%
4.0mA/cm^2	12.5%	1.5%	86.0%

It is obvious that anodizing increased the roughness of the electrode. However, the degree of roughness is strongly dependent on the anodizing conditions. This is further evidence of the porous nature of the product formed in alkaline electrolytes.

Although data on zinc are scarce, measurements are available for the reflectivity of anodized aluminium.[17] Reflectivities above 90% are associated with a continuous oxide of aluminium formed in non-aggressive electrolyte. The oxide grown in acidic and very alkaline electrolytes (aggressive) gives specular reflectivities in the range 20-50%. Values of specular reflectivity of up to 80% were also reported[16,18] for zinc electrodes subjected to much less oxidation than in the present study.

Impedance measurements. The impedance Z of the anodized electrode is the sum of two components, the resistive component and the capacitive component. In some experiments each component was studied

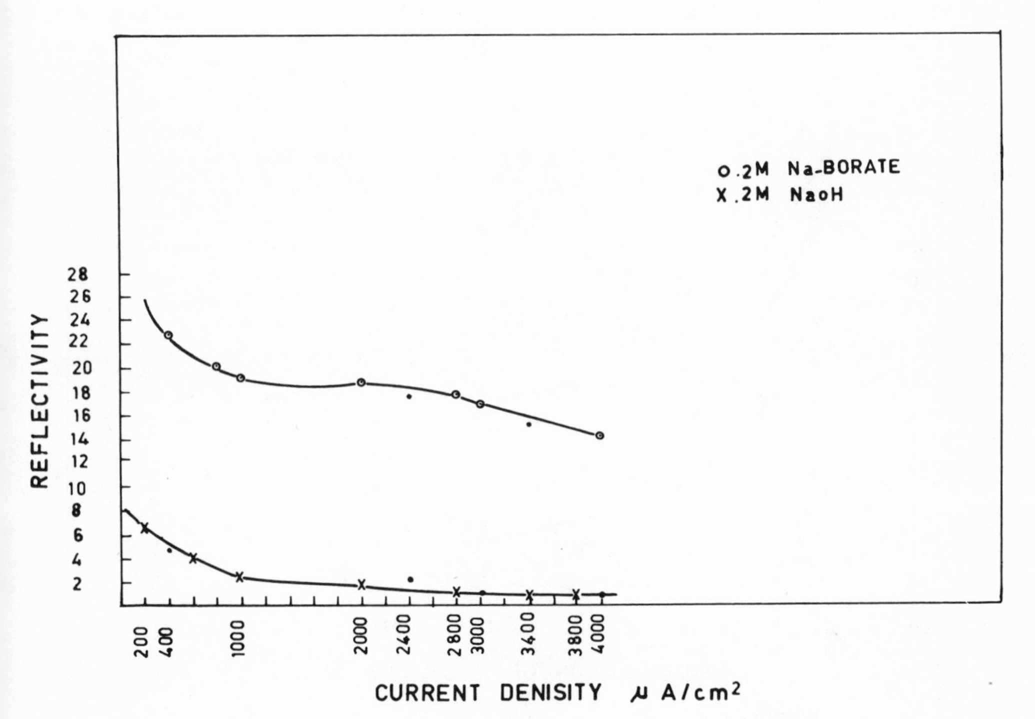

Fig. 5: Effect of C.D. on the reflectivity of anodized zinc.

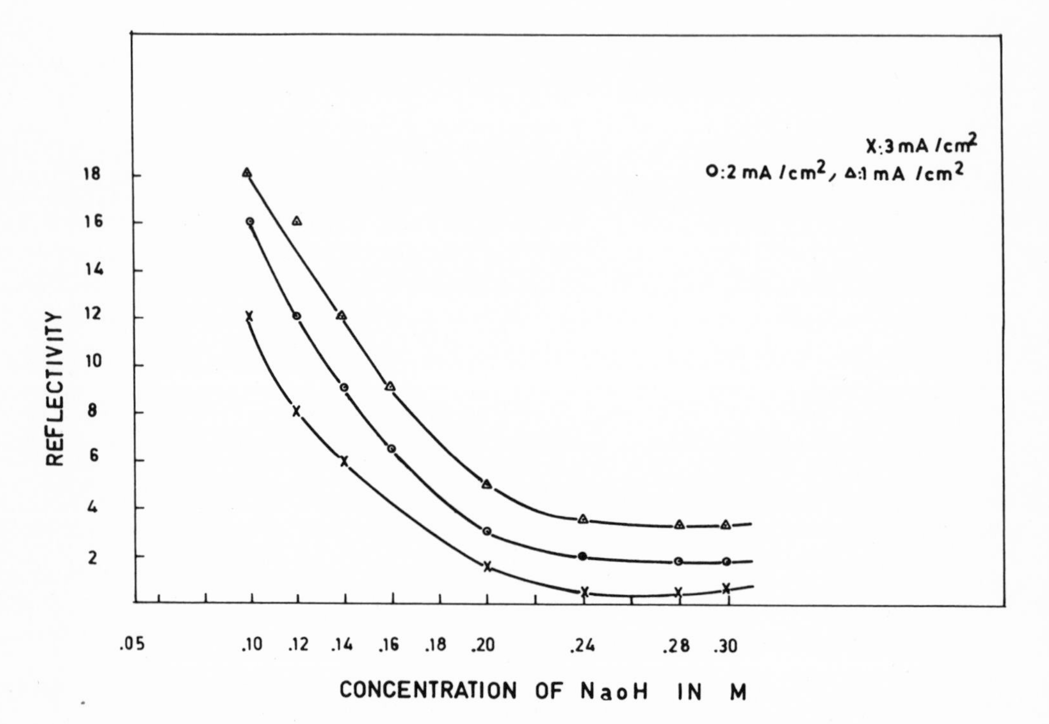

Fig. 6: Effect of concentration of sodium hydroxide on reflectivity of anodized zinc.

separately.

In the borate solution of pH 9.5, zinc behaved under constant current almost in a similar way as valve (Ta, Zr) metals. The continuous growth by ionic migration in the insulating film increases the electrode potential steadily. Figure 7 shows that the resistance R and inverse capacitance 1/C increase linearly with time. It is possible to compute the thickness of the oxide film using the relationship

$$D = (A\ \varepsilon/4\ \pi).\ 1/C$$

where D is the film thickness, ε is the dielectric constant of the anodic film, C is the measured capacitance, and A is the interfacial area.

This relationship is applicable if the dielectric properties of the film ($A\varepsilon$) do not change as the film thickness, but this is not the case, as will be demonstrated later.

In Figs. 8 and 9 both the resistance R and overall impedance Z are plotted against the time of anodizing at 1 mA/cm^2 in borate and hydroxide electrolytes respectively. In both cases, the resistance increases linearly. The slope of the line is a function of the specific resistivity of the continuously deposited film. This slope is almost doubled by increasing the alkalinity of the electrolyte from pH 10.5 to pH 13. Vermilyea[19] and others[20] have shown that the resistance of anodic oxide film on valve metals is strongly dependent on the forming electrolyte. This was explained in some cases by the porous nature of the film. The incorporation of impurities may be the basis of understanding this phenomenon.

The steeper increase in overall impedance with time of anodizing in 0.1M borate electrolyte is an indication of the increase in the capacitive component 1/C. However, the overall impedance of anodized zinc in 0.1 M hydroxide decreases with time despite the increase of the resistive component. This indicates a decrease in the capacitive component or an increase in the capacitance of the anode. This is further clarified by Fig. 10, which summarizes the effect of current density of anodizing on the capacitance of zinc anodes in both 0.1 M borate and 0.1 M hydroxide electrolytes.

In the equation relating film thickness to its capacitance, it is difficult to assume that the dielectric properties of the film are independent of anodizing conditions. The increase in the capacitance of the anodic product with increasing pH may be explained by the effect of the electrolyte on either the surface area or the permititivity ε_o of the formed film. The roughness of zinc anodized in more alkaline solutions was greater. This was shown by

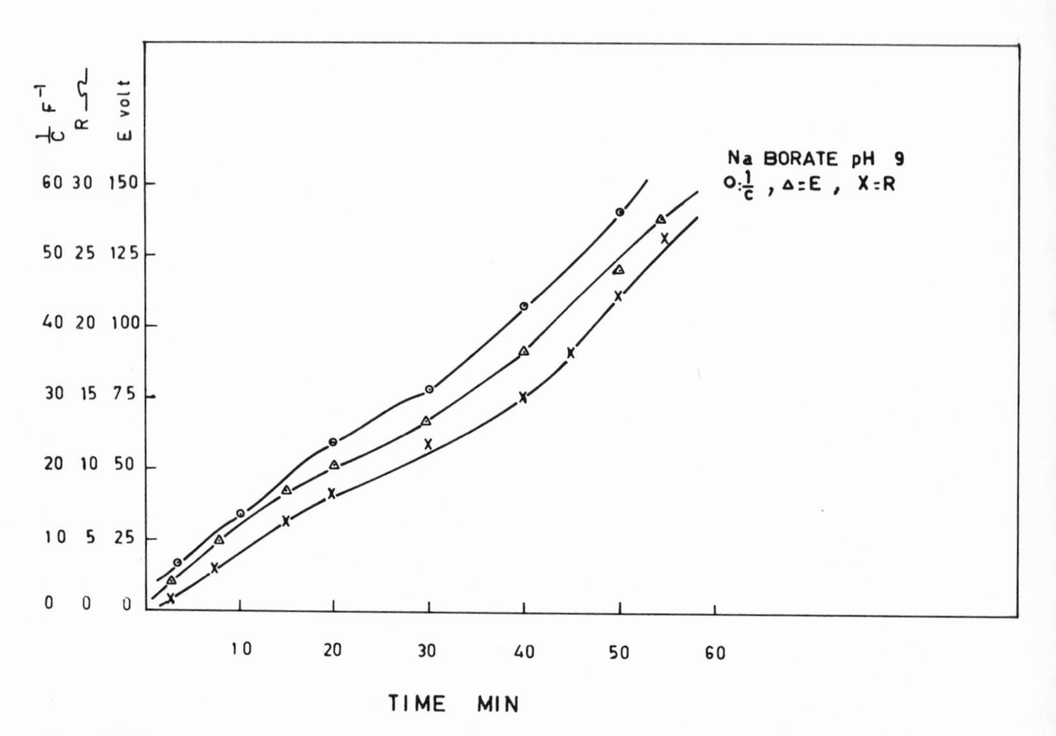

Fig. 7: Effect of time on E, R and 1/C of zinc anodized.

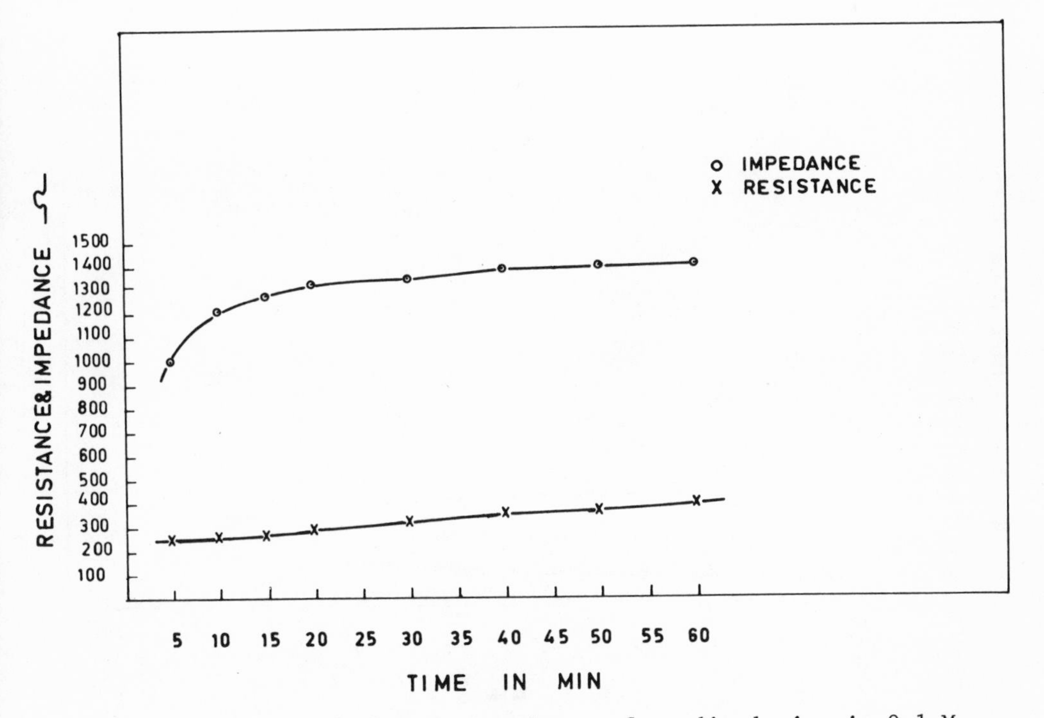

Fig. 8: Effect of time on impedance of anodized zinc in 0.1 M sodium borate electrolyte.

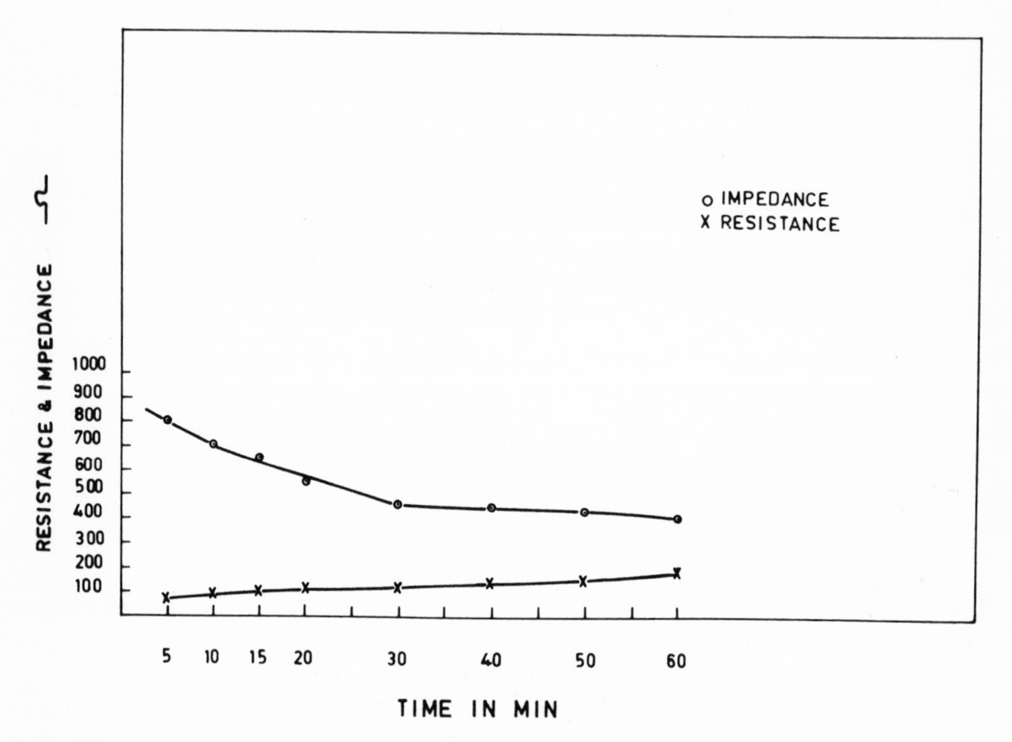

Fig. 9: Effect of time on the resistance and impedance of anodized zinc in 0.1 M sodium hydroxide electrolyte.

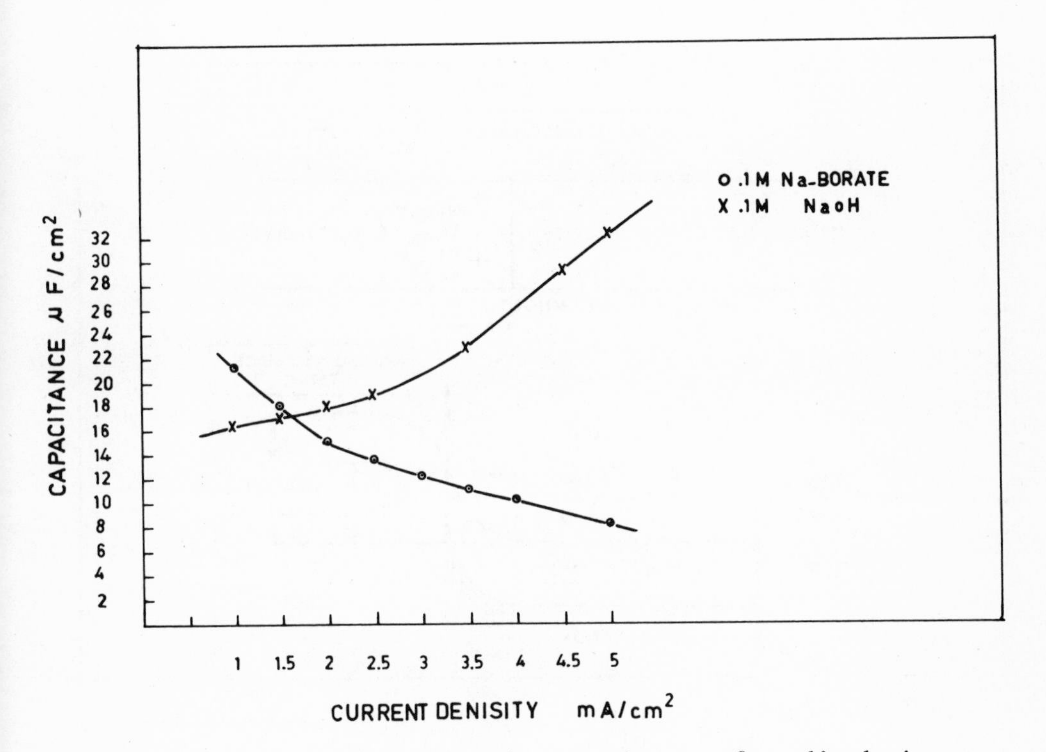

Fig. 10: Effect of C.D. on the capacitance of anodized zinc.

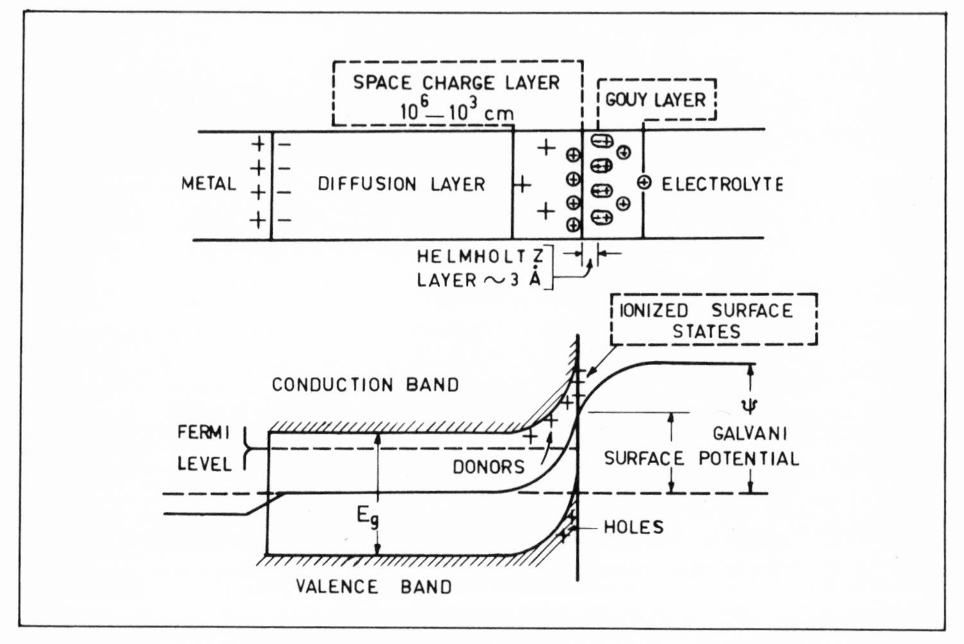

Fig. 11: Structure and energy levels of semiconductor zinc oxide/electrolyte.

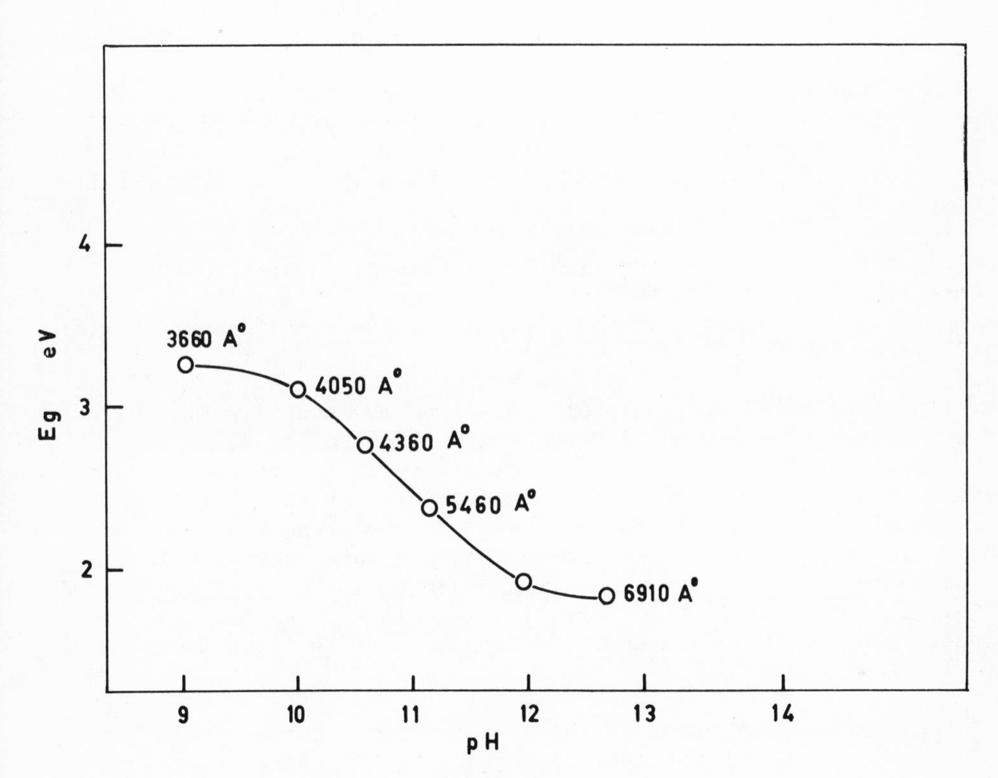

Fig. 12: Threshold wavelength-pH anodic polarized ($0.5\ mA/cm^2$) zinc.

electrochemical measurements and supported by the reflectivity results of this study. Reported values of ε_o of zinc oxide vary between 8 and 40 depending upon the impurities and the technique of formation.[7]

DISCUSSION AND CONCLUSION

From the results of this study, one can assume that the anodic film has semiconducting properties which are a function of the electrochemical forming conditions. Figure 11 gives an idealized schematic structure and energy levels of the semiconducting anodic film. Impurity levels are introduced due to incorporated anions from the electrolyte.[17,19] The anodic film formed in borate solution is expected to contain BO_3^{---} in its crystal, replacing O^{--}. Some OH^- ions are introduced in the anodic film formed in most alkaline electrolytes.

An n-type semiconductor with lower valency impurity anions should yield higher electronic conductivity. The opposite would be the case with higher valency impurity anions[21,22] (compare Fig.8 for borate with Fig. 9 for hydroxide).

The intrinsic energy gap of zinc oxide is about 3.2 eV[7] which is equivalent to a threshold wave length equal to 366 Å. A pure and perfect stoichiometric oxide does not react to light of longer wave lengths. Introducing defects and impurities into the oxide would fill some levels in the energy gap. The present photoeffects and reflectivity studies have shown that the threshold wave length λ_t of the anodic film was strongly dependent on the electrolyte used (Fig. 12). The anodic film formed on zinc in 1.0M sodium hydroxide electrolyte was very dark and sensitive to visible light (even to red light with λ = 6910 Å).

Optical sensitization of the anodic product is thus possible by introducing OH^- in the crystal, replacing O^{--}. This technique is as efficient as solid state sensitization by incorporating organic dyes and inorganic impurities.[23]

The system studied proved to have rather specific capacitance properties yet to be fully understood. However, porosity of the film leading to an increase in the interfacial area should be considered as a possible explanation.

Finally, it would appear valuable to examine the possibilities for practical application of this type of phenomena. For example, an electrolytic photo cell could be constructed.

REFERENCES

1. H. Fry, M. Whitaker, J. Electrochem. Soc. 106 606 (1959).

2. J. Sarghi, M. Fleichmann, Electrochim. Acta 1 161 (1958).

3. T.P. Dirkse, D. De Wite, R. Shoemaker, J. Electrochem. Soc. 115 442 (1968).

4. R.F. Ashton, M.T. Hepworth, Corrosion 24 50 (1968).

5. T.P. Dirkse, J. Appl. Electrochem. 1 27 (1971).

6. T.P. Dirkse, L.A. Van der Lugi, N. Hampson, J. Electrochem. Soc. 118 1606 (1971); Electrochim. Acta 17 32 (1972).

7. G. Heiland, E. Mollwo, F. Stockmann, Solid State Physics, vol. 8, Academic Press, New York (1959) 191.

8. L. Azaroff, J. Modern Physics 87 813 (1962).

9. J.E. Dewald, Bell System Tech. J. 39 615 (1960).

10. S.R. Morrison, T. Freund, Electrochim. Acta 13 1343 (1968).

11. R.A. Mickelson, W.D. Kingery, J. Appl. Physics 37 3541 (1966).

12. K. Huber, Helv. Chim. Acta 26 1253 (1943); Z. Electrochem. 62 675 (1958).

13. M.W. Breiter, W. Vedder, Trans. Farad. Soc. 63 1042 (1967).

14. N. De Zoubov, M. Pourbaix, Atlas of Electrochemical Equilibria in Aqueous Solutions, M. Pourbaix, ed., Pergamon, Oxford (1966) 406.

15. T.A.T. Cowell, J. Woods, Brit. J. Appl. Phys. 2 1053 (1969).

16. S.M. El Raghy, Ph.D. Thesis, University of London (1969).

17. L. Young, Anodic Oxide Films, Academic Press, London (1961) 78.

18. M.E. Ismail, Ph.D. Thesis, University of Alexandria, Egypt (1971).

19. D.A. Vermilyea, Report No. 65-RL-4033M, General Electric Research Information, Schenectady, New York (1965).

20. W. McNeil, L.L. Gruss, D.G. Husted, J. Electrochem. Soc. 112 713 (1965).

21. A.D. Wadsley, Non-stoichiometric Compounds, L. Mandelcorn, ed., Academic Press, New York (1963) 23.

22. G. Mueller, R. Helbig, J. Phys. Chem. Solids 32 1971 (1971).

23. H. Kokado, T. Nakayama, B. Inone, J. Phys. Chem. Solids 34 1 (1973).

CONTENTS OF VOLUME 2

POLYMERS

GLASS

COMPOSITES AND CERAMICS

METALLURGY

CONTENTS OF VOLUME 3

ARCHEOLOGY

INDEX